HISTORY OF STRENGTH OF MATERIALS

HISTORY OF STRENGTH OF MATERIALS

With a brief account of the history
of theory of elasticity and
theory of structures

STEPHEN P. TIMOSHENKO

Professor of Engineering Mechanics
Stanford University

DOVER PUBLICATIONS, INC., NEW YORK

Published in Canada by General Publishing Company, Ltd., 30 Lesmill Road, Don Mills, Toronto, Ontario.
Published in the United Kingdom by Constable and Company, Ltd., 10 Orange Street, London WC2H 7EG.

This Dover edition, first published in 1983, is an unabridged and unaltered republication of the work originally published in 1953 by McGraw-Hill Book Company, Inc., N.Y.

International Standard Book Number: 0-486-61187-6

Manufactured in the United States of America
Dover Publications, Inc., 180 Varick Street, New York, N.Y. 10014

Library of Congress Cataloging in Publication Data

Timoshenko, Stephen, 1878-1972.
   History of strength of materials.

   Reprint. Originally published: New York : McGraw-Hill, 1953.
   1. Strength of materials—History. I. Title.
TA405.T53   1983   620.1'12   82-17713

Preface

This book was written on the basis of lectures on the history of strength of materials which I have given during the last twenty-five years to students in engineering mechanics who already had knowledge of strength of materials and theory of structures. During the preparation of the book for publication, considerable material was added to the initial contents of the lectures, but the general character of the course remained unchanged. In writing the book, I had in mind principally those students who, after a required course in strength of materials, would like to go deeper into the subject and learn something about the history of the development of strength of materials. Having this in mind, I did not try to prepare a repertorium of elasticity and give a complete bibliography of the subject. Such a bibliography can be found in existing books such as "A History of the Elasticity and Strength of Materials" by Todhunter and Pearson and the articles in volume 4_4 of the "Encyklopädie der Mathematischen Wissenschaften," edited by F. Klein and C. Müller.

I wanted rather to follow the example of Saint-Venant's "Historique Abrégé"* and give to a larger circle of readers a historical review of the principal steps in the development of our science without going into too much detail. In doing this, I considered it desirable to include in the history brief biographies of the most prominent workers in this subject and also to discuss the relation of the progress in strength of material to the state of engineering education and to the industrial development in various countries. There is no doubt, for example, that the development of railroad transportation and the introduction of steel as structural material brought many new problems, dealing with strength of structures, and had a great influence on the development of strength of materials. The development of combustion engines and light airplane structures has had a similar effect in recent times.

Progress in strength of materials cannot be satisfactorily discussed without considering the development of the adjacent sciences such as theory of elasticity and theory of structures. There exists a close interrelation in the development of those sciences, and it was necessary to include some of their history in the book. In doing so I have taken

* The historical introduction which Saint-Venant added to his edition of Navier's book: "Résumé des Leçons."

from the history of theory of elasticity only those portions closely related to the development of strength of materials and omitted all material related to the purely theoretical and mathematical progress of that science. In the same way, in dealing with the development of theory of structures, the portions having only technical interest were not included in this book.

In this writing, I have tried to follow the chronological form of presentation and divided the history of the subject into several periods. For each of those periods I have discussed the progress made in strength of materials and in adjacent sciences. This order was not always strictly followed, and in some discussions of the works of a particular author I have found it more expedient to put in the same place the review of all his publications, although some of them did not belong to the period under discussion.

In the preparation of the book, the existing publications on the history of sciences were very helpful. In addition to the books already mentioned I had in my hands the third edition of Navier's book, "Résumé des Leçons . . . ," edited by Saint-Venant and containing his "Historique Abrégé . . . " and his numerous notes which are now of great historical interest. I consulted also Saint-Venant's translation of Clebsch's book on Elasticity, which in itself contains the history of elasticity in its earlier periods. Among the biographies I found the following very useful: "Histoire des Sciences Mathématiques et Physiques" by M. Marie, "Geschichte der Technischen Mechanik," by M. Rühlmann, several biographies in English, and collections of "Éloges Academiques" by François Arago and by Joseph Bertrand. For the review of newer publications it was necessary to go through many periodicals in various tongues. This took a considerable amount of time, but the writer will feel completely rewarded if his work will save some labor for other workers in history of strength of materials.

I am thankful to my colleagues at Stanford University—to Prof. Alfred S. Niles for his comments on the portions of the manuscript dealing with the early history of trusses and the Maxwell-Mohr method of analyzing statically indeterminate trusses; and to Prof. Donovan H. Young who gave much constructive advice at the time of preparation of the manuscript. I am also very grateful to Dr. R. E. D. Bishop for his reading of the entire manuscript and his numerous important comments, and to our graduate student, James Gere, who checked the proofs.

Stephen P. Timoshenko

Stanford, Calif.
December, 1952

Contents

Preface v

Introduction 1

I. THE STRENGTH OF MATERIALS IN THE SEVENTEENTH CENTURY
   1. Galileo 7
   2. Galileo's work on strength of materials 11
   3. Organization of the national academies of science . . 15
   4. Robert Hooke 17
   5. Mariotte 21

II. ELASTIC CURVES
   6. The mathematicians Bernoulli 25
   7. Euler 28
   8. Euler's contribution to strength of materials . . . 30
   9. Lagrange 37

III. STRENGTH OF MATERIALS IN THE EIGHTEENTH CENTURY
   10. Engineering applications of strength of materials . . 41
   11. Parent 43
   12. Coulomb 47
   13. Experimental study of the mechanical properties of structural materials in the eighteenth century . . 54
   14. Theory of retaining walls in the eighteenth century . . 60
   15. Theory of arches in the eighteenth century 62

IV. STRENGTH OF MATERIALS BETWEEN 1800 AND 1833
   16. L'Ecole Polytechnique 67
   17. Navier 70
   18. Navier's book on strength of materials 73
   19. The experimental work of French engineers between 1800 and 1833 80
   20. The theories of arches and suspension bridges between 1800 and 1833 83
   21. Poncelet 87
   22. Thomas Young 90

23. Strength of materials in England between 1800 and 1833 . . 98
24. Other notable European contributions to strength of materials 100

V. THE BEGINNING OF THE MATHEMATICAL THEORY OF ELASTICITY
25. Equations of equilibrium in the theory of elasticity. . 104
26. Cauchy 107
27. Poisson 111
28. G. Lamé and B. P. E. Clapeyron 114
29. The theory of plates. 119

VI. STRENGTH OF MATERIALS BETWEEN 1833 AND 1867
30. Fairbairn and Hodgkinson 123
31. The growth of German engineering schools 129
32. Saint-Venant's contributions to the theory of bending of beams 135
33. Jourawski's analysis of shearing stresses in beams . . 141
34. Continuous beams 144
35. Bresse 146
36. E. Winkler 152

VII. STRENGTH OF MATERIALS IN THE EVOLUTION OF RAILWAY ENGINEERING
37. Tubular bridges 156
38. Early investigations on fatigue of metals 162
39. The work of Wöhler 167
40. Moving loads 173
41. Impact 178
42. The early stages in the theory of trusses 181
43. K. Culmann 190
44. W. J. Macquorn Rankine 197
45. J. C. Maxwell's contributions to the theory of structures 202
46. Problems of elastic stability. Column formulas. . . 208
47. Theory of retaining walls and arches between 1833 and 1867 210

VIII. THE MATHEMATICAL THEORY OF ELASTICITY BETWEEN 1833 AND 1867
48. The physical elasticity and "the elastic constant controversy" 216
49. Early work in elasticity at Cambridge University . . 222
50. Stokes. 225
50a. Barré de Saint-Venant 229

Contents

- 51. The semi-inverse method 233
- 52. The later work of Saint-Venant 238
- 53. Duhamel and Phillips 242
- 54. Franz Neumann 246
- 55. G. R. Kirchhoff 252
- 56. A. Clebsch 255
- 57. Lord Kelvin 260
- 58. James Clerk Maxwell 268

IX. STRENGTH OF MATERIALS IN THE PERIOD 1867–1900
- 59. Mechanical Testing Laboratories 276
- 60. The work of O. Mohr 283
- 61. Strain energy and Castigliano's theorem 288
- 62. Elastic stability problems 293
- 63. August Föppl 299

X. THEORY OF STRUCTURES IN THE PERIOD 1867–1900
- 64. Statically determinate trusses 304
- 65. Deflection of trusses 311
- 66. Statically indeterminate trusses 316
- 67. Arches and retaining walls 323

XI. THEORY OF ELASTICITY BETWEEN 1867 AND 1900
- 68. The work of Saint-Venant's pupils 328
- 69. Lord Rayleigh 334
- 70. Theory of elasticity in England between 1867 and 1900 339
- 71. Theory of elasticity in Germany between 1867 and 1900 344
- 71a. Solutions of two-dimensional problems between 1867 and 1900 350

XII. PROGRESS IN STRENGTH OF MATERIALS DURING THE TWENTIETH CENTURY
- 72. Properties of materials within the elastic limit . . . 355
- 73. Fracture of brittle materials 358
- 74. Testing of ductile materials 362
- 75. Strength theories 368
- 76. Creep of metals at elevated temperatures 372
- 77. Fatigue of metals 377
- 78. Experimental stress analysis 383

XIII. THEORY OF ELASTICITY DURING THE PERIOD 1900–1950
- 79. Felix Klein 389
- 80. Ludwig Prandtl 392
- 81. Approximate methods of solving elasticity problems . 397

- 82. Three-dimensional problems of elasticity 401
- 83. Two-dimensional problems of elasticity 405
- 84. Bending of plates and shells 408
- 85. Elastic stability 412
- 86. Vibrations and impact 417

XIV. Theory of Structures during the Period 1900–1950

- 87. New methods of solving statically indeterminate systems 422
- 88. Arches and suspension bridges 426
- 89. Stresses in railway tracks 430
- 90. Theory of ship structures 434

Name Index 441

Subject Index 449

Introduction

From the earliest times when people started to build, it was found necessary to have information regarding the strength of structural materials so that rules for determining safe dimensions of members could be drawn up. No doubt the Egyptians had some empirical rules of this kind, for without them it would have been impossible to erect their great monuments, temples, pyramids, and obelisks, some of which still exist. The Greeks further advanced the art of building. They developed statics, which underlies the mechanics of materials. Archimedes (287–212 B.C.) gave a rigorous proof of the conditions of equilibrium of a lever and outlined methods of determining centers of gravity of bodies. He used his theory in the construction of various hoisting devices. The methods used by the Greeks in transporting the columns and architraves of the temple of Diana of Ephesus are shown in Figs. 1 to 3.

The Romans were great builders. Not only some of their monuments and temples remain, but also roads, bridges, and fortifications. We know something of their building methods from the book by Vitruvius,[1] a famous Roman architect and engineer of the time of Emperor Augustus. In this book, their structural materials and types of construction are described. Figure 4 shows a type of hoist used by the Romans for lifting heavy stones. The Romans often used arches in their buildings. Figure 5 shows the arches in the famous Pont du Gard, a bridge which is in service to this day in southern France. A comparison[2] of the proportions of Roman arches with those of the present time indicates that nowadays much lighter structures are built. The Romans had not the advantages provided by stress analysis. They did not know how to select the proper shape and usually took semicircular arches of comparatively small span.

Most of the knowledge that the Greeks and Romans accumulated in the way of structural engineering was lost during the Middle Ages and only since the Renaissance has it been recovered. Thus when the famous Italian architect Fontana (1543–1607) erected the Vatican obelisk at the order of Pope Sixtus V, (Fig. 6), this work attracted wide attention from

[1] Vitruvius, "Architecture," French translation by De Bioul, Brussels, 1816.

[2] For such a comparison, see Alfred Leger, "Les Travaux Publics aux temps des Romains," p. 135, Paris, 1875.

European engineers. But we know that the Egyptians had raised several such obelisks thousands of years previously, after cutting stone from the quarries of Syene and transporting it on the Nile. Indeed, the Romans had carried some of the Egyptian obelisks from their original sites and erected them in Rome; thus it seems that the engineers of the six-

FIGS. 1 to 4. Bottom, Greeks' methods of transporting columns. Top, type of hoist used by the Romans.

teenth century were not as well equipped for such difficult tasks as their predecessors.

During the Renaissance there was a revival of interest in science, and art leaders appeared in the field of architecture and engineering. Leonardo da Vinci (1452–1519) was a most outstanding man of that period. He was not only the leading artist of his time but also a great scientist and engineer. He did not write books, but much information

Introduction

was found in his notebooks[1] regarding his great discoveries in various branches of science. Leonardo da Vinci was greatly interested in mechanics and in one of his notes he states: "Mechanics is the paradise of mathematical science because here we come to the fruits of mathematics." Leonardo da Vinci uses the method of moments to get the correct solutions of such problems as those shown in Figs. 8a and 8b. He applies the notion of the principle of virtual displacements to analyze various systems of pulleys and levers such as are used in hoisting devices. It seems that

Fig. 5. The famous Pont du Gard.

Leonardo da Vinci had a correct idea of the thrust produced by an arch. In one of his manuscripts there is a sketch (Fig. 9) of two members on which a vertical load Q is acting and the question is asked: What forces are needed at a and b to have equilibrium? From the dotted-line parallelogram, in the sketch, it can be concluded that Leonardo da Vinci had the correct answer in this case.

Leonardo da Vinci studied the strength of structural materials experimentally. In his note "Testing the Strength of Iron Wires of Various Lengths" he gives the sketch shown in Fig. 10, and makes the following remark: "The object of this test is to find the load an iron wire can carry. Attach an iron wire 2 braccia long to something that will firmly support

---

[1] A bibliography of Leonardo da Vinci's work is given in the Encyclopaedia Britannica. Also a selection of passages from the manuscripts will be found in the book by Edward McCurdy, "Leonardo da Vinci's Note-books." See also the book by W. B. Parsons, "Engineers and Engineering in the Renaissance," 1939. From the latter book Fig. 10 and the quotations given in this article are taken.

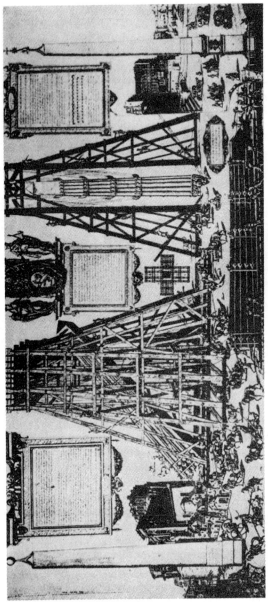

Fig. 6. The erection of the Vatican obelisk.

Introduction

it, then attach a basket or any similar container to the wire and feed into the basket some fine sand through a small hole placed at the end of a hopper. A spring is fixed so that it will close the hole as soon as the wire breaks. The basket is not upset while falling, since it falls through a very short distance. The weight of sand and the location of the fracture of the wire are to be recorded. The test is repeated several times to check the results. Then a wire of one-half the previous length is tested and the additional weight it carries is recorded; then a wire of one-fourth length is tested and so forth, noting each time the ultimate strength and the location of the fracture."[1]

Leonardo da Vinci also considered the strength of beams and stated a general principle as follows: "In every article that is supported, but is free to bend, and is of uniform cross section and material, the part that is farthest from the supports will bend the most." He recommends that a series of tests be made, starting with a beam that could carry a definite weight when supported at both ends, and then taking successively longer

Fig. 7. Leonardo da Vinci.

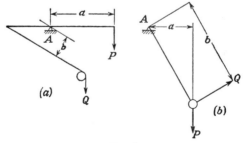

Fig. 8.

beams of the same depth and width, and recording what weight these would carry. His conclusion was that the strength of beams supported at both ends varies inversely as the length and directly as the width. He also made some investigation of beams having one end fixed and the other free and states: "If a beam 2 braccia long supports 100 libbre, a beam 1 braccia long will support 200. As many times as the shorter length is

[1] See Parsons, "Engineers and Engineering in the Renaissance," p. 72.

contained in the longer, so many times more weight will it support than the longer one." Regarding the effect of depth upon the strength of a beam there is no definite statement in Leonardo da Vinci's notes.

Apparently Leonardo da Vinci made some investigations of the strength of columns. He states that this varies inversely as their lengths, but directly as some ratio of their cross sections.

These briefly discussed accomplishments of da Vinci represent perhaps the first attempt to apply statics in finding the forces acting in members of structures and also the first experiments for determining the strength

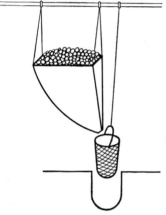

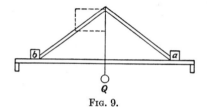

FIG. 9.

FIG. 10. Tensile test of wire by Leonardo da Vinci.

of structural materials. However, these important advances were buried in da Vinci's notes and engineers of the fifteenth and sixteenth centuries continued, as in the Roman era, to fix the dimensions of structural elements by relying only on experience and judgment.

The first attempts to find the safe dimensions of structural elements analytically were made in the seventeenth century. Galileo's famous book "Two New Sciences"[1] shows the writer's efforts to put the methods applicable in stress analysis into a logical sequence. It represents the beginning of the science of *strength of materials*.

[1] See English translation by Henry Crew and Alfonso de Salvio, New York, 1933.

CHAPTER I

The Strength of Materials in the Seventeenth Century[1]

1. *Galileo* (1564–1642)

Galileo was born in Pisa[2] and was a descendant of a noble Florentine house. Galileo received his preliminary education in Latin, Greek, and logic in the monastery of Vallombrosa, near Florence. In 1581, he was placed in Pisa University, where he was to study medicine. But very soon the lectures on mathematics began to attract his attention, and he threw all his energy into studying the work of Euclid and Archimedes. It seems that, through Cardan's books,[3] he became acquainted with Leonardo da Vinci's discoveries in mechanics. In 1585, Galileo had to withdraw from the University, due to lack of means, without taking the degree and he returned to his home in Florence. There, Galileo gave private lessons in mathematics and mechanics and continued his own scientific work.

Fig. 11. Galileo.

In 1586, he made a hydrostatic balance for measuring the density of various substances and he carried

---

[1] The history of mechanics of materials during the seventeenth and eighteenth centuries is discussed in the preface to the book "Traité Analytique de la résistance des Solides" by P. S. Girard, Paris, 1798.

[2] See J. J. Fahie, "Galileo, His Life and Work," New York, 1903. See also the novel "The Star-gazer" by Zsolt de Harsanyi, English translation by P. Tabor, New York, 1939.

[3] See P. Duhem, "Les Origines de la Statique," p. 39, Paris, 1905. Cardan (1501–1576) discusses mechanics in some of his mathematical publications. His presentation of this science is very similar to that of Leonardo da Vinci, and it is usually assumed that Cardan had access to the latter's manuscripts and notebooks.

7

out investigations of the centers of gravity in solid bodies. This work made him known, and in the middle of 1589 he was given the mathematical professorship at Pisa when he was twenty-five and a half years old.

During his time in Pisa (1589–1592), Galileo continued his work in mathematics and mechanics and made his famous experiments on falling bodies. On the basis of these experiments the treatise "De Motu Gravium" was prepared in 1590, and it represents the beginning of dynamics as we know it today. The principal conclusions of this work were (1) all bodies fall from the same height in equal times; (2) in falling, the final velocities are proportional to the times; (3) the spaces fallen through are proportional to the squares of the times. These conclusions were in complete disagreement with those of Aristotelian mechanics, but Galileo did not hesitate to use them in his disputes with the representatives of the Aristotelian school. This produced feelings of animosity against the young Galileo, and finally he had to leave Pisa and return to Florence. At this difficult time some friends helped him to get the professorship in the University of Padua. In connection with the official appointment there the following statement[1] was made at that time: "Owing to the death of Signor Moletti, who formerly lectured on Mathematics at Padua, the Chair has been for a long time vacant, and, being a most important one, it was thought proper to defer electing any one to fill it till such time as a fit and capable candidate should appear. Now Domino Galileo Galilei has been found, who lectured at Pisa with very great honour and success, and who may be styled the first in his profession, and who, being ready to come at once to our said university, and there to give the said lectures, it is proper to accept him."

On Dec. 7, 1592, Galileo embarked upon his new duties with a discourse "which has won the greatest admiration, not only for its profound knowledge, but for its eloquence and elegance of diction" During his first years at Padua, Galileo was extraordinarily active. His lectures became so well known that students from other European countries came to Padua. A room capable of containing 2,000 students had to be used eventually for these lectures. In 1594 the famous treatise on mechanics ("Della Scienza Meccanica") was written. In this treatise various problems of statics were treated by using the principle of virtual displacements. The treatise attained a wide circulation in the form of manuscript copies. At about the same time, in connection with some problems in shipbuilding, Galileo became interested also in the strength of materials. Very soon astronomy attracted Galileo's attention. It is known that during his first years in Padua, Galileo taught the Ptolemaic system, as was the custom in those days. But in a letter to Kepler, as early as

[1] See the book by J. J. Fahie, p. 35.

1597, he states: "Many years ago I became a convert to the opinions of Copernicus, and by this theory have succeeded in explaining many phenomena which on the contrary hypothesis are altogether inexplicable." A rumor of the invention of a telescope reached Padua in 1609, and, on the strength of meager information, Galileo succeeded in building one of his own with a magnifying power of 32. With this instrument, he made

FIG. 12. The living room in Galileo's villa at Arcerti.

a series of outstanding astronomical discoveries. He showed that the Milky Way consists of lesser stars, described the mountainous nature of the moon, and, in January, 1610, he saw Jupiter's satellites for the first time. This last discovery had a great effect on the further development of astronomy, for the visible motion of this system became a very powerful argument in favor of the Copernican theory. All these discoveries made Galileo famous. He was nominated "philosopher and mathematician extraordinary" to the grand duke of Tuscany and in September,

1610, abandoned Padua for Florence. In his new position Galileo had no duties other than to continue his scientific work and he put all his energy into astronomy. He discovered the peculiar shape of Saturn, observed the phases of Venus, and described the spots upon the sun.

All these brilliant discoveries and Galileo's enthusiastic writing in favor of the Copernican theory attracted the attention of the Church. The

Fig. 13. The title page of Galileo's book, "Two New Sciences."

discrepancy between the new view of the planetary system and that of the Scriptures was brought before the Inquisition and, in 1615, Galileo received a semiofficial warning to avoid theology and limit himself to physical reasoning. In 1616, the great work of Copernicus was condemned by the Church, and, during the seven succeeding years, Galileo stopped publishing his controversial work in astronomy. In 1623, Maffeo Barberini, a friend and admirer of Galileo, was elected to the pontifical throne, and Galileo, expecting a more favorable treatment of his astronomical publications, started writing his famous book dealing with the two ways of regarding the universe, which appeared in print in 1632. Since the book definitely favored the Copernican theory, its sale was prohibited by the Church and Galileo was called to Rome by the Inquisition. There he was condemned and had to read his recantation. After his return to Florence, he had to live in his villa at Arcerti in strict seclusion, which he did during the remaining eight years of his life. It was then that he wrote his famous book "Two New Sciences,"[1] in which he recapitulated the results of all his previous work in the various fields of mechanics. The book was printed by the Elzevirs at Leiden in 1638 (Fig. 13). A portion of the book, dealing with the mechanical properties of structural materials and with the strength of beams, constitutes the first publication in the field of strength of materials, and from that date the history of mechanics of elastic bodies begins.

2. *Galileo's Work on Strength of Materials*

All Galileo's work on the mechanics of materials is included in the first two dialogues of his book "Two New Sciences." He begins with several observations made during his visits to a Venetian arsenal and discusses geometrically similar structures. He states that if we make structures geometrically similar, then, with increase of the dimensions, they become weaker and weaker. In illustration he states: "A small obelisk or column or other solid figure can certainly be laid down or set up without danger of breaking, while very large ones will go to pieces under the slightest provocation, and that purely on account of their own weight." To prove this, he starts with a consideration of the strength of materials in simple tension (Fig. 14) and states that the strength of a bar is proportional to its cross-sectional area and is independent of its length. This strength of the bar Galileo calls the "absolute resistance to fracture" and he gives some figures relating to the ultimate strength of copper. Having the absolute resistance of a bar, Galileo investigates the resistance to fracture of the same bar if it is used as a cantilever with the load at the end (Fig. 15). He states: "It is clear that, if the cylinder breaks, fracture will occur at the point *B* where the edge of the mortise acts as a fulcrum for the lever *BC*, to which the force is applied; the thickness of the solid *BA* is the other arm of the lever along which is located

FIG. 14. Galileo's illustration of tensile test.

[1] Galileo, "Two New Sciences," English translation by Henry Crew and Alfonso de Salvio, The Macmillan Company, New York, 1933.

the resistance. This resistance opposes the separation of the part BD, lying outside the wall, from that portion lying inside. From the preceding, it follows that the magnitude of the force applied at C bears to the magnitude of the resistance, found in the thickness of the prism, *i.e.*, in the attachment of the base BA to its contiguous parts, the same ratio

Fig. 15. Galileo's illustration of bending test.

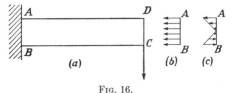

Fig. 16.

which half the length BA bears to the length BC."[1] We see that Galileo assumes that when fracture occurs the "resistance" is uniformly distributed over the cross section BA (Fig. 16b). Assuming that the bar has a rectangular cross section and that the material follows Hooke's law up to fracture, we obtain the stress distribution shown in Fig. 16c. The resisting couple corresponding to this stress distribution is equal only to one-third of the moment assumed by Galileo. Thus for such a material

[1] See "Two New Sciences," English translation, p. 115.

The Strength of Materials in the Seventeenth Century   13

Galileo's theory gives a value three times larger than the actual breaking load for the load at *C*. Actual materials do not follow Hooke's law until they fail, and the stress distribution at fracture is different from that shown in Fig. 16c in such a way as to diminish the discrepancy between the prediction of Galileo's theory and the true value of the breaking load.

On the basis of his theory Galileo draws several important conclusions. Considering a rectangular beam, he puts the question: "How and in what proportion does a rod, or rather a prism whose width is greater than its thickness, offer more resistance to fracture when the force is applied in the direction of its breadth than in the direction of its thickness." Using his assumption (Fig. 16b), he gives the correct answer: "Any given ruler or prism, whose width exceeds its thickness, will offer greater resistance to fracture when standing on edge than when lying flat, and this in the ratio of the width to the thickness."[1]

Continuing the discussion of the cantilever-beam problem and keeping a constant cross section, Galileo concludes that the bending moment, due to the weight of the beam, increases as the square of the length. Keeping the length of a circular cylinder constant and varying its radius, Galileo finds that the resisting moment increases as the cube of the radius. This result follows from the fact that the "absolute" resistance is proportional to the cross-sectional area of the cylinder and that the arm of the resisting couple is equal to the radius of the cylinder.

Considering geometrically similar cantilever beams under the action of their weights Galileo concludes that, while the bending moment at the built-in end increases as the fourth power of the length, the resisting moment is proportional to the cube of the linear dimensions. This indicates that the geometrically similar beams are not equally strong. The beams become weaker with increase in dimensions and finally, when they are large, they may fail under the action of their weight only. He also observes that, to keep the strength constant, the cross-sectional dimensions must be increased at a greater rate than the length.

With these considerations in mind, Galileo makes the following important general remark: "You can plainly see the impossibility of increasing the size of structures to vast dimensions either in art or in nature; likewise the impossibility of building ships, palaces, or temples of enormous size in such a way that their oars, yards, beams, iron-bolts, and, in short, all their other parts will hold together; nor can nature produce trees of extraordinary size because the branches would break down under their own weight; so also it would be impossible to build up the bony structures of men, horses, or other animals so as to hold together and perform their normal functions if these animals were to be increased enormously in

[1] See "Two New Sciences," English translation, p. 118.

height; for this increase in height can be accomplished only by employing a material which is harder and stronger than usual, or by enlarging the size of the bones, thus changing their shape until the form and appearance of the animals suggest a monstrosity. . . . If the size of a body be diminished, the strength of that body is not diminished in the same proportion; indeed the smaller the body the greater its relative strength. Thus a small dog could probably carry on his back two or three dogs of his own size; but I believe that a horse could not carry even one of his own size."[1]

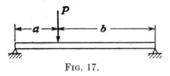

Fig. 17.

Galileo also considers a beam on two supports (Fig. 17) and finds that the bending moment is greatest under the load and is proportional to the product *ab*, so that to produce fracture with the smallest load this load must be put at the middle of the span. He observes that the possibility exists of economizing material by reducing the size of the cross section near the supports.

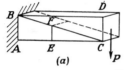 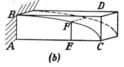

Fig. 18.

Galileo gives a complete derivation of the form of a cantilever beam of equal strength, the cross section of which is rectangular. Considering first a prismatical beam $ABCD$ (Fig. 18a), he observes that part of the material can be removed without affecting the strength of the beam. He shows also that if we remove half of the material and take the beam in the form of the wedge ABC, the strength at any cross section EF will be insufficient since, while the ratio of the bending moment at EF to that at AB is in the ratio $EC:AC$, the resisting moments, proportional to the square of the depth, will be in the ratio $(EC)^2:(AC)^2$ at these cross sections. To have the resisting moment varying in the same proportion as the bending moment, we must take the parabolic curve BFC (Fig. 18b). This satisfies the requirement of equal strength, since for a parabola we have

$$\frac{(EF)^2}{(AB)^2} = \frac{EC}{AC}$$

Finally, Galileo discusses the strength of hollow beams and states[2] that such beams "are employed in art—and still more often in nature—in a

---

[1] See "Two New Sciences," English translation, p. 130.
[2] See "Two New Sciences," English translation, p. 150.

The Strength of Materials in the Seventeenth Century 15

thousand operations for the purpose of greatly increasing strength without adding to weight; examples of these are seen in the bones of birds and in many kinds of reeds which are light and highly resistant both to bending and breaking. For if a stem of straw which carries a head of wheat heavier than the entire stalk were made up of the same amount of material in solid form, it would offer less resistance to bending and breaking. This is an experience which has been verified and confirmed in practice where it is found that a hollow lance or a tube of wood or metal is much stronger than would be a solid one of the same length and weight. . . . " Comparing a hollow cylinder with a solid one of the same cross-sectional area, Galileo observes that their absolute strengths are the same and, since the resisting moments are equal to their absolute strength multiplied by the outer radius, the bending strength of the tube exceeds that of the solid cylinder in the same proportion as that by which the diameter of the tube exceeds the diameter of the cylinder.

3. Organization of the National Academies of Science

During the seventeenth century there was a rapid development in mathematics, astronomy, and in the natural sciences. Many learned men became interested in the sciences and experimental work in particular received much attention. Many of the universities were controlled by the Church, and since this was not favorable for scientific progress, learned societies were organized in several European countries. The purpose of these was to bring men with scientific interests together and to facilitate experimental work. This movement started in Italy where, in 1560, the Accademia Secretorum Naturae was organized in Naples. The famous Accademia dei Lincei was founded in Rome in 1603, and Galileo was one of its members. After Galileo's death, the Accademia del Cimento was organized in Florence with the support of the grand duke Ferdinand de' Medici and his brother Leopold. Galileo's pupils Viviani and Torricelli participated in the work of that academy. In the volume of the academy publications, considerable space is devoted to such problems as those of the thermometer, barometer, and pendulum and also to various experiments relating to vacua.[1]

In England at about the same time, scientific interest drew a group of men together, and they met whenever suitable opportunities presented themselves. The mathematician Wallis describes these informal meetings as follows: "About the year 1645, while I lived in London, beside the conversation of divers eminent divines as to matters theological, I had the opportunity of being acquainted with divers worthy persons, inquisitive into natural philosophy, and other parts of human learning; and particularly of what hath been called the 'New Philosophy' or 'Experimental

[1] "Saggi di Naturali Esperienze," 2d ed., Florence, 1691.

16   *History of Strength of Materials*

Philosophy.' We did by agreements, divers of us, meet weekly in London on a certain day and hour under a certain penalty, and a weekly contribution for the charge of experiments, with certain rules agreed upon amongst us to treat and discourse of such affairs. . . . Our business was (precluding matters of theology and state affairs) to discourse and consider of *Philosophical Enquiries,* and such as related thereunto; as Physick, Anatomy, Geometry, Astronomy, Navigation, Staticks, Magnetics, Chymicks, Mechanicks, and Natural Experiments; with the state of these studies, as then cultivated at home and abroad. We then discoursed on the circulation of the blood, the valves in the veins, the Copernican Hypothesis, the Nature of Comets and New Stars, the Satellites of Jupiter, the oval shape [as it then appeared] of Saturn, the spots in the Sun, and its turning on its own Axis, the Inequalities and Selenography of the Moon, the several Phases of Venus and Mercury, the Improvement of Telescopes, and grinding of Glasses for that purpose, the Weight of Air, the Possibility or Impossibility of Vacuities and Nature's Abhorrence thereof, the Torricellian Experiment in Quicksilver, the Descent of heavy Bodies, and the degrees of Acceleration therein; and divers other things of like nature. Some of which were then but New Discoveries, and others not so generally known and embraced as now they are, with other things appertaining to what hath been called the 'New Philosophy' which from the times of Galileo at Florence, and Sir Francis Bacon (Lord Verulam) in England, hath been much cultivated in Italy, France, Germany, and other parts abroad, as well as with us in England. Those meetings in London continued . . . and were afterwards incorporated by the name of the Royal Society, etc., and so continue to this day."[1]

The date upon which the First Charter was sealed (July 15, 1662) is usually taken as being that of the Royal Society's foundation. In the list of those invited to become members of the Society we find the names of Robert Boyle, physicist and chemist; Christopher Wren, architect and mathematician, and John Wallis, mathematician. As the curator, whose duty would be "to furnish the Society every day they meet, with three or four considerable experiments," Robert Hooke was appointed.

The French Academy of Sciences also had its origin in the informal meetings of scientists. Father Mersenne (1588–1648) established, and fostered until his death, a series of conferences which were attended by such men as Gassendi, Descartes, and Pascal. Later on these private gatherings of scientists were continued in the house of Habert de Montmor. In 1666, Louis XIV's minister Colbert took official steps to organize the Academy of Sciences which was to have as its members specialists in various scientific fields. The mathematician Roberval, the astronomer

[1] Sir Henry Lyons, "The Royal Society 1660–1940," 1944.

Cassini, the Danish physicist Römer (who measured the velocity of light), and the French physicist Mariotte appear in the first membership list of the Academy.[1] Somewhat later (in 1770) the Berlin Academy of Sciences was organized, and in 1725 the Russian Academy of Sciences was opened in St. Petersburg.

All these academies published their transactions and these had a great influence on the development of science in the eighteenth and nineteenth centuries.

4. Robert Hooke (1635–1703)[2]

Robert Hooke was born in 1635, the son of a parish minister who lived on the Isle of Wight. As a child he was weak and infirm, but very early he showed great interest in making mechanical toys and in drawing. When he was thirteen years old, he entered Westminster School and lived in the house of Dr. Busby, master. There he learned Latin, Greek, and some Hebrew and became acquainted with the elements of Euclid and with other mathematical topics. In 1653, Hooke was sent to Christ Church, Oxford, where he was a chorister, and this gave him an opportunity to continue his study so that, in 1662, he took the degree of Master of Arts. In Oxford he came into contact with several scientists and, being a skilled mechanic, he helped them in their research work. About 1658 he worked with Boyle and perfected an air pump. He writes: "About the same time having an opportunity of acquainting myself with Astronomy by the kindness of Dr. Ward, I applied myself to the improving of the pendulum for such observations and I contriv'd a way to continue the motion of the pendulum . . . . I made some trials for this end, which I found to succeed to my wish. The success of these made me farther think of improving it for finding the Longitude and *the Method I had made for myself for Mechanick Inventions*, quickly led me to the use of Springs, instead of Gravity, for the making a body vibrate in any Posture." This marks the beginning of his experiments with springs.

In 1662, on the recommendation of Robert Boyle, Hooke became curator of the experiments of the Royal Society and his knowledge of mechanics and inventive ability were put to good use by the Society. He was always ready to devise apparatus to demonstrate his own ideas or to illustrate and clarify any point arising in the discussions of the Fellows.

[1] J. L. F. Bertrand, "L'Académie des Sciences et les Académiciens de 1666 a 1793," Paris, 1869.
[2] See "The Life and Work of Robert Hooke" by R. T. Gunther in "Early Science in Oxford," vols. VI–VIII. See also the paper by E. N. Da C. Andrade, *Proc. Roy. Soc. (London)*, vol. 201, p. 439, 1950. The quotations given in this article are taken from these sources.

Between the years 1663–1664 Robert Hooke became interested in microscopy and in 1665 his book "Micrographia" was published.[1] There we find not only information about Hooke's microscope but also descriptions of his important new discoveries. Hooke conceived the idea that "light is a very short vibrative motion transverse to straight lines of propagation." He explained the interference colours of soap bubbles and the phenomenon of Newton's rings.

In 1664, Hooke became professor of geometry in Gresham College, but he continued to present his experiments, inventions and descriptions of new instruments to the Royal Society and to read his Cutlerian Lectures.[2]

At a meeting of the Royal Society on May 3, 1666, Hooke said: "I will explain a system of the world very different from any yet conceived and it is founded on the three following positions:

"I. That all the heavenly bodies have not only a gravitation of their parts to their own proper centre, but that they also mutually attract each other within their spheres of action.

"II. That all bodies having a simple motion, will continue to move in a straight line, unless continually deflected from it by some extraneous force, causing them to describe a circle, an ellipse, or some other curve.

"III. That this attraction is so much the greater as the bodies are nearer. As to the proportion in which those forces diminish by an increase of distance, I own [says he] I have not discovered it although I have made some experiments to this purpose. I leave this to others, who have time and knowledge sufficient for the task."[3]

We see that Hooke had a clear picture of universal gravitation, but, it seems, he lacked the mathematical knowledge to prove Kepler's laws.

After the Great Fire of London in September, 1666, Hooke made a model incorporating his proposals for the rebuilding and the magistrates of the City made him a surveyor. He was very active in this reconstruction work and designed several buildings.

In 1678, the paper "De Potentiâ Restitutiva," or "Of Spring," was published. It contains the results of Hooke's experiments with elastic bodies. This is the first published paper in which the elastic properties of materials are discussed. Regarding the experiments, he says: "Take a wire string [Fig. 19][4] of 20, or 30, or 40 ft long, and fasten the upper part thereof to a nail, and to the other end fasten a Scale to receive the weights: Then with a pair of Compasses take the distance of the bottom of the scale from the ground or floor underneath, and set down the said distance,

---

[1] See "Early Science in Oxford," vol. XIII.
[2] See "Early Science in Oxford," vol. VIII.
[3] See John Robison, "Elements of Mechanical Philosophy," p. 284, Edinburgh, 1804.
[4] Figure 19 is taken from Hooke's paper.

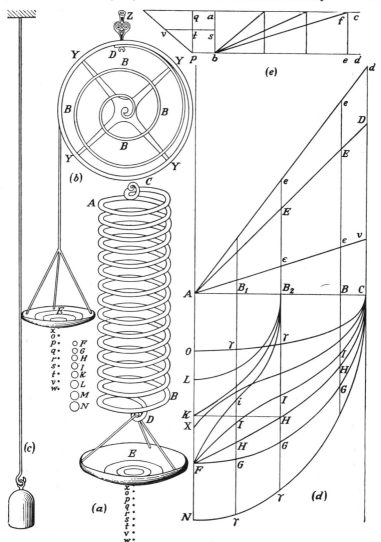

Fig. 19. Devices used in Hooke's experiments.

then put in weights into the said scale and measure the several stretchings of the said string, and set them down. Then compare the several stretchings of the said string, and you will find that they will always bear the same proportions one to the other that the weights do that made them." Hooke also describes experiments with helical springs, with watch springs coiled into spirals, and with "a piece of dry wood that will bend and return, if one end thereof be fixed in a horizontal posture, and to the other end be hanged weights to make it bend downwards." He not only discusses the deflection of this beam, but also considers the deformations of longitudinal fibers and makes the important statement that the fibers on the convex side are extended during bending while the fibers on the concave side are compressed. From all these experiments Hooke draws the following conclusion: "It is very evident that the Rule or Law of Nature in every springing body is, that the force or power thereof to restore itself to its natural position is always proportionate to the distance or space it is removed therefrom, whether it be by rarefaction, or the separation of its parts the one from the other, or by a Condensation, or crowding of those parts nearer together. Nor is it observable in these bodies only, but in all other springy bodies whatsoever, whether metal, wood, stones, baked earth, hair, horns, silk, bones, sinews, glass, and the like. Respect being had to the particular figures of the bodies bended, and the advantagious or disadvantagious ways of bending them. From this principle it will be easy to calculate the several strength of Bows . . . as also of the Balistae or Catapultae used by the Ancients . . . . It will be easy to calculate the proportionate strength of the spring of a watch. . . . From the same also it will be easy to give the reason of the Isochrone motion of a Spring or extended string, and of the uniform sound produced by those whose vibrations are quick enough to produce an audible sound. From this appears the reason why a spring applied to the balance of a watch doth make the vibrations thereof equal, whether they be greater or smaller . . . . From this it will be easy to make a Philosophical Scale to examine the weight of any body without putting in weights . . . . This Scale I contrived in order to examine the gravitation of bodies towards the Center of the Earth, *viz.* to examine whether bodies at a further distance from the center of the earth did not loose somewhat of their power or tendency towards it . . . ."

We see that Robert Hooke not only established the relation between the magnitude of forces and the deformations that they produce, but also suggested several experiments in which this relation can be used in solving some very important problems. This linear relation between the force and the deformation is the so-called *Hooke's law*, which later on was used as the foundation upon which further development of the mechanics of elastic bodies was built.

5. Mariotte

Mariotte (1620–1684) spent most of his life in Dijon where he was the prior of St.-Martin-sous-Beaune. He became one of the first members of the French Academy of Sciences, in 1666, and was largely responsible for the introduction of experimental methods into French science. His experiments with air resulted in the well-known Boyle-Mariotte law which states that, at constant temperature, the pressure of a fixed mass of gas multiplied by its volume remains constant.

In the mechanics of solid bodies, Mariotte established the laws of impact, for, using balls suspended by threads, he was able to demonstrate the conservation of momentum. He invented the ballistic pendulum.

Mariotte's investigations in elasticity are included in a paper on the motion of fluids.[1] Mariotte had to design the pipe lines for supplying water to the Palace of Versailles and, as a result of this, became interested in the bending strength of beams. Experimenting with wooden and glass rods, he found that Galileo's theory gives exaggerated values for the breaking load and thus he developed his own theory of bending in which the elastic properties of materials was taken into consideration.

He starts with simple tensile tests. Figure 20a[2] shows the arrangement used in his tensile tests of wood. In Fig. 20b, the tensile test of paper is shown. Mariotte was not only interested in the absolute strength of materials but also in their elastic properties and found that, in all the materials tested, the elongations were proportional to the applied forces. He states that fracture occurs when the elongation exceeds a certain limit. In his discussion of the bending of a cantilever (see Fig. 20c), he begins with a consideration of the equilibrium of a lever AB (Fig. 20d), supported at C. On the left-hand arm of the lever, three equal weights $G = H = I = 12$ lb are suspended at the distances $AC = 4$ ft, $DC = 2$ ft, $EC = 1$ ft. To balance them by a load F applied at the distance $BC = 12$ ft, we must take $F = 7$ lb. If now the load F is increased somewhat, the lever begins to rotate about point C. The displacements of points A, D, and E are in proportion to their distances from C, but the forces applied at those points continue to be equal to 12 lb. Consider now the same lever, but assume that the loads G, H, I are replaced by the three identical wires DI, GL, HM (Fig. 20e), the absolute strength of which is equal to 12 lb. In calculating the load R which is required to produce fracture of the wires, Mariotte observes that, when the force in the wire DI reaches its ultimate value 12 lb, the forces in wires GL and HM, proportional to their elongations, will be 6 lb and 3 lb, respectively,

---

[1] This paper was edited by M. de la Hire in 1686, after Mariotte's death. See also the second volume of Mariotte's collected works (2d ed., The Hague, 1740).

[2] Figure 20 is taken from Mariotte's collected works.

and the ultimate load R will be only $5\frac{1}{4}$ lb and not 7 lb, as it was in the preceding case.

Mariotte uses similar reasoning in considering the bending of a cantilever beam (Fig. 20f). Assuming that, at the instance of fracture, the right-hand portion of the beam rotates with respect to point D, he concludes that the forces in its longitudinal fibers will be in the same proportion as their distances from D. From this it follows that in the case of

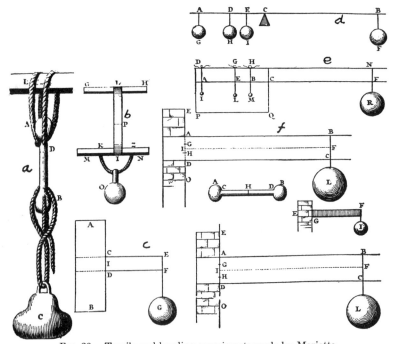

FIG. 20. Tensile and bending experiments made by Mariotte.

a rectangular beam, the sum of these forces will be equal to $S/2$ (that is, only half of the absolute strength of the beam in tension) and that their moment with respect to D will be $S/2 \times \frac{2}{3}h = Sh/3$, where h is the depth of the beam. Equating this to the moment Ll of the applied load L, we find that the ultimate load is

$$L = \frac{Sh}{3l} \qquad (a)$$

Thus, by taking the deformation of fibers into consideration and using the same point of rotation D, as Galileo did, Mariotte finds that the ultimate load in bending L is to the absolute strength S as $(h/3):l$. This means that the ultimate load is equal to only two-thirds of the value calculated by Galileo.

Now, Mariotte goes further with his analysis and, referring again to the rectangular beam (Fig. 20f), he observes that the fibers in the lower portion ID of the cross section are in compression, while the fibers in the upper portion IA are in tension. To calculate the load L, which is required to overcome the resistance of the fibers in tension, he uses Eq. (a) by putting $h/2$, instead of h, which gives

$$L_1 = \frac{Sh}{6l} \quad (b)$$

Considering the compressed fibers in the lower portion ID of the cross section, Mariotte assumes that the same law of force distribution holds as in the case of tension and that the ultimate strength is the same. Hence the contribution to the strength of the beam made by the compressed fibers will also be equal to L_1, and is given by Eq. (b). The total strength will be given by the previously established equation (a). We see that in his analysis Mariotte used a theory of stress distribution in elastic beams which is satisfactory. His assumption regarding the force distribution in fibers is correct, but in calculating the moment of the tensile forces with respect to point I it is necessary not only to substitute $h/2$ for h in Eq. (a), but also to use $S/2$ instead of S. This error prevented Mariotte from arriving at the correct formula for the failure of beams, the material of which follows Hooke's law up to fracture.

To verify his theory, Mariotte experimented with wooden cylindrical bars of $\frac{1}{4}$ in. in diameter. The tensile test gave the absolute strength as $S = 330$ lb. Testing the bar as a cantilever beam[1] of length $l = 4$ in., he found the ultimate load equal to $L = 6$ lb, which gives $S:L = 55$, while Eq. (a) gives[2] $S:L = 48$ and Galileo's theory gives $S:L = 32$. Mariotte tries to explain the discrepancy between his experimental results and those predicted by Eq. (a) as being due to a "time effect." He says that the specimen in tension might well fracture under a load of 300 lb if the load acted for a sufficiently long time. When he repeated the experiments with rods of glass, Mariotte again found that his formula [Eq. (a)] gives a more accurate forecast than does that of Galileo.

This French physicist also conducted experiments with beams supported at both ends, and he found that a beam with built-in ends can carry, at its center, twice the ultimate load for a simply supported beam of the same dimensions.

[1] Mariotte mentions in his paper that the tests were made in the presence of Roberval and Huyghens.
[2] Note that Eq. (a) is derived for a rectangular cross section but is used by Mariotte for a circle also.

By means of a very interesting series of tests, Mariotte found the bursting strength of pipes under internal hydrostatic pressure. For this purpose he used a cylindrical drum AB (Fig. 21) to which a long vertical pipe was attached. By filling the drum and the pipe with water and increasing the height of water level[1] in the latter, he was able to burst the drum. In this way he deduced that the required thickness of pipes must be proportional to their internal pressure and to the diameter of the pipe.

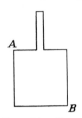

FIG. 21. Cylindrical drum used in Mariotte's bursting tests.

Dealing with the bending of uniformly loaded square plates, Mariotte correctly establishes, on the strength of similarity considerations, that the total ultimate load on the plate remains constant and independent of the size of the plate, if the thickness is always the same.

We see that Mariotte considerably enhanced the theory of mechanics of elastic bodies. By introducing considerations of elastic deformation, he improved the theory of bending of beams and then used experiments to check his hypothesis. Experimentally, he checked some of Galileo's conclusions regarding the way in which the strength of a beam varies with the span. He investigated the effects on the strength of a beam brought about by clamping its ends and gave a formula for the bursting strength of pipes.

[1] The height of water in some experiments approached 100 ft.

CHAPTER II

Elastic Curves

6. The Mathematicians Bernoulli[1]

The Bernoulli family originally lived in Antwerp, but, because of the religious persecution of the Duke of Alba, they left Holland and, toward the end of the sixteenth century, settled in Basel. Starting near the end of the seventeenth century this family produced outstanding mathematicians for more than a hundred years. In 1699, the French Academy of Sciences elected the two brothers Jacob and John Bernoulli as foreign members, and up to 1790 there were always representatives of the Bernoulli family in that institution.

During the last quarter of the seventeenth and the beginning of the eighteenth centuries a rapid development of the infinitesimal calculus took place. Started on the Continent by Leibnitz (1646–1716),[2] it progressed principally by the work of Jacob and John Bernoulli. In

FIG. 22. Jacob Bernoulli.

trying to expand the field of application of this new mathematical tool, they discussed several examples from mechanics and physics. One such example treated[3] by Jacob Bernoulli (1654–1705) concerned the shape of the deflection curve of an elastic bar and in this way he began an

---

[1] For biographies, see "Die Mathematiker Bernoulli" by Peter Merian, Basel, 1860.

[2] Newton developed the fundamentals of calculus independently in England, but on the Continent Leibnitz's method of presentation and his notation were adopted and used in the quick growth of that branch of mathematics.

[3] After some preliminary discussion of the problem, printed in Leibnitz's publication "Acta Eruditorum Lipsiae," 1694, he put his final version of the problem in the "Histoire de l'Académie des Sciences de Paris," 1705. See also "Collected Works of J. Bernoulli," vol. 2, p. 976, Geneva, 1744.

important chapter in the mechanics of elastic bodies. Whereas Galileo and Mariotte investigated the strength of beams, Jacob Bernoulli made calculations of their deflection; he did not contribute to our knowledge of the physical properties of materials. Following Mariotte's assumption regarding the position of the neutral axis, he took the tangent to the boundary of the cross section on the concave side perpendicular to the plane of action of the external loads. Considering a rectangular beam built-in at one end and loaded at the other by a force P, he takes the deflection curve as shown in Fig. 23. Let $ABFD$ represent an element of the beam the axial length of which is ds. If, during bending, the cross

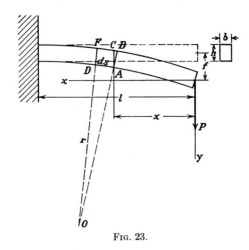

Fig. 23.

section AB rotates with respect to the cross section FD about the axis A, the elongation of the fibers between the two adjacent cross sections is proportional to the distance from axis A. Assuming Hooke's law and denoting the elongation of the outermost fiber on the convex side by Δds, we find that the resultant of the tensile forces in all fibers of the cross section AB is

$$\frac{1}{2} \frac{m \, \Delta ds}{ds} bh \tag{a}$$

where bh is the cross-sectional area and m is a constant depending on the elastic properties of the material of the beam. The moment of this resultant with respect to the axis A must be equal to the moment Px of the applied load with respect to the same axis, and we obtain the equation

$$\frac{1}{2} \frac{m \, \Delta ds}{ds} bh \cdot \frac{2}{3} h = Px \tag{b}$$

Observing now that

$$\frac{\Delta ds}{ds} = \frac{h}{r}$$

we put Eq. (b) in the form

$$\frac{C}{r} = Px \qquad (c)$$

where

$$C = \frac{mbh^3}{3}$$

Due to Jacob Bernoulli's erroneous assumption regarding the axis of rotation of the cross section AB, we have found his incorrect value for the constant C. However, the general form of Eq. (c), stating that the curvature of the deflection curve at each point is proportional to the bending moment at that point, is correct and it was used later on by other mathematicians (principally Euler) in their investigations of elastic curves.

John Bernoulli (1667–1748), the younger brother of Jacob, was considered to be the greatest mathematician of his time. As a result of his teaching, the first book on calculus was written by the Marquis de l'Hôpital in 1696. The original lectures of John Bernoulli on differential calculus were published by Naturforschende Gesellschaft of Basel in 1922, on the occasion of the three hundredth anniversary of the Bernoullis having attained citizenship of Basel. It was John Bernoulli who formulated the principle of virtual displacements in his letter to Varignon.[1] Although he was interested in the elastic properties of materials, his contribution in that field was of little importance.[2] Much more important contributions to strength of materials were made by John Bernoulli's son Daniel and his pupil L. Euler.

Daniel Bernoulli (1700–1782) is best known for his famous book "Hydrodynamica," but he also contributed to the theory of elastic curves. He suggested to Euler that he should apply the variational calculus in deriving the equations of elastic curves by remarking in a letter, "Since no one is so completely the master of the isoperimetric method (the calculus of variations) as you are, you will very easily solve the following problem in which it is required that $\int ds/r^2$ shall be a minimum."[3] This integral, as we know now, represents the strain energy of a bent bar neglecting a constant factor. Euler's work, based upon this suggestion, will be discussed later (see page 32).

[1] See "Nouvelle Mécanique" by Varignon, vol. 2, p. 174, Paris, 1725.

[2] Elasticity is discussed in the first three chapters of John Bernoulli's book, "Discours sur les loix de la communication du Mouvement," Paris, 1727.

[3] See P. H. Fuss, "Correspondance Mathématique et Physique," letter 26, vol. II, St. Petersburg, 1843.

Daniel Bernoulli was the first to derive the differential equation governing lateral vibrations of prismatical bars and he used it to study particular modes of this motion. The integration of this equation was done by Euler and it will be discussed later (see page 35), but Daniel Bernoulli made a series of verifying experiments and, about his results, he writes to Euler: "These oscillations arise freely, and I have determined various conditions, and have performed a great many beautiful experiments on the position of the knot points and the pitch of the tone, which agree beautifully with the theory."[1] Thus Daniel Bernoulli was not only a mathematician but also an experimenter. Some of his experiments furnished new mathematical problems for Euler.

7. Euler (1707–1783)

Leonard Euler[2] was born in the vicinity of Basel. His father was the pastor of the neighboring village of Riechen. In 1720, Euler entered the University of Basel, which at that time was a very important center of mathematical research, since the lectures of John Bernoulli attracted young mathematicians from all parts of Europe. The young student's mathematical talents were soon noticed, and John Bernoulli, in addition to his usual lectures, gave him private lessons weekly. At sixteen, Euler obtained his master's degree, and before he was twenty he had participated in an international competition for a prize offered by the French Academy of Sciences and had published his first scientific paper.

Fig. 24. Leonard Euler.

The Russian Academy of Sciences was opened in 1725 at St. Petersburg. The two sons of John Bernoulli, Nicholas and Daniel, accepted invitations to become members of the new institute. After settling in Russia, they helped Euler to obtain a position as an associate there. In the summer of 1727, Euler moved to St. Petersburg and, free from any other duties, he was able to put all his energy into mathematical research. He became a member of the

---

[1] See P. H. Fuss, letter 30, vol. 2.
[2] See "Leonard Euler" by Otto Spiess, Leipzig. See also Condorcet's eulogy, printed in "Lettres de L. Euler à une Princesse D'Allemagne," Paris, 1842.

Academy in the department of physics in 1730 and in 1733, when Daniel
Bernoulli left St. Petersburg after his brother's death (in 1726) and
returned to Basel, Euler took his place as the head of the department of
mathematics. During the time he was at the Russian Academy at St.
Petersburg, Euler wrote his famous book on mechanics,[1] and in it, instead of employing the geometrical methods used by Newton and his pupils, Euler introduced analytical methods. He showed how the differential equations of motion of a particle can be derived and how the motion of the body can be found by integrating these differential equations. This method simplified the solution of problems, and the book had a great influence on subsequent developments in mechanics. Lagrange states in his "Mécanique analytique" (1788) that Euler's book was the first treatise on mechanics in which the calculus was applied to the science of moving bodies.

About the time of the publication of his book, Euler became interested in elastic curves. Also, from the correspondence between Euler and Daniel Bernoulli, it can be seen that the latter drew Euler's attention to the problem of the lateral vibration of elastic bars and to the investigation of the corresponding differential equation.

Frederick II (Frederick the Great) became the king of Prussia in 1740. He was interested in science and philosophy and wanted to have the best scientists at the Prussian Academy. By that time Euler was recognized as an outstanding mathematician and the new king invited him to become a member of the Berlin Academy. Since there was political unrest in Russia at that time, Euler accepted the offer and in the summer of 1741 he moved to Berlin. He retained some connection with the Russian Academy and continued to publish many of his memoirs in the *Commentarii Academiae Petropolitanae*.[2] In Berlin, Euler continued his mathematical research and his papers appeared in the annual publications of the Prussian and Russian academies.

In 1744 his book "Methodus inveniendi lineas curvas . . . " appeared. This was the first book on variational calculus and it also contained the first systematic treatment of elastic curves. This will be discussed later.

While at Berlin, Euler wrote his "Introduction to Calculus" (1748), "Differential Calculus" (2 vols., 1755), and "Integral Calculus" (3 vols.), the last of these being published in St. Petersburg (1768–1770). All these books guided mathematicians for many years, and it can be said that all

---

[1] "Mechanica sive motus scientia analytice exposita," 2 vols., St. Petersburg, 1736, German translation by J. P. Wolfers, Greiswald, 1848 and 1850.

[2] Memoirs of the Russian Academy.

the eminent men of mathematics living toward the end of the eighteen century and at the beginning of the nineteenth were Euler's pupils.[1]

After the death of Maupertuis (1759), Euler was in charge of the academy and, because of this, he had a considerable amount of executive work. He had also to look for money for the upkeep of the academy during the difficult period of the Seven Years' War. In 1760, Berlin was occupied by the invading Russian army, and Euler's house was pillaged. When the Russian commander, General Totleben, was informed about this, he immediately apologized to Euler and ordered that he be payed compensation. The Russian Empress Elizabeth sent an additional sum which more than recompensed the mathematician.

Catherine II became the empress of Russia in 1762. She favored scientific research and progress and wanted to improve the Russian Academy of Sciences. Very soon negotiations were started with Euler with a view to his returning to St. Petersburg. Catherine II was able to make a better offer than Frederick II, and thus, in 1766, after twenty-five years work in Berlin, Euler went back to St. Petersburg. He was given a very kind reception at the imperial court and Catherine II presented him with a house. He was once more freed from financial difficulties and was able to devote all his time to scientific work. He was approaching sixty and his eyesight was very poor. He had lost one eye in 1735 and, since a cataract was starting to form in the other, he was approaching complete blindness. However, this did not stop his activity, and the number of papers he wrote each year in this latter period of his life was greater than ever. To accomplish this, Euler had assistants to whom he gave complete explanations of new problems and the methods of solution which would have to be used. With this information the assistants were able to go ahead with the work and, after further discussion with the master, they would put the results into writing for Euler's final approval. More than 400 papers were produced by the old man during this last period (1766-1783) of his life, and more than forty years after his death the Russian Academy of Sciences was still publishing his papers in its annual memoirs.

8. Euler's Contribution to Strength of Materials

Euler, as a mathematician, was interested principally in the geometrical forms of elastic curves. He accepted Jacob Bernoulli's theory that the curvature of an elastic beam at any point is proportional to the bending moment at that point, without much discussion. On the basis of that assumption, he investigated the shapes of the curves which a slender

[1] Condorcet in his eulogy states: "Tous les mathématiciens célèbres qui existent aujourdhui sont ses éleves: il n'en est aucun qui ne se soit formé par la lecture de ses ouvrages. . . . "

elastic bar will take up under various loading conditions. The principal results of Euler's work in this direction are to be found in his book "Methodus inveniendi lineas curvas . . . ,"[1] mentioned before. He approaches the problem from the point of view of variational calculus, the subject with which he deals in that book. Introducing this method, Euler observes: "Since the fabric of the universe is most perfect, and is the work of a most wise Creator, nothing whatsoever takes place in the universe in which some relation of maximum and minimum does not appear. Wherefore there is absolutely no doubt that every effect in the universe can be explained as satisfactorily from final causes, by the aid of the method of maxima and minima, as it can from the effective causes themselves . . . . Therefore, two methods of studying effects in Nature lie open to us, one by means of effective causes, which is commonly called the direct method,

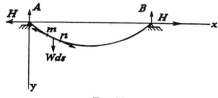

Fig. 25.

the other by means of final causes . . . . One ought to make a special effort to see that both ways of approach to the solution of the problem be laid open; for thus not only is one solution greatly strengthened by the other, but, more than that, from the agreement between the two solutions we secure the very highest satisfaction." To illustrate the two methods, Euler mentions the problem of a catenary. If a chain is suspended at A and B (Fig. 25), we can obtain the curve of equilibrium by using the "direct method." We then consider the forces acting on one infinitesimal element mn of the curve and write equations of equilibrium of those forces. From these equations the required differential equation of the catenary follows. But for the same purpose we can also use the "method of final causes" and attack the problem with a consideration of the potential energy of the gravity forces. From all those geometrically possible curves the required one is that which gives this potential energy a minimum value, or what is equivalent, the curve of equilibrium is that in which the chain's center of gravity occupies the lowest position. In this way the problem reduces to that of finding the maximum of the integral

---

[1] An English translation of the Appendix to this book, containing the study of elastic curves, was made by W. A. Oldfather, C. A. Ellis, and D. M. Brown. See *Isis*, vol. XX, p. 1, 1933 (reprinted in Bruges, Belgium). See also the German translation in "Ostwald's Klassiker," nr. 175.

$\int_0^s wy\, ds$ when the length s of the curve is given, w being the weight per unit length. Applying the rules of variational calculus we arrive at the same differential equation as before.

Taking the case of an elastic bar, Euler mentions that the "direct method" of establishing the equation of elastic curves was used by Jacob Bernoulli (see page 27). To use "the method of final causes," Euler needs the expression for strain energy, and here he uses the information given to him by Daniel Bernoulli. He states: "The most illustrious and, in this sublime fashion of studying nature, most perspicacious man, Daniel Bernoulli, had pointed out to me that he could express in a single formula, which he calls the potential force, the whole force which inheres in a curved elastic strip, and that this expression must be a minimum in the elastic curve," and then continues (according to Bernoulli) "if the strip be of uniform cross section and elasticity, and if it be straight, when in its natural position, the character of the curve will be such that in this case the expression $\int_0^s ds/R^2$ is an absolute minimum." Using his variational calculus, Euler obtains Jacob Bernoulli's differential equation for elastic curves which, in the case shown in Fig. 23, becomes

$$C \frac{y''}{(1 + y'^2)^{\frac{3}{2}}} = Px \qquad (a)$$

Since Euler does not limit his discussion to the consideration of small deflections, the term y'^2 in the denominator cannot be neglected and the equation is a complicated one. Euler integrates it by series and shows that if the deflection f (see Fig. 23) is small, Eq. (a) gives

$$C = \frac{Pl^2(2l - 3f)}{6f} \qquad (b)$$

If we neglect the term $3f$ in the numerator, the usual formula for the deflection of the end of a cantilever is obtained, viz.,

$$f = \frac{Pl^3}{3C} \qquad (c)$$

The neglected term allows for the fact that, due to deflection, the length l is somewhat smaller than the initial length of the bar.

Euler does not discuss the physical significance of the constant C which he calls the "absolute elasticity," merely stating that it depends on the elastic properties of the material and that, in the case of rectangular beams, it is proportional to the width and to the square of the depth h. We see that Euler was mistaken in assuming that C is proportional to h^2, instead of h^3. His recommendation that Eq. (b) should be used for an experimental determination of C was followed up by many experimenters.[1]

Euler considers the various cases of bending shown in Fig. 26,[2] and classifies the corresponding elastic curves according to values of the angle

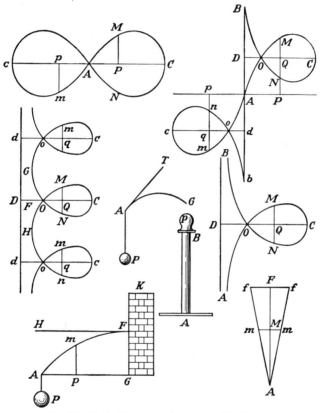

FIG. 26. Deflection curves investigated by Euler.

between the direction of the force P and the tangent at the point of application of the load. When this angle is very small, we have the important case of columns buckling under the action of an axial compressive force. Euler shows (see the column AB in Fig. 26) that here the equation of the elastic curve can readily be solved and that the load at

---

[1] See, for example, P. S. Girard's work (p. 7).
[2] This figure is taken from Euler's original work.

which the buckling occurs is given by the equation

$$P = \frac{C\pi^2}{4l^2} \quad (d)$$

He states: "Therefore, unless the load P to be borne be greater than $C\pi^2/4l^2$, there will be absolutely no fear of bending; on the other hand, if the weight P be greater, the column will be unable to resist bending. Now when the elasticity of the column and likewise its thickness remain the same, the weight P which it can carry without danger will be inversely proportional to the square of the height of the column; and a column twice as high will be able to bear only one-fourth of the load." We see that, two hundred years ago, Euler established the formula for the buckling of columns which now has such a wide application in analyzing the elastic stability of engineering structures.

Euler also considers bars of variable cross section and, as an example, discusses the deflection of a cantilever (Fig. 26), the rigidity of which is proportional to the distance x.

Again, Euler treats the bending of bars which have some initial curvature $1/R_0$ and states that, in this case, Eq. (a) must be replaced by

$$C\left(\frac{1}{R} - \frac{1}{R_0}\right) = Px \quad (e)$$

that is, for initially curved bars, the change of curvature at each point is proportional to the bending moment. He shows that if the initial curvature $1/R_0$ is constant along the length of the bar, Eq. (e) can be treated in the same manner as that for straight bars considered previously. He discusses also the following interesting problem: What must be the initial form of a cantilever if a load, applied at its end, is to make it straight?

When a distributed load acts on a beam, such as the weight of the beam or a hydrostatic pressure, Euler shows that the differential equation of the elastic curve will be of the fourth order. He succeeds in solving this equation for the case of hydrostatic pressure and obtains the deflection curve in an algebraic form.

Further on in Euler's book, we find a treatment of the lateral vibration of bars. By limiting his discussion to the case of small deflections, he states that he is able to take d^2y/dx^2 as the curvature of the deflected beam and writes the differential equation of the curve in the same form as it is used now. To eliminate the effect of gravity forces he assumes that the vibrating bar AB is vertically built-in at the end A (Fig. 27a) and considers the motion of an infinitesimal element mn of weight $w\,dx$. He observes that this motion is the same as that of a simple isochronous

Elastic Curves

pendulum (Fig. 27b), and the force pulling the element toward the x axis must be the same as that in the case of the pendulum, i.e., for small oscillations, it is equal to $wy\ dx/l$. Euler reasons: "It follows that, if to the single element mn of the strip, equal forces $wy\ dx/l$ should be applied in the opposite direction (in the direction of y axis) the strip in the position $AmnB$ would be in a state of equilibrium. Hence the strip, while oscillating, will assume the same curvature which it would take on when at rest, if at the individual elements mn it should be acted upon by the forces $wy\ dx/l$ in the direction y." In this way Euler

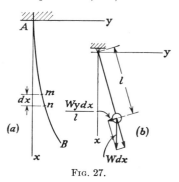

Fig. 27.

arrives at the conclusions which we now reach by applying D'Alembert's principle.[1] Having the distributed load Euler now obtains the required differential equation of motion by twice differentiating the equation

$$C\frac{d^2y}{dx^2} = M$$

and observing that the second derivative of M is equal to the intensity of lateral load. This gives

$$C\frac{d^4y}{dx^4} = \frac{wy}{l} \qquad (f)$$

and Euler concludes: "By this equation, therefore, the character of the curves $AmnB$ is expressed, and from that, if it be adapted to the case presented, the length l (of the equivalent pendulum) will be determined. That being known, the oscillatory motion itself will become known." After this he integrates Eq. (f) and, using the assumed conditions at the ends of the strip, finds the frequency equation from which the frequencies of the consecutive modes of vibration can be calculated.

Euler does not limit his analysis to the case of a cantilever but also discusses the transverse motion of bars (1) with simply supported ends, (2) with built-in ends, and (3) with both ends entirely free. For all these cases he established a formula for the frequency f of the type

$$f = m\sqrt{\frac{\pi^4 Cg}{wl^4}} \qquad (g)$$

---

[1] D'Alembert's book "Traité de Dynamique" was published in 1743, but it was unknown to Euler when he was writing the appendix "De Curvis Elasticis" to his book "Methodus inveniendi . . . ."

where m is a number depending upon the conditions at the ends of the bar and on the mode of vibration. In conclusion Euler states that Eq. (g) can be used not only for an experimental check of his theory but also that it offers a practical means of determining the "absolute elasticity" C of the strip.

In 1757, Euler again published work concerning the problem[1] of the buckling of columns. Here he gives a simple derivation of the formula for the critical load by using the simplified differential equation

$$C \frac{d^2y}{dx^2} = -Py$$

He gives here a more satisfactory discussion of the quantity C and concludes that it must have the dimensions of a force multiplied by the square of a length. In subsequent papers Euler extends his analysis to cover columns of variable cross section and also treats the problem of a column with an axial load distributed along its length, but he does not arrive at correct solutions for these more complicated problems.

Jacques Bernoulli (1759-1789)[2] refers to Euler's paper "Von dem Drucke eines mit einem Gewichte beschwerten Tisches auf eine Flaeche" in which we have the first treatment of a statically indeterminate problem. It was solved under the assumption that the table top remains plane.

Euler also explored the theory of the deflection and vibration of a perfectly flexible membrane. Considering a membrane as being composed of two systems of strings which are perpendicular to each other, he derived the partial differential equation[3]

$$\frac{\partial^2 z}{\partial t^2} = a^2 \frac{\partial^2 z}{\partial x^2} + b^2 \frac{\partial^2 z}{\partial y^2}$$

This notion was used later by Jacques Bernoulli who, in his investigation of bending and vibration of a rectangular plate, considered the plate as a lattice of two systems of beams and obtained an equation of the form[4]

$$k \left(\frac{\partial^4 z}{\partial x^4} + \frac{\partial^4 z}{\partial y^4} \right) = Z$$

This he used to explain Chladni's experimental results for the vibration of plates (see page 119).

[1] Sur la force des colonnes, *Mem. Acad. Berlin*, vol. 13, 1759.
[2] Nephew of Daniel Bernoulli. He perished in the River Neva.
[3] *Novi Comm. Acad. Petrop.*, vol. 10, p. 243, 1767. Euler used this idea again in studying the vibrations of bells (see the same publication, p. 261).
[4] *Nova acta*, vol. 5, 1789, St. Petersburg.

9. Lagrange (1736–1813)

Lagrange was born in Turin[1] and his father was a rich man who lost his property through some poor speculations. Young Lagrange later considered that this loss had some redeeming features, since, had he been rich, he might not have taken up mathematics. Early on, he showed exceptional mathematical ability and at the age of nineteen he had already become a professor of mathematics at the Royal Artillery School in Turin.

With a group of pupils he founded a society which later became the Turin Academy of Sciences. In the first volume of the publications of that institute (which appeared in 1759) several of Lagrange's memoirs were printed, and between them they constitute his important work in variational calculus. This common interest brought him into correspondence with Euler. Euler's very high opinion of Lagrange's work led him to nominate the latter as a foreign member of the Berlin Academy and Lagrange was elected in 1759.

FIG. 28. J. L. Lagrange.

In 1766, he was invited to replace Euler at that academy on the recommendation of Euler and D'Alembert and moved to Berlin. Here he found excellent working conditions and soon published a long series of important papers.

At that time, he also prepared his famous "Mécanique analytique." In it, using D'Alembert's principle, and the principle of virtual displacements, Lagrange introduced the notions of "generalized coordinates" and "generalized forces" and reduced the theory of mechanics to certain general formulas from which the necessary equations in any particular problem could be derived. In the preface Lagrange states that there are no figures in his book because the methods which he uses do not require geometrical or mechanical considerations, but only algebraic operations which have to follow a prescribed order. In Lagrange's hands, mechanics became a branch of analysis which he called the "geometry of four dimensions." There were few people at that time who could appreciate such presentation of mechanics, and Lagrange had difficulty in finding a pub-

[1] See the biography of Lagrange by Delambre, "Oeuvres de Lagrange," vol. I, Paris, 1867.

lisher for that work. Eventually, it was published in Paris in 1788, a hundred years after Newton's "Principia" appeared.

After the death of Frederick the Great the conditions for scientific work in Berlin deteriorated and, no longer finding the same appreciation as before, Lagrange moved to Paris in 1787. He found a warm reception awaiting him in the French capital, was lodged in the Louvre, and received a grant equal to that which he had hitherto enjoyed. However, as a result of overwork, Lagrange completely lost interest in mathematics and for two years his printed volume of the "Mécanique" lay unopened. During this period he showed some interest in other sciences, especially in chemistry, and also participated in the work of the commission which was debating the introduction of the metric system into France.

This was the time of the French Revolution and the revolutionary government began purging the membership of the commission. Several great scientists, such as the chemist Lavoisier and the astronomer Bailly, were executed, and Lagrange planned to leave the country. But at that time a new school, the École Polytechnique, was opened and Lagrange was asked to lecture on the calculus at the new institute. This activity revived his interest in mathematics and his lectures began to attract not only students but also teachers and professors. As a result of these lectures he wrote two books, "Fonctions analytiques" and "Traité de la Résolution des équations numériques." During the last years of his life Lagrange was occupied with a revision of his book on mechanics, but he died in 1813, when only about two-thirds of this work was done. The second volume of the revised edition appeared after his death.

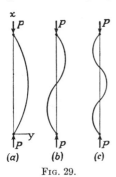

Fig. 29.

The most important contribution made by Lagrange to the theory of elastic curves is his memoir "Sur la figure des colonnes."[1] He begins with a discussion of a prismatic bar having hinges at the ends (Fig. 29) and assumes there to be a small deflection under the action of the axial compressive force P. This leads him to the equation

$$C \frac{d^2y}{dx^2} = -Py \qquad (a)$$

which had already been discussed by Euler (see page 36). He shows that the solution of this equation

$$y = f \sin \sqrt{\frac{P}{C}} x$$

[1] See "Oeuvres de Lagrange," vol. 2, p. 125.

satisfied the end conditions only if

$$\sqrt{\frac{P}{C}} \, l = m\pi$$

where m is an integer. From this it follows that the load at which a very small bending of the column may occur is given by the equation

$$P = \frac{m^2\pi^2 C}{l^2} \qquad (b)$$

Thus it is possible to have an infinite number of buckling curves. To produce the curve with one half wave, as in Fig. 29a, we have to apply a load four times larger than that calculated by Euler for the case where one end is built-in. To give the curve in Fig. 29b, the required load is sixteen times Euler's load, and so on. Lagrange does not limit himself to a calculation of the critical values of the load P but goes on to investigate the deflections which will exist if the load P exceeds the critical value. For this purpose he uses the equation containing the exact expression for the curvature instead of the approximate one in Eq. (a) and integrates by series obtaining

$$l = \frac{m\pi}{\sqrt{P/C}}\left[1 + \frac{Pf^2}{4(4C)} + \frac{9P^2f^4}{4\cdot 16(16C^2)} + \frac{9\cdot 25 P^3 f^6}{4\cdot 16 \cdot 36(64C^3)} + \cdots\right]$$

For $f = 0$, this equation gives formula (b). For small values of f, the series converges rapidly and the load corresponding to a given deflection can readily be calculated.

Continuing, Lagrange considers columns of variable cross section (the columns being solids of revolution) and inquires how to find that curve which, by revolution about an axis, generates the column of greatest efficiency. As a measure of efficiency Lagrange takes the ratio of the critical load P to the square of the volume V of the column. For curves which have the same curvature at both ends of the column and have tangents at the ends parallel to the axis of the column, Lagrange concludes that the column of greatest efficiency has a cylindrical form. He arrives at the same conclusion by considering curves which pass through four points taken at equal distances from the axis. Thus Lagrange did not succeed in getting a satisfactory solution to the problem of the form of the column of maximum efficiency. Later on the same problem was discussed by several other authors.[1]

In a second memoir,[2] "Sur la force des ressorts pliés," Lagrange dis-

[1] See Clausen, *Bull. phys.-math. acad. Sci.* (*St. Petersburg*), vol. IX, 1851; E. Nicolai, *Bull. Polytech. Inst. St. Petersburg*, vol. 8, 1907; Blasius, *Z. Math. u. Physik*, vol. 62.

[2] "Oeuvres de Lagrange," vol. 3, p. 77.

cusses the bending of a uniform strip built-in at one end and loaded at the other. He makes the usual assumption that the curvature is proportional to the bending moment and discusses several instances which might be of some interest in studying flat springs such as are used in watches. The form of Lagrange's solution is too complicated for practical application.

Although Lagrange's contributions to strength of materials are of more theoretical than practical interest, his method of generalized coordinates and generalized forces later found applications in strength of materials and proved of great value in solving problems of practical importance.

CHAPTER III

Strength of Materials in the Eighteenth Century

10. Engineering Applications of Strength of Materials

During the seventeenth century, scientific investigation developed principally in the hands of men working in academies of science. Few people were interested in the mechanics of elastic bodies and, although Galileo, Hooke, and Mariotte considered some questions of elasticity and of the strength of structures raised by practical problems, scientific curiosity was the principal motive power in their work. During the eighteenth century, the scientific results of the preceding hundred years found practical applications and scientific methods were gradually introduced in various fields of engineering. New developments in military and structural engineering required not only experience and practical knowledge, but also the ability to analyze new problems rationally. The first engineering schools were founded, and the first books on structural engineering were published. France was ahead of other countries in this development, and the study of the mechanics of elastic bodies progressed during the eighteenth century principally due to scientific activity in that country.

In 1720, several military schools were opened in France for the training of experts in fortifications and artillery, and, in 1735, Belidor (1697–1761) published a textbook on mathematics[1] for use in those schools. Not only mathematics but also its applications in mechanics, geodesy, and artillery are discussed by the writer. Although Belidor includes only elementary mathematics in this work, he recommends those of his pupils who are mathematically inclined, to study calculus also and mentions the book "Analyse des Infiniment Petits" by the Marquis de L'Hôpital, the first book on calculus ever to be published. To appreciate how rapidly the application of mathematics was developing, we have only to think that, at the end of the seventeenth century, only four men (Leibnitz, Newton, and the two brothers Bernoulli) were working on the calculus and were familiar with this new branch of mathematics.

[1] Belidor, "Nouveau Cours de Mathématique a l'Usage de L'Artillerie et du Génie," Paris, 1735.

42   History of Strength of Materials

In 1729, Belidor's book "La Science des Ingénieurs" was published. This book enjoyed great popularity amongst structural engineers and was reprinted many times. The last edition, with notes added by Navier, appeared in 1830. In this book there is a chapter dealing with strength of materials. The theory here does not go beyond the results obtained by Galileo and Mariotte, but Belidor applies it to his experiments with wooden beams and gives the rules for determining the safe dimensions of beams. In these calculations, Belidor shows that the then established practice of choosing the sizes of beams is not satisfactory and he recommends a more rational method of approach in solving the problem. For that purpose he uses Galileo's statement that the strength of a rectangular beam is proportional to the width and to the square of the depth of the cross section. In conclusion, he expresses the opinion that not only simple beams but also more complicated systems of bars such as are used in the construction of roofs and bridge trusses can be analyzed and that methods of selecting safe dimensions can be established.

In 1720, the Corps of Engineers of Ways of Communications was established by the French government and, in 1747, the famous school École des Ponts et Chaussées was founded in Paris for training engineers in construction work on highways, channels, and bridges. This school, as we shall see later, played a great part in the development of our science.[1] The first director of this school, Jean-Rodolphe Perronet (1708–1794), was a famous engineer who designed and constructed several large arch bridges, the channel of Bourgogne, and many important structures in Paris. His memoirs[2] were widely read by structural engineers. The chapter of these memoirs which deals with the testing of structural materials will be discussed later (see page 58).

Toward the end of the eighteenth century (in 1798) the first book on strength of materials by Girard was published.[3] A historical introduction in this book is of great interest, for it contains a discussion of the principal investigations on the mechanics of elastic bodies which were made in the seventeenth and eighteenth centuries. In discussing bending of beams, Girard considers Galileo's and Mariotte's methods of analysis and it seems that, at that time, both theories were used. In the case of brittle materials, such as stone, engineers used Galileo's hypothesis which assumes that, at the instance of fracture, the internal forces are uniformaly distributed over the cross section. For wooden beams, Mariotte's hypoth-

[1] The history of this famous school can be found in *Ann. ponts et chaussées*, 1906; see the article by de Dartein.

[2] Déscription des projets de la construction des ponts, etc., "Oeuvres de Perronet," Paris, 1788.

[3] "Traité Analytique de la Résistance des Solides" by P. S. Girard, ingenieur des Ponts et Chaussées, Paris, 1798.

esis was preferred, by which it was assumed that the intensity of internal forces varies from zero at the concave side to the maximum tension at the most remote fiber (on the convex side). However, Girard clearly states that the fibers on the concave side are in compression and that on the convex side they are in tension. He sees that the axis with respect to which the moment of internal forces should be calculated lies in the cross section, but at the same time he upholds Mariotte's mistaken impression that the position of the neutral axis is immaterial. Thus, in the theory of strength of beams, Girard's book does not improve upon Mariotte's work. Girard follows Euler's work very closely in connection with beam deflections and does not limit his derivations to small deflections. Consequently he arrives at complicated formulas which are not satisfactory for practical application. In his earlier work, Euler assumed that the flexural rigidity of a beam is proportional to the cube of its linear dimensions. Later on, he corrected this assumption and took flexural rigidity as being proportional to the fourth power. Girard does not investigate this point and used Euler's incorrect initial assumption.

In the second part of the book, Girard deals with beams of equal strength. For such structures he has to satisfy the equation

$$\frac{Mh}{I} = \text{const}$$

at every cross section, where M is the bending moment, I is the moment of inertia of the cross section with respect to the neutral axis, and h is the depth of the beam. Girard shows that, for a given load, we can find a variety of shapes for these beams of equal strength, depending on the manner in which we vary the cross-sectional dimensions. In the eighteenth century this problem was very popular and there are many papers dealing with it. But this kind of work did not contribute much to the more practically important theory of the strength of beams.

The third part of Girard's book deals with experiments on the bending and buckling of wooden columns. It represents Girard's original work in the bending of beams and will be discussed later (see page 58).

From this brief discussion, it will be appreciated that in solving problems on the strength of beams, engineers of the eighteenth century used the theories of seventeenth century. But, at the same time, a more satisfactory theory of bending of beams took shape during this period and this will be discussed in the following articles.

11. *Parent*

In the last chapter we saw how the mathematicians of the eighteenth century developed the theory of elastic curves by taking the curvature proportional to bending moment. They did this work without inquiring how the stress is distributed in the beam. While this mathematical work progressed, investigations were made dealing with the physical aspect of the bending problem and these resulted in a better understanding of the distribution of stresses. Concerning these investigations we shall take the work of Parent (1666–1716). Parent was born in Paris.[1] His parents wanted him to study law, but his interest was in mathematics and the physical sciences. After his graduation, Parent never practiced law. He spent most of his time studying mathematics and lived by giving lessons in this subject. In 1699, Des Billettes was elected to membership of the French Academy and he brought Parent with him as an élève (assistant). This position at the Academy gave Parent the opportunity to meet French scientists and to participate in the meetings of the academy. Here he had a chance to show his great knowledge in various fields of science and he had several memoirs published in the Academy volumes. Not all of his memoirs were accepted for these publications and, in 1705, Parent started his own journal in which he printed his papers and also reviewed the work of other mathematicians. In 1713 these papers were reedited in three volumes and printed as his "Recherches de Mathématique et de Physique."

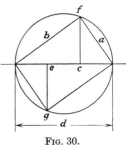

Fig. 30.

In his first memoirs[2] dealing with the bending of beams, Parent follows Mariotte's assumption and takes, as the neutral axis, the tangent to the boundary at the concave side perpendicular to the plane of action of the loads. He uses this theory for finding various forms of beams of equal strength. He also discusses the interesting problem of how a rectangular beam must be cut from a circular trunk in order to have the maximum strength and shows that, for a given diameter d (Fig. 30), the product ab^2 must have the maximum value. This is obtained by dividing the diameter into three equal parts and erecting the perpendiculars cf and eg (as shown).

In 1713, Parent published two memoirs[3] on the bending of beams, which represent an important step forward. In the first of these, he shows that the equation

$$L = \frac{Sh}{3l} \quad (a)$$

[derived by Mariotte (see page 22) for rectangular beams and used later for other shapes of cross sections] cannot be applied to circular tubes or to

---

[1] See the article in "Histoire de l'Académie des Sciences," Paris, 1716.

[2] See "Histoire de l'Académie des Sciences," 1704, 1707, 1708, 1710, Paris.

[3] "Essais et Recherches de Mathématique et de Physique," vol. 2, p. 567; vol. 3, p. 187.

solid circular beams. Assuming, with Mariotte, that the cross section rotates about the tangent nn (Fig. 31) and that the forces in the fibers are proportional to the distances from this tangent, Parent finds that, for a solid circular cross section, the ultimate moment of these forces with respect to the axis nn is $5Sd/16$, where S is the "absolute strength" of the beam. This indicates that, in circular beams, $5d/16$ (instead of $h/3$) must be substituted in the above equation (a).

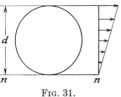

Fig. 31.

In the second memoir, Parent gives a very important discussion regarding the position of the axis about which the moment of resisting fiber forces must be calculated. Considering a rectangular beam built-in at the cross section AB (Fig. 32a) and carrying the load L, he assumes first that rotation occurs about the axis B. The corresponding stresses, shown in Fig. 32b, can be replaced by the resultant F. Continuing this force to its intersection at E with the line of action of the load L, he concludes, from equilibrium considerations, that the resultant R of the forces F and L must pass through the axis of rotation B. Now he states that a single point B has not sufficient resistance to serve as a support for this

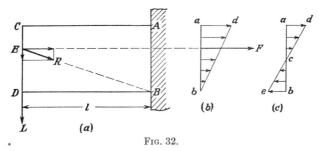

Fig. 32.

resultant and concludes that a considerable portion of the cross section AB must serve as a support for R and work in compression. Without doubt, he says, this consideration forced Mariotte to use the variation shown in the two equal triangles of Fig. 32c instead of the stress distribution depicted by the triangle abd (Fig. 32b). Taking this second mode of stress distribution and observing that at the instant of fracture the tension in the most remote fiber must always be equal to σ_{ult}, he concludes that the resisting moment, represented by the two triangles, is only half as great as that represented by one triangle of Fig. 32b. Thus he corrects the mistake made by Mariotte and upheld later by such men as Jacob Bernoulli and Varignon.[1] We know that the distribution of stresses

[1] See Varignon's article in "Histoire de l'Académie des Sciences," Paris, 1702.

represented by two equal triangles (Fig. 32c) is correct only so long as the material of the beam follows Hooke's law and it cannot be used for calculating the ultimate value of the bending load L. From Mariotte's experiments, Parent knows that the ultimate resisting couple has a magnitude between the predictions of the stress distributions of Figs. 32b and 32c. He overcomes this difficulty by assuming that, at the instant of fracture, the axis of rotation (neutral line) does not pass through the center of the cross section and the stress distribution is like that shown in Fig. 33. The maximum tension ad is always the same and represents the ultimate tensile strength of fibers. Replacing the stresses acting on the cross section ab by their resultants and adding the tensile force F and the external load L together as shown in the figure, Parent concludes from the condition of equilibrium that the resultant compressive force acting on the portion bc of the cross section ab must be equal to the resultant tensile force F acting on the upper portion of the same cross section. He also notes that in addition to normal forces, a shearing force of magnitude L will act on the cross section ab. We see that the statical problem in the bending of beams is completely solved by Parent, and he clearly shows that the resisting forces distributed over the built-in cross section ab must constitute a system of forces balancing the external load. In

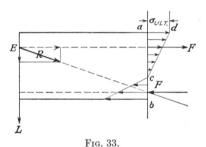

Fig. 33.

one of his previous papers,[1] Parent remarks that the neutral line may shift during the increase of load and approach the tangent to the boundary at the concave side at the instant of fracture. Now, having the stress distribution shown in Fig. 33, Parent uses it in Mariotte's experimental results and shows that his (Parent's) ultimate resisting couple coincides with that found experimentally if the neutral axis is placed so that $ac:ab = 9:11$.

In taking a triangular stress distribution as shown in Fig. 33, Parent assumes that Hooke's law holds and that the modulus in tension is different from the modulus in compression. He also considers the case where the material does not follow that law and correctly observes that the ultimate resisting couple becomes smaller than that represented by the two triangles in Fig. 33 if the material is such that strain increases in a smaller proportion than the stress. Having no experimental information regarding the mechanical properties of materials, Parent was naturally unable

[1] See his memoirs, vol. 2, pp. 588–589.

Strength of Materials in the Eighteenth Century

to develop further his fundamentally correct ideas regarding this problem. From this brief discussion of Parent's work it can be seen that he had much clearer ideas regarding the distribution of stresses in beams than his predecessors. However, his work remained unnoticed, and during the eighteenth century the majority of engineers continued to use formulas based on Mariotte's theory. The cause of this may perhaps lie in the fact that Parent's principal results were not published by the Academy and appeared in the volumes of his collected papers which were poorly edited and contain many misprints. Moreover, Parent was not a clear writer, and it is difficult to follow his derivations. In his writing he was very critical of the work of other investigators, and doubtless this made him unpopular with scientists of his time. Sixty years passed after the appearance of Parent's work before further progress was made in the science of mechanics of elastic bodies. At the end of this period, we find the outstanding work of Coulomb.

12. Coulomb

C. A. Coulomb (1736–1806) was born at Angoulême.[1] After obtaining his preliminary education in Paris, he entered the military corps of engineers. He was sent to the island of Martinique where, for nine years, he was in charge of the various works of construction which led him to study the mechanical properties of materials and various problems of structural engineering. While on this island, he wrote his famous paper "Sur une Application des Régles de maximis et minimis à quelques problèmes de statique relatifs à l'architecture" which was presented in 1773 to the French Academy of Sciences.[2] In the preface to this work Coulomb says: "This memoir written some years ago, was at first meant for my own individual use in the work upon which I was engaged in my profession. If I dare to present it to this Academy it is only because the feeblest endeavors are kindly welcomed by it when they have a useful objective. Besides, the Sciences are monuments consecrated to the public good.

FIG. 34. C. A. Coulomb.

---

[1] See J. B. J. Delambre, Éloge historique de Coulomb, *Mém. inst. natl. France*, vol. 7, p. 210, 1806. See also S. C. Hollister, "The Life and Work of C. A. Coulomb," *Mech. Eng.*, 1936, p. 615.

[2] Published in *Mém. acad. sci. savants étrangers*," vol. VII, Paris, 1776.

Each citizen ought to contribute to them according to his talents. While great men will be carried to the top of the edifice where they can mark out and construct the upper storeys, ordinary artisans who are scattered through the lower storeys or are hidden in the obscurity of the foundations should seek only to perfect that which cleverer hands have created."

After his return to France, Coulomb worked as an engineer at Rochelle, the Iste of Aix, and Cherbourg. In 1779, he shared (with Van Swinden) the prize awarded by the Academy for a paper on the best way to construct a compass; 1781 saw him win the Academy prize for his memoir[1] "Théorie des machines simples," in which the results of his experiments on the friction of different bodies slipping on one another (dry, or coated with greasy substances) were presented. After 1781, Coulomb was stationed permanently at Paris, where he was elected to membership of the Academy and was able to find better facilities for scientific work. He turned his attention to researches in electricity and magnetism. For the measurement of small electric and magnetic forces, he developed a very sensitive torsion balance and in connection with this work he investigated the resistance of wire to torsion.[2]

At the outbreak of the French Revolution in 1789, Coulomb retired to a small estate which he possessed at Blois. In 1793, the Academy was closed, but two years later it reappeared under the new name of L'Institut National des Sciences et des Arts. Coulomb was elected one of the first members of this new institution and his last papers dealing with the viscosity of fluids and with magnetism were published in the *Mémoires de l'Institut* (1801, 1806). Coulomb was appointed one of the general inspectors of studies in 1802 and devoted much of his energy to the betterment of public education. This activity required much traveling, which was too strenuous for his age and weak health, and he died in 1806. His work remains, and we are still using his theories of friction, strength of structural materials, and of torsion.

No other scientist of the eighteenth century contributed as much as Coulomb to the science of mechanics of elastic bodies. The principal advances are included in his paper of 1773.

The paper begins with a discussion of experiments which Coulomb made for the purpose of establishing the strength of some kind of sandstone. For his tensile tests, Coulomb uses square plates 1 ft by 1 ft by 1 in. thick and gives the test specimens the form shown in Fig. 35a.[3]

---

[1] See Mém. présentés par savants étrangers, vol. X, p. 161. This memoir, together with the previously mentioned one relating to the theory of structures and several others of interest to mechanical engineers, was reedited in the form of a book, "Théorie des machines simples," 1821, Paris.

[2] "Recherches théoriques et experimentales sur la force de torsion et sur l'élasticité des fils de métal," *Mém. acad. sci.*, 1784.

[3] This figure is taken from Coulomb's work.

Strength of Materials in the Eighteenth Century

In this way he finds the ultimate strength in tension to be 215 lb per in.[2] For testing the same material in shear, he uses rectangular bars 1 by 2 in. and applies a shearing force P at the built-in cross section ge (see Fig. 35b). He finds that the ultimate strength in shear is equal to the ultimate strength in tension in this case. Finally, he carries out bending

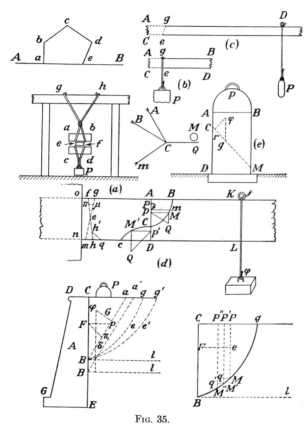

Fig. 35.

tests (Fig. 35c), using bars 1 in. in depth, 2 in. in width, and 9 in. in length. He finds that the ultimate load P is equal to 20 lb.

After this, Coulomb gives a theoretical discussion of the bending of beams (Fig. 35d). Taking a rectangular cantilever and considering a cross section AD, he concludes that the fibers of the upper portion AC of the cross section will work in tension while the fibers of lower part will be compressed. Resolving the forces in the fibers into horizontal and vertical components (as shown by vectors PQ and $P'Q$) and applying the three equations of statics, he concludes that the sum of horizontal forces, the distribution of which is given along AD by the curve BCc, must vanish. The sum of vertical components must be equal to the applied load ϕ and the moment with respect to the axis C of all forces in the fibers must be equal to the moment of the applied load ϕ with respect to the same axis. He notes that these equations are independent of the law governing the forces and elongations of the fibers. Taking a perfectly elastic material which follows Hooke's law up to fracture, and considering an element $ofhn$ at the built-in end, he assumes that, due to the bending action of the load ϕ, the plane fh will take up a position gm and the small triangles fge and emh will represent the deformations and the stresses of the fibers. If σ is the maximum tension at point f, he finds that the moment of internal forces is $\sigma(\overline{fh})^2/6$.[1] Denoting the ultimate strength of the bar in tension by S and the arm of the load ϕ and the depth of the beam by l and h, the equation for calculating the ultimate load becomes

$$\frac{Sh}{6} = \phi l \qquad (a)$$

He notes that the effect of shearing forces on the strength of the beam can be neglected if its depth is small in comparison with its length.

Considering now materials which are absolutely rigid and assuming rotation about point h and that a uniform distribution of stresses exists over cross sections, Coulomb concludes that the equation for calculating the ultimate load is

$$\frac{Sh}{2} = \phi l \qquad (b)$$

Applying this equation to the previously mentioned experiments, he finds that the calculated ultimate load is somewhat larger than that given by experiment. He concludes that the point of rotation cannot be at h but must be at some point h' (see Fig. 35d) so that the portion hh' of the cross section will be in compression.

Thus we see that, in his theory of bending, Coulomb used equations of statics correctly in analyzing internal forces and had clear ideas regarding the distribution of these forces over the cross section of the beam. It seems that Parent's work was unknown to him. For in his analysis, Coulomb refers only to C. Bossut[2] who, in his work "La Construction la plus avantageuse des digues," recommends that wooden beams be treated as elastic and stone beams as if absolutely rigid.

[1] The width of the beam is taken equal to unity.
[2] C. Bossut (1730–1814) was professor of mathematics at the military engineering school at Mézières and a member of the Academy. He published a three-volume course in mathematics (1765), a two-volume course of hydrodynamics (1771), and a history of mathematics (1810) in two volumes which are of great interest.

As his next problem, Coulomb contemplates the compression of a prism by an axial force P (Fig. 35c). He assumes that fracture is due to sliding along a certain plane CM and that it occurs when the component of force P along this plane becomes larger than the cohesive resistance in shear along the same plane. On the strength of his previously found experimental results, Coulomb thinks that the ultimate strength in shear, τ_{ult}, is equal to the strength σ_{ult} in tension. Denoting then the cross-sectional area of the prism by A and the angle between the axis of the prism and the normal to the plane CM by α, he obtains an equation for calculating the ultimate load. It is

$$P \sin \alpha = \frac{\sigma_{ult} A}{\cos \alpha} \tag{c}$$

from which

$$P = \frac{\sigma_{ult} A}{\sin \alpha \cos \alpha} \tag{d}$$

This equation shows that the smallest value of P is obtained when $\alpha = 45°$ and that this value is twice as large as the ultimate strength of the column in tension. To bring the theory into better agreement with experimental results, Coulomb proposes that, not only should cohesive resistance along the CM plane be considered, but also friction. Thus, he takes, instead of Eq. (c), the equation

$$P \sin \alpha = \frac{\sigma_{ult} A}{\cos \alpha} + \frac{P \cos \alpha}{n} \tag{e}$$

where $1/n$ is the coefficient of friction. From this equation,

$$P = \frac{\sigma_{ult} A}{\cos \alpha [\sin \alpha - (\cos \alpha / n)]} \tag{f}$$

Selecting α so as to make P a minimum, he finds

$$\tan \alpha = \frac{1}{\sqrt{(1 + 1/n^2)} - 1/n} \tag{g}$$

Assuming that for brick $1/n = \frac{3}{4}$, Coulomb finds that $\tan \alpha = 2$, and thus obtains $P = 4\sigma_{ult} A$ from Eq. (f). This agrees well with his experimental results.

In the second part of the same paper (of 1773), Coulomb considers the stability of retaining walls and arches, but we shall postpone our discussion of this till later.

In 1784, Coulomb produced his memoir on torsion, and we now pass on to an examination of its contents. Coulomb determines the torsional rigidity of a wire by observing the torsional oscillations of a metal cylinder suspended by it (Fig. 36a).[1] Assuming that the resisting torque of a twisted wire is proportional to the angle of twist ϕ, he obtains the

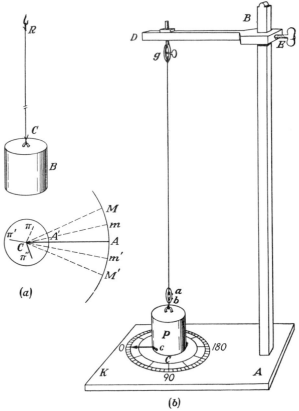

Fig. 36. Coulomb's device for torsional vibration tests.

differential equation

$$n\phi = -I\ddot{\phi}$$

and, by integration, finds the formula for the period of vibration, viz.,

$$T = 2\pi \sqrt{I/n} \qquad (h)$$

By experiment, Coulomb finds that this period is independent of the angle of twist, if the angle is not too large, and concludes that the assumption of the proportionality of the torque to the angle of twist is correct. Now Coulomb proceeds with his experiments, taking wires of the same

[1] This figure is taken from Coulomb's memoir.

material but of different lengths and diameters. In this way, he establishes the following formula for the torque M:

$$M = \frac{\mu d^4}{l} \phi \qquad (i)$$

where l is the length of the wire, d its diameter, and μ a constant of the material. Comparing steel and brass wires, Coulomb finds that the ratio of these constants is $3\frac{1}{3}:1$, and hence concludes that where we need great rigidity, as in pivots, the use of steel is preferable.

After establishing the fundamental equation (i), Coulomb extends his investigation of the mechanical properties of the materials of which the wires are made. For each kind of wire he finds the torsional limit of elasticity, beyond which some permanent set occurs. Also, he shows that, if the wire is twisted initially far beyond the elastic limit, the material becomes harder and its elastic limit is raised whereas the quantity μ in Eq. (i) remains unchanged. Again, by annealing, he can remove the hardness produced by the plastic deformation. On the strength of these experiments Coulomb asserts that, in order to specify the mechanical characteristics of a material, we need two quantities, viz, μ defining the elastic property of the material and the elastic limit which depends upon the magnitude of the cohesive forces. By cold work or quenching we can increase these cohesive forces and so raise the elastic limit, but we cannot change the elastic property of the material defined by the constant μ. To demonstrate that this conclusion is applicable to other kinds of deformation, Coulomb makes bending tests with steel bars which differ only in their heat-treatment and shows that for small loads they give the same deflection (irrespective of their thermal history) but that the elastic limit of the annealed bars is much lower than that of the quenched ones. Thus at larger loads the annealed bars develop considerable permanent set while the heat-treated metal continues to be perfectly elastic, *i.e.*, the heat-treatment changes the elastic limit but leaves the elastic property of the material unaltered. Coulomb frames the hypothesis that each elastic material has a certain characteristic arrangement of molecules which is not disturbed by small elastic deformations. Beyond the elastic limit some permanent sliding of molecules takes place, resulting in an increase of the cohesive forces though the elastic property remains unaffected.

Coulomb also discusses the damping of torsional oscillations and shows experimentally that the principal cause of it is not air resistance, but some imperfection of the material of his wires. Where the motion is small, he finds that the amount by which the amplitude is diminished per cycle is approximately proportional to the amplitude. But where the swings are larger (the elastic limit being raised by cold work), the damping effect increases at a higher rate than the amplitude and the test results become more erratic.

13. Experimental Study of the Mechanical Properties of Structural Materials in the Eighteenth Century

Mention has already been made of the experimental work of such scientists as Hooke, Mariotte, and Coulomb. Their experiments were carried out chiefly in order to verify certain theories of mechanics of materials. But independently of these, a considerable amount of experimental work of a more practical nature was completed during the eight-

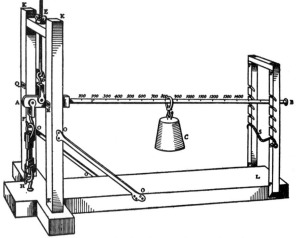

FIG. 37. Musschenbroek's tensile-test machine.

eenth century, since information was needed regarding the strength of the materials used in structural engineering.

Réaumur (1683–1757) relied upon mechanical testing in studying various technological processes involved in steel manufacture.[1] He made tensile tests of wire to evaluate the effects of various kinds of heat-treatment and developed a method of measuring hardness by studying the indentations produced by pressing the corners of two perpendicular triangular prisms together.

An extensive series of mechanical tests of various materials was made by Petrus van Musschenbroek (1692–1761) who was professor of physics at Utrecht and later at Leiden. In his work " Physicae experimentales et

[1] During the years 1720–1722, Réaumur presented several memoirs to the French Academy of Sciences. See also his book "L'art de convertir le fer forgé en acier," Paris, 1722.

geometricae" (1729), he describes his methods of testing and the apparatus which he designed for this purpose. Figure 37 shows his tensile testing machine while Fig. 38 indicates the types of specimen he used and the methods of grasping the ends. Although the lever system gave him a considerable magnification of the force, he could produce fracture of wood and steel with his machine only when the cross sections of the test pieces were small. The results of his tests were incorporated in his book on physics[1] and they were widely used by engineers. Figure 39 represents Musschenbroek's machine for making bending tests. By using wooden rectangular bars this physicist verified Galileo's theory that the strength of beams in bending is proportional to bh^2. Using the results which he obtained with small test specimens, Musschenbroek shows how the ultimate load for large beams, such as are used in structures, can be calculated.

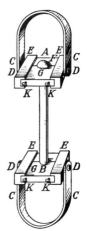

Figure 40 shows Musschenbroek's apparatus for carrying out compression tests of struts. This was the first time that a practical investigation of the phenomenon of lateral buckling was undertaken, and, as a result of his research, that writer makes the important statement that the buckling load is inversely proportional to the square of the length of the strut. As we have already seen, Euler came to the same conclusion using his mathematical analysis of elastic curves.

FIG. 38. Method of grasping the ends of tensile-test specimens.

The experiments made by Musschenbroek were criticized by Buffon (1707–1788). The latter is known principally for his work in the natural sciences, but his interests seem to have embraced many subjects. Although he was the director of the botanical garden in Paris and the organizer of the museum of natural sciences, Buffon translated Newton's "Methods of Fluxions" into French and made very extensive studies of the mechanical properties of wood.[2] His investigations showed that the strength of wood specimens taken from the same trunk varies very considerably with the location from which they are taken, both as regards distance from the axis of the trunk and also the distance along it. Thus experiments with small specimens such as those used by Musschenbroek

[1] See the French translation of the book, "Essai de Physique," Leiden, 1751. From the interesting preface of this book, it may be seen that problems of physics attracted the interest of greatly different groups of people in those days and that scientific societies were organized in several cities of Holland.

[2] See Buffon, "Histoire Naturelle," vol. 2, p. 111, Paris, 1775.

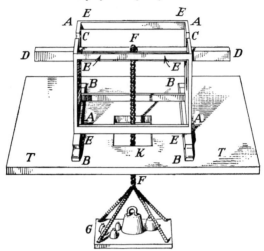

Fig. 39. Musschenbroek's machine for bending tests.

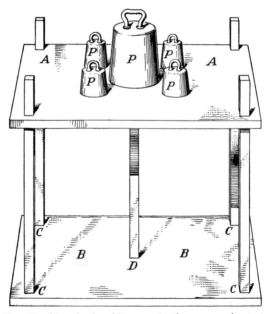

Fig. 40. Musschenbroek's apparatus for compression tests.

do not give the structural engineer all the information that he needs and tests on full size beams are necessary. Buffon made many of these tests using square beams of 4-by-4 in. to 8-by-8 in. cross sections and with spans up to 28 ft. The beams were simply supported at the ends and loaded at their centers, and the results confirmed Galileo's assertion that the strength of a beam is proportional to the width of the cross section and to the square of its height. They showed also that the strength of beams can be assumed proportional to the density of the material.

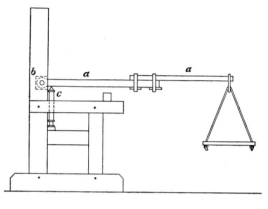

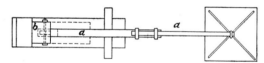

FIG. 41. Gauthey's testing machine.

In connection with the construction of the church of Sainte Geneviève in Paris, questions as to the proper cross-sectional dimensions of pillars arose, and the opinions of the leading French architects and engineers were divided. It became imperative that tests on the compression strength of various kinds of stone be made. To make the necessary tests, the French engineer Gauthey (1732–1807), who was the author of the well-known books on bridges,[1] designed and constructed a special machine, shown in Fig. 41. The contrivance uses the lever principle and is somewhat similar to that used by Musschenbroek for tensile tests (see Fig. 37). Specimens were used in the form of cubes usually having

[1] Gauthey, "Traité de la Construction des Ponts," 1809–1813. This treatise was edited, after Gauthey's death, by his nephew Navier and will be discussed later, together with Navier's work (see page 71).

sides 5 cm long. Comparing his test results with the compressive loads which stones had to carry in some existing structures, Gauthey found that (assuming central action of compressive force) the factor of safety was usually not smaller than 10. He explains this comparatively low load on stone structures by the fact that, in actual buildings, the load may act with some eccentricity and may not be normal to the plane on which it acts. He also mentions some experiments which showed that the compressive strength of elongated specimens is usually much smaller than that obtained from cubes.

Rondelet (1734–1829) improved Gauthey's machine by replacing the axis b (Fig. 41) with a knife-edge, thus reducing the friction in the apparatus. A description of this machine appears in Rondelet's work, "L'Art de Batir," Paris, 1802.

A considerable amount of experimental work dealing with the strength of stones was undertaken by Perronet. This was brought about by the construction of the arch bridge at Neuilly-sur-Seine. He designed and installed a machine similar to that of Gauthey at the École des Ponts et Chaussées and added a means for making tensile tests. In his capacity of professor and structural engineer at the school, Perronet made a large number of tests.

During the closing years of the eighteenth century, a series of investigations with wooden struts was made. This work was started by J. E. Lamblardie (1747–1797) who succeeded Perronet as director of the École des Ponts et Chaussées and who helped to organize the famous École Polytechnique. It was continued by P. S. Girard, the author of the first book on strength of materials (see page 42). The machine constructed for these experiments is shown in Fig. 42. The vertical strut is compressed by the use of the lever XY rotating about the axis V and loaded at y. The circular disks at the upper end of the specimen slide between the verticals AB and EF and prevent the end from lateral movement. The lower end of the specimen is supported at R. The same machine was also used for lateral bending tests when the lower end R' of the strut produces pressure on the center of the horizontally placed beam. In order to make a comparison of strut-test results with the predictions of Euler's column formula, the required flexural rigidity of the struts was obtained experimentally, as Euler suggested (see page 33). These experiments showed that wooden struts were very far from being perfectly elastic. The deflections during lateral bending were not proportional to the load and were not constant for a given load but increased with the time of action of the force. The means of fastening the ends of the struts and the method of applying the load were open to criticism, so that no satisfactory agreement between the test results and Euler's theory was obtained.

In St. Petersburg, G. B. Bülfinger did some experiments to check Galileo's and Mariotte's theories of bending.[1] He finds that Mariotte's theory is the better for explaining the experimental results. Bülfinger

FIG. 42. Girard's experiments with wooden struts.

thinks that Hooke's law is not borne out by the experiments and suggests a parabolic relation

$$\epsilon = a\sigma^m$$

where m is a constant to be determined experimentally. The experimental work that we have discussed was done primarily by French engineers and they amassed a considerable volume of test results which were of practical importance. For instance, in Gauthey's treatise on bridges

[1] *Comm. Acad. Petrop.*, vol. 4, p. 164, 1735.

14. Theory of Retaining Walls in the Eighteenth Century

The practice of using retaining walls to prevent earth from sliding is a very old one. But in the old days the problem of how to select the proper dimensions of these structures was attacked by rule-of-thumb methods. Toward finding some rational basis for a design procedure, various hypotheses were proposed by engineers during the eighteenth century for calculating the earth pressure acting on the wall. Also, some laboratory experiments were made.[1]

Belidor contributed considerably to the solution of this problem and included a chapter on retaining walls in his book (see page 42) "La Science des Ingénieurs" (1729). He notes that the horizontal force Q required

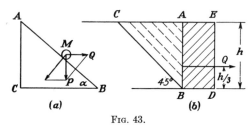

Fig. 43.

to keep a ball of weight P (Fig. 43a) in equilibrium on the inclined plane AB is equal to $P \tan \alpha$. Considering now the earth behind the retaining wall $ABDE$ (Fig. 43b) and assuming that, in the absence of the wall the unsupported earth will have the slope BC of about 45 deg, he concludes that the triangular prism of earth ABC has a tendency to slide along the plane BC. If that sliding were frictionless, as in the case of the ball, the horizontal reaction of the wall necessary to keep the prism of earth in equilibrium must be equal to the weight of that prism. But due to friction a much smaller force will be required, and Belidor proposes that the reaction of the wall may safely be taken equal to one-half of the weight of the prism. Denoting the height of the wall by h and the weight of unit volume of the earth by γ, he finds that the horizontal pressure per unit length exerted by the earth upon the wall is equal to $\gamma h^2/4$. Repeating the same reasoning for any plane parallel to BC, as shown in the figure by dotted lines, Belidor concludes that the earth pressure on the plane AB follows the triangular law and the resultant pressure Q acts at a dis-

---

[1] The complete story of the theory of retaining walls and of these experiments in the eighteenth century is told in the book by K. Mayniel, "Traité expérimental, analytique et pratique de la Poussée des Terres et de Murs de Revêtements," Paris, 1808.

tance $h/3$ from the base BD of the wall. The turning moment of this pressure with respect to the edge D, which must be considered in choosing the thickness of the wall, is then equal to $\gamma h^3/12$. Using this value of the moment in calculating the thickness, Belidor arrives at proportions which are in satisfactory agreement with the then established practice.

Further progress in the theory of retaining walls was made by Coulomb. In the memoir of 1773 (mentioned on page 47), he considers the earth pressure on the portion BC of the vertical side CE of the wall (Fig. 35) and assumes that the earth has a tendency to slide along some plane aB. Neglecting friction on CB, he concludes that the reaction H of the wall is horizontal. The weight W of the prism CBa is $(\gamma h^2 \tan \alpha)/2$, where h is the depth of the point B and α is the angle CBa. The resultant R of the reactions along the sliding plane will form the friction angle ψ with the normal to the plane Ba, that is, R will lie at an angle $\alpha + \psi$ with the horizontal. The triangle in Fig. 44 will then represent the condition of equilibrium of the wedge CBa in Fig. 35 and from it he concludes that

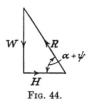

Fig. 44.

$$H = W \cot (\alpha + \psi) = \frac{\gamma h^2}{2} \tan \alpha \cot (\alpha + \psi)$$

$$= \frac{\gamma h^2}{2} \frac{1 - \mu \tan \alpha}{1 + \mu \cot \alpha} \quad (a)$$

where $\mu = \tan \psi$ is the coefficient of friction.

The next step is to select such a value for the angle α as to make the reaction H a maximum. The derivative of H [given by expression (a)] with respect to α is equated to zero, and he finds that

from which
$$\frac{1 - \mu \tan \alpha}{1 + \mu \cot \alpha} = \tan^2 \alpha \quad (b)$$

$$\tan \alpha = -\mu + \sqrt{1 + \mu^2} \quad (c)$$

Substituting this value into expression (a) he obtains the required maximum pressure on the retaining wall. In his derivation Coulomb also accounts for cohesive forces acting on the plane Ba and shows that Eq. (c) still holds. Again, the case of an extra load P (Fig. 35) is considered in Coulomb's paper.

As a second problem, Coulomb envisages the wall producing a horizontal pressure Q' on the prism CBa such that the prism moves upward. He calculates that value of the angle α which makes Q' a minimum. Finally Coulomb considers the case where sliding of the earth takes place along a curved surface such as that indicated by the line Beg in Fig. 35 and he briefly discusses the shape of the curve which would produce the maximum pressure on the wall.

Some simplification of Coulomb's method was introduced by Prony (1755–1839) to aid its practical application. Prony was born in Chamelet near Lyons and he entered l'École des Ponts et Chaussées in 1776. After graduating in 1780, he worked under Perronet on the construction of the bridge at Neuilly. In 1785, he assisted Perronet in the reconstruction of the harbor of Dunkirk and accompanied Perronet on a visit to England. At the time of the French Revolution Prony became a member of the French commission on the metric system and, when the decimal system of dividing the circle was adopted and demanded the preparation of new trigonometric tables, he was put in charge of the work. Prony was one of the founders of the famous École Polytechnique (1794) and he became the first professor of mechanics at that school. In 1798 he became the director of l'École des Ponts et Chaussées. His books on hydraulics ["Architecture hydrolique" (1790–1796, 2 vols.)] and on mechanics ["Leçons de mécanique analytique" (1810)] were widely used in French engineering schools.

To return to Coulomb's analysis, observing that $\mu = \tan \psi$, Prony threw Eq. (b) into a simpler form, viz.,

$$\cot (\alpha + \psi) = \tan \alpha$$

from which it follows that

$$\alpha = \tfrac{1}{2}(90° - \psi)$$

This indicates that the plane Ba, corresponding to the maximum earth pressure in Fig. 35, bisects the angle between the vertical BC and the line of natural slope. In this way, Prony developed a practical method of selecting the proper dimensions of retaining walls.[1]

15. Theory of Arches in the Eighteenth Century

The art of building arches is very old. The Romans used arches extensively (with semicircular spans) in the construction of their bridges and aqueducts. In fact, they were so skilled in this work that some of their structures remain intact to this day,[2] and we can study the proportions and the methods of construction that they used. There was apparently no theory for determining the safe dimensions of arches, and the Roman builders were guided only by empirical rules. In mediaeval times there

---

[1] R. Prony, "Instruction-Pratique pour déterminer les dimensions des murs de revêtement," Germinal, An X, Paris.

[2] Drawings of Roman bridges are given in Gauthey's work (previously mentioned) "Traité de la Construction des Ponts," vol. 1, 1809.

was little or no road or bridge construction, and the Roman methods became forgotten.

The advent of the Renaissance, with its resuscitation of European economic life, made it necessary for builders to learn anew the art of building arches. At first, the proportions of these structures were again taken in a purely empirical way, but toward the end of the seventeenth century and at the beginning of the eighteenth, some attempts were made by French engineers to draw up a theory of arches. France was at that time ahead of other countries in the development of its road system and, there, arch bridges were widely used in conjunction with highways.

The first application of statics to the solution of arch problems is due to Lahire[1] (1640–1718), a member of the French Academy. He learned his mathematics from Desargues and developed an interest in geometry

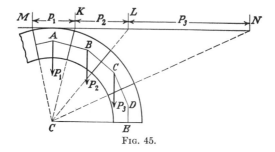

Fig. 45.

under the influence of that great mathematician. In Lahire's book "Traité de Méchanique" (1695) we find the funicular polygon used for the first time in arch analysis. He considers a semicircular arch (Fig. 45) and assumes that the surfaces of contact of the wedges forming it are perfectly smooth so that only normal pressures can be transmitted through them. With this assumption he investigates the following problem: what weights P_1, P_2, P_3, . . . must the wedges of the arch have in order to ensure stability of the structure? Lahire solves this question geometrically, as shown in Fig. 45. Drawing a horizontal tangent MN to the outer surface and extending the radii which subdivide the arch into wedges to the intersection points M, K, L, N, he finds that the requirement of stability is met if the weights P_1, P_2, . . . are in the same ratio as the lengths MK, KL, LN. He also finds that the pressures acting between the wedges of the arch are represented in such a case by the lengths CK, CL, CN. Using the language of today, we should state that the figure $ABCDE$ is the funicular polygon constructed for the

[1] The history of the development of the theory of arches is given in a very interesting memoir by Poncelet; see *Comp. Rend.*, vol. 35, p. 493, 1852.

system of vertical forces P_1, P_2, P_3, ... and that the figure $MKLNC$ represents the polygon of forces rotated by 90 deg about the pole C.

Proceeding as above, Lahire finds that the weight of the last wedge supported by the horizontal plane CE must be infinitely large, since the radius CE is parallel to the tangent MN. This showed Lahire that, under the assumption of perfect smoothness at the wedge surfaces, stability of a semicircular arch cannot exist. He remarks that in an actual arch cement will prevent the wedges from sliding, so contributing to the stability. Hence, the established ratio for $P_1:P_2:P_3$ ... need not be complied with rigorously.

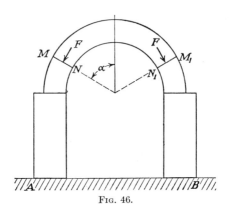

Fig. 46.

Later on, Lahire returns to the arch problem[1] and applies his theory to the determination of proper dimensions for the pillars supporting a semicircular arch (Fig. 46). Assuming that fracture occurs at the most unfavorably located positions of the arch, say at the sections MN and M_1N_1, he calculates the forces F representing the actions of the upper portion of the arch on the lower part, again neglecting friction. Then he calculates the weight of the pillars that is required to counteract the forces F and prevent rotation of pillars about points A and B.

This method of analysis was first put to practical use by Belidor who described Lahire's method in his book "La Science des Ingénieurs" and suggested that the angle α in Fig. 46 should be taken equal to 45 deg. Lahire's method was used later by Perronet and Chezy in the preparation of a table for use in calculations of the thickness of arches.[2]

Coulomb made further progress in the theory of arches. In his time,

[1] See his memoir in "Histoire de L'Académie des Sciences," Paris, 1712.
[2] M. Lesage, "Recueil des Mémoires extraits de la bibliothèque des Ponts et Chaussées," Paris, 1810.

Strength of Materials in the Eighteenth Century

it was known from experiments with models[1] that typical failures of arches are such as is shown in Fig. 47. We can conclude from this figure that in investigating arch stability, it is not sufficient only to consider relative sliding of the wedges, but also the possibility of relative rotation must be checked. In his memoir of 1773, Coulomb contemplates

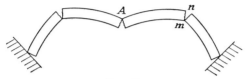

Fig. 47.

both of these types of failure. Let $ABDE$ in Fig. 48 represent one-half of a symmetrical arch which is symmetrically loaded. Denoting the horizontal thrust at the cross section AB by H and the weight of the portion $ABmn$ of the arch by Q, he finds that the normal and the tangential components of the resultant force acting on the plane mn are

$$H \cos \alpha + Q \sin \alpha$$

and

$$Q \cos \alpha - H \sin \alpha$$

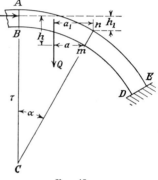

Fig. 48.

where α is the angle between this plane and the vertical AC. The smallest value of H required to prevent the portion $ABmn$ of the arch from sliding down along the plane mn is given by the equation

$$Q \cos \alpha - H \sin \alpha = \mu(H \cos \alpha + Q \sin \alpha) + \tau A \quad (a)$$

where μ is the coefficient of friction and τA the total resistance of the arch to shearing along the plane mn. From this equation he finds that

$$H = \frac{Q \cos \alpha - \mu Q \sin \alpha - \tau A}{\sin \alpha + \mu \cos \alpha} \quad (b)$$

From all possible values of the angle α it is now necessary to find that which makes expression (b) a maximum. Suppose that H is the corre-

[1] Experiments of this kind, made in 1732 by Danizy at the Academy of Montpellier, are described by Frézier in his book "Traité de la Coupe des Pierres," vol. 3, Strasbourg, 1739. Similar experiments were made on a larger scale by L. C. Boistard. See Lesage, "Recueil de divers mémoires . . . ," vol. 2, p. 171, Paris, 1808.

sponding value of the thrust. Evidently this is the smallest force required to prevent the upper portion of the arch from sliding down along the mn plane. In a similar manner he can find the value H' of the thrust which will start sliding of the portion $ABmn$ of the arch in the upward direction. This value is

$$H' = \frac{Q \cos \alpha + \mu Q \sin \alpha + \tau A}{\sin \alpha - \mu \cos \alpha} \qquad (c)$$

He now selects that value of α which makes expression (c) a minimum. Suppose that H' has the corresponding value. It is clear that to eliminate any possibility of sliding the actual thrust must be larger than H and smaller than H'.

Turning now to possible rotation about the point m, such as that illustrated in Fig. 47, and using the notation of Fig. 48, he finds the limiting value H_1 of the thrust. It is

$$H_1 = \frac{Qa}{h} \qquad (d)$$

Similarly for rotation about the point n he obtains

$$H_1' = \frac{Qa_1}{h_1} \qquad (e)$$

The maximum value of expression (d) and the minimum of expression (e) give the limits between which the value of the actual thrust must be to obviate rotation. Such is Coulomb's method of attack. From his study of practical cases, Coulomb concludes that usually the designer of an arch need concern himself only with limits (d) and (e). He suggests that the point of application of the thrust should be moved to A (Fig. 48) in order to make H_1 in expression (d) as small as possible.

We find that Coulomb does not give definite rules for designing arches but only determines limits for the thrust which is necessary for stability. For this reason, the value of Coulomb's work was not appreciated by the engineers of his time[1] but later, in the nineteenth century, when graphic methods were developed for calculating the limits (d) and (e), his ideas were used extensively by arch builders.

At the end of the eighteenth century a considerable number of tests were performed by Gauthey,[2] Boistard,[3] and Rondelet.[4] All these experiments showed the kind of failure illustrated by Fig. 47 and upheld the assumptions which were made by Coulomb in developing his theory.

[1] Practical engineers preferred at that time to use the empirical formulas of Perronet based on Lahire's theory for calculating the necessary thickness of arches.
[2] Gauthey, "Dissertation sur les dégradations du Panthéon Français," Paris, 1800.
[3] Boistard, "Recueil d'experiences et observations," Paris, 1800.
[4] Rondelet, "Art de batir," vol. 3, p. 236.

CHAPTER IV

Strength of Materials between 1800 and 1833

16. L'École Polytechnique[1]

As mentioned before (see Art. 10), several engineering schools were organized in France during the eighteenth century. During the French Revolution, instruction in those schools and also in the universities was discontinued, since the students and their professors were regarded with some suspicion by the revolutionary government. But at the same time France was at war with the European coalition and badly needed engineers to push forward the construction of fortifications, roads, and bridges and for the development of artillery. A group of scientists and engineers under the leadership of the great mathematician Gaspard Monge proposed (to the new government) that an engineering school of a new type should be organized to replace all those

Fig. 49. Gaspard Monge.

which had existed under the old régime. The proposition was officially approved in 1794, and at the end of the same year instruction started in the new school. In 1795, the school was given its present name, École Polytechnique.

The organization of the new school differed greatly from that of the schools it superseded. All the old privileges were abolished and boys of all social classes could enter the school. In order to select the best pupils,

---

[1] A very complete history of this famous school, "École Polytechnique, Livre du Centenaire 1794–1894," was published in 1895 by the alumni. See also Pinet, "l'Histoire de l'École Polytechnique," 1886.

competitive entrance examinations were introduced.[1] When the school was started, there were many idle scientists and professors in Paris who wanted to teach and a most outstanding group of men formed the teaching staff of the new school. Such famous men as Lagrange, Monge, and Prony took part by teaching mathematics and mechanics. Very soon Fourier and Poisson joined them.

The teaching system of the new school, prepared by Monge, differed substantially from previously established practice. There had been no uniformity of entrance requirements in the old schools and no joint lectures for large groups of pupils were given. The method of teaching had had an atmosphere of apprenticeship since practical engineers explained to individual students (or to small groups) how this or that type of structure must be designed and constructed. If some theoretical information in mathematics or mechanics was required, which was not familiar to the students, the necessary additional instruction would be given by the professor of engineering or by one of the students with a better mathematical training. No joint lectures in fundamental subjects such as mathematics, mechanics, or physics were given in those old schools.

The administration of the new school was based upon completely different ideas. It was assumed that the various branches of engineering require the same preparation in such general subjects as mathematics, mechanics, physics, and chemistry. It was assumed also that if a student received a good training in those fundamental sciences, it would not be difficult for him to acquire the necessary special knowledge in any particular branch of engineering. In accordance with this general notion, the first two years of the program were devoted exclusively to the fundamental sciences and condensed engineering courses were taken in the third year. Subsequently, training in engineering was discontinued and the École Polytechnique became a school for the fundamental sciences only giving the necessary preparation for students planning later to enter one of the engineering schools such as the École des Ponts et Chaussées, École de Mines, military academies, and so on.

The prime mover in the new school was Gaspard Monge (1746–1818) whose enthusiasm and capability as a teacher contributed much to the success of the school. He was born in Beaune in a family of very limited means and studied in the local school. He was so outstanding in his work that before he was sixteen he became the physics teacher. He was always interested in geometry and in making drawings. A military engineer happened to notice the boy's unusual ability and arranged for him to

---

[1] This method of selection has been continued up to our time. Each year the entrance examination of the École Polytechnique attracts the best pupils of French high schools to Paris. Only a small fraction of the number of students competing is admitted.

study at the military school of Mézières which was at that time perhaps the most renowned school in Europe. Here he came into contact with Professors Camus and Bossut, who very soon noticed his talents in mathematics and helped him to progress so rapidly in the school that, in 1768, he became a professor of mathematics at that school. It was there that he developed his flair for teaching. His lectures aroused great interest in the students amongst whom were Lazare Carnot, and Prieur. These two men became very influential at the time of the French Revolution and proved very helpful in the creation of the École Polytechnique.

At Mézières Monge evolved the completely new branch of mathematics known as descriptive geometry, which has since become a very important subject in engineering education. In 1780, Monge was elected to membership of the Academy of Sciences and, in 1783, he left the school at Mézières and moved to Paris where he became a professor at l'École de Marine. For use in this school he wrote the treatise on statics[1] which was used for many years as the textbook in several engineering schools in France.

As a progressive thinker, Monge enthusiastically supported the revolution and, at one time, he held office in the government as the secretary of the navy. Very soon, however, he became disgusted with administrative work, returned to science, and took up the organization of the École Polytechnique. Monge was not only a great scientist but also an outstanding teacher and a powerful lecturer. He believed in a lecture system whereby students could meet the outstanding scientists of their day and which allowed them to obtain the fundamentals of their subjects from men who themselves participated in the creation and development of science. The lecture system at the École Polytechnique was a great success. The greatest mathematicians of that time lectured at the school and they succeeded in imparting an interest in their subjects to the students. The enthusiasm of the teachers and students resulted in great success, and the first classes of the school gave the country a series of famous scientists and engineers including Poinsot, Biot, Malus, Poisson, Gay-Lussac, Arago, Cauchy, and Navier.

The lectures for large groups of students were interspersed with problem classes and periods set aside for drawing and working in the physics and chemistry laboratories.[2] For this kind of work the big classes were divided into groups (*brigades*) of twenty and a special instructor was attached to each group. The most outstanding of these young instructors were promoted in time to professorship, so that the school not only trained future engineers, but also professors and scientists. The

---
[1] "Traité Élémentaire de Statique," 1st ed., 1786.
[2] It seems that this was the first time that laboratory work was introduced as part of a teaching program.

advances made in mechanics of elastic bodies during the first half of the nineteenth century were accomplished principally by men who graduated from this school.

From the very beginning, the school published the *Journal Polytechnique* and it became a leading journal in mathematics. In it appeared not only the original papers of the teaching staff but also the courses of lectures which were given. This was the first time that calculus, mechanics, and physics were taught to large groups of students and for the success of this innovation it was necessary to have textbooks. Thus, the professors were obliged to prepare their lectures for publication. In this way a series of important books was written such as "Géométrie descriptive" and "Cours d'Analyse appliquée a la Géométrie" by Monge, "Leçons de Mécanique Analytique" by Prony, "Traité de Mécanique" by Poisson, "Calcul différentiel et integral" by Lacroix, and "Traité de Physique" by d'Haüy. These books were widely read not only in France but also in other countries and exerted great influence in the advancement of the fundamental sciences.

Other countries modeled their systems of engineering education upon that adopted at the École Polytechnique. For instance, its program was used as a guide by the organizers of the Polytechnical Institute in Vienna and the Polytechnical Institute in Zürich was founded by General Dufour, a former pupil of the École Polytechnique. The French system of technical education was copied in Russia and it had an influence on the establishment of the United States Military Academy at West Point.

France was the first country in which engineers were given a thorough training in the fundamental sciences. With their broad scientific education, these men proved very successful and contributed much to the development of the engineering sciences. The prestige of French engineers at that time was very high and often they were asked by foreign countries for assistance in the solution of difficult engineering problems.

17. Navier

The famous memoir of Coulomb, presented in 1773, contained the correct solutions of several important problems in mechanics of materials, but it took engineers more than forty years to understand them satisfactorily and to use them in practical applications.[1] Further progress in our science was made by Navier[2] (1785–1836) but, although he worked after Coulomb's death, his first publications did not contain the advances made

[1] Poncelet in his historical review of various theories of arches (previously mentioned) remarks: "Coulomb's memoir contains in few pages so many things that during forty years the attention of engineers and scientists was not fixed on any of them."

[2] Navier's biography and a complete list of his publications are given in the third edition of his book on strength of materials edited by Saint-Venant, Paris, 1864.

by his predecessor. Navier was born in Dijon in the family of a well-to-do lawyer. When fourteen years old, he lost his father and was taken into the house of his uncle, who was the famous French engineer Gauthey, and who paid great attention to the boy's education. In 1802, Navier passed the competitive entrance examination to enter the École Polytechnique and, in 1804, after graduating from that school, he was admitted to the École des Ponts et Chaussées, where his uncle had previously studied and taught mathematics. Gauthey used every opportunity to temper his nephew's theoretical studies with practical knowledge regarding bridge and channel construction. Thus Navier, when he graduated in 1808, was well prepared to carry out theoretical investigations of practical problems.

FIG. 50. Navier.

He immediately had an opportunity to use his knowledge and ability in important work. Gauthey, who died in 1807, was occupied during his last years with the preparation of a treatise on bridges and channels. The work was not completed, and it was Navier who did the final editing and who published the three volumes. The first volume, containing a history of bridge construction and also a description of the most important new bridges, appeared in 1809. The second was published in 1813 and the last one, dealing with the construction of channels, appeared in 1816. To bring the work up to date, Navier added editorial notes in many places. These have great historical interest since they describe the state of the mechanics of elastic bodies at the beginning of the nineteenth century. Comparing these notes with Navier's later work, we can see the progress made by our science in his time principally due to his own efforts. The note on page 18, volume 2, is especially interesting in that respect. In it, the complete theory of bending of prismatical bars is given. It may be seen that Navier has no knowledge of Parent's important memoir (see page 45) or of Coulomb's work. With Mariotte and Jacob Bernoulli, he thinks that the position of the neutral axis is immaterial and takes the tangent to the cross section on the concave side for this axis. He also thinks that Mariotte's formula (see page 22) is accurate enough for calculating the strength of beams and goes into the analysis of beam deflections. By means of some unsatisfactory assumptions, he derives an expression for the flexural rigidity consisting of two terms and suggests that bending and compression tests of bars should be used for determining the two constants appearing in his formula.

Navier did not retain these erroneous ideas for long and, in 1819, when he started his lectures on strength of materials at the École des Ponts et Chaussées, some mistakes in his theory had already been eliminated.[1] But the method of finding the position of the neutral axis remained incorrect, *i.e.*, Navier assumed that this axis must so divide the cross section that the moment of tensile stresses about it must be equal to the moment of the compressive stresses. Only in the first printed edition (1826) of his lectures is this remedied, and it is proved there that, for materials following Hooke's law, the neutral axis must pass through the centroid of the cross section.

Navier published a new edition of Belidor's "La Science des ingénieurs" in 1813 and, in 1819, he reedited the first volume of Belidor's "l'Architecture Hydrolique." In both of these books we see numerous important notes made by Navier who wanted the contents to satisfy the requirements of his time.

In 1820, Navier presented a memoir on the bending of plates to the Academy of Sciences and, in 1821, his famous paper giving the fundamental equations of the mathematical theory of elasticity appeared.

Although occupied with theoretical work and with editing books, Navier always had some practical work in hand, usually in connection with bridge engineering. Great changes in that field took place during the end of the eighteenth and the beginning of the nineteenth centuries. Up to that time stone was the chief material used for the construction of important bridges but now the use of metals was becoming more common. England was industrially the most advanced country at that time, and wide application of metals in industry was started there. John Smeaton (1724–1792)[2] was the first outstanding engineer to use cast iron to any great extent and he utilized it in his construction of windmills, water wheels, and pumps. The first cast-iron bridge was built (in 1776–1779) over the River Severn by Abraham Darby. Other countries followed England and, at the turn of the century, cast-iron bridges appeared in Germany and France. These bridges were made in the form of arches so that the material worked chiefly in compression. These new structures were not always strong enough and a number of them failed.

Engineers concluded that cast-iron bridges were not safe for large spans and started to build suspension bridges, the basic principle of which is very old. Some very old ones have been found in China and South

---

[1] See Saint-Venant's remark on p. 104 of his history of strength of materials and theory of elasticity included in his famous edition of Navier's "Résumé des Leçons . . . de la résistance des corps solides," Paris, 1864.

[2] Smeaton's biography is given in the first volume of "Reports of John Smeaton, F. R. S.," London, 1812.

Strength of Materials between 1800 and 1833

America.[1] But the first suspension bridges capable of withstanding the rigors of more modern times were erected in North America at the end of the eighteenth century. James Finley built his first suspension bridge in Pennsylvania in 1796. At the beginning of the nineteenth century there already existed a considerable number of such bridges in that state. The most important was the one over the Schuylkill River near Philadelphia. British engineers followed the Americans, and, during the first quarter of the nineteenth century, a large number of these bridges were constructed in England. The most important of them, the Menai Bridge with a center span of 550 ft, was designed and built (1822–1826) by Telford[2] (1757–1834).

The French government was very interested in the new development, and Navier was sent to England to study the art of building suspension bridges. After two visits (in 1821 and 1823) he presented his "Rapport et Mémoire sur les Ponts Suspendus" (1823), which contains not only a historical review of the field and a description of the most important existing bridges but also the theoretical methods of analyzing such structures.[3] For fifty years, this report was one of the most important books covering the design of suspension bridges and to this day it has retained some of its importance.

In 1824 Navier was elected a member of the Academy, and in 1830 he became professor of calculus and mechanics at the École Polytechnique. His lectures on these subjects, "Résumé des Leçons de Mécanique," were published, and for many years they enjoyed great popularity among French engineers.

18. Navier's Book on Strength of Materials

In 1826, the first printed edition of Navier's book on strength of materials[4] appeared, and his main achievements in that field were incorporated

---

[1] Interesting historical information regarding old metal bridges can be found in G. C. Mehrtens' lectures on metal bridges, vol. 1, Leipzig, 1908. See also A. A. Jakkula, "A History of Suspension Bridges in Bibliographical Form," Texas Agricultural and Mechanical College, 1941.

[2] See "The Story of Telford" by A. Gibb, London, 1935. See also Smiles, "Lives of the Engineers," vol. 2. Telford organized the London Institute of Civil Engineers (1821) and was the president of that institute to the end of his life.

[3] The English engineers were not much interested in theory at that time. Thus Brewster writes about Telford that "he had a singular distaste for mathematical studies, and never even made himself acquainted with the elements of geometry; so remarkable indeed was this peculiarity, that when we had occasion to recommend to him a young friend as a neophyte in his office, and founded our recommendation on his having distinguished himself in mathematics, he did not hesitate to say that he considered such acquirements as rather disqualifying than fitting him for the situation."

[4] Copies of Navier's lectures had been distributed amongst his students since 1819. Saint-Venant, who was one of Navier's students, refers to this material when he discusses some errors in Navier's early work.

in it. If we compare this book with those of the eighteenth century, we clearly see the great progress made in mechanics of materials during the first quarter of the nineteenth. Engineers of the eighteenth century used experiments and theory to establish formulas for the calculation of *ultimate* loads. Navier, from the very beginning, states that it is very important to know the limit up to which structures behave perfectly elastically and suffer no permanent set. Within the elastic range, deformation can be assumed proportional to force and comparatively simple formulas can be established for calculating these quantities. Beyond the elastic limit the relation between forces and deformations becomes very complicated and no simple formulas can be derived for estimating ultimate loads. Navier suggests that the formulas derived for the elastic condition should be applied to existing structures, which had proved to be sufficiently strong, so that safe stresses for various materials could be determined which later could be used in selecting proper dimensions for new structures.

In the first two articles of the book the author discusses simple compression and simple tension of prismatical bars and indicates that, to specify the characteristics of a material, it is not sufficient to give its ultimate strength, but that it is necessary also to state its modulus of elasticity E, which he defines as the ratio of the load per unit of cross-sectional area to the unit elongation produced.[1] Since measurements of very small elongations in the elastic range are required for determining E, Navier could draw very little information from previous experiments. Consequently he made his own tests with the iron that he used in the construction of the Pont des Invalides in Paris.[2] He determined the modulus E for that material.

In the third article, Navier discusses bending of prismatical bars and assumes from the very beginning that bending occurs in the same plane as that in which the forces act so that his analysis is valid only for beams having a plane of symmetry and which are loaded in that plane. Assuming that cross sections remain plane during bending and using three equations of statics he concludes that the neutral axis passes through the centroid of the cross section and that the curvature is given by the equation

$$\frac{EI}{\rho} = M \qquad (a)$$

in which I denotes the moment of inertia of the cross section with respect to the neutral axis. Assuming that deflections are small and

---

[1] The modulus of elasticity was first introduced in mechanics of elastic bodies by Thomas Young (see p. 92). But he defined it in a different way. Navier's definition is now generally accepted.

[2] See "Rapport et Mémoire sur les ponts suspendus," 2d ed., p. 293.

taking the x axis in the direction of the axis of the beam, he finds the relationship

$$EI \frac{d^2y}{dx^2} = M \qquad (b)$$

Since Euler's time, this equation had been used for calculating the deflection of cantilevers and of symmetrically loaded simply supported beams. Navier uses it for any kind of lateral loading of a simple beam, in which case the deflection curve is represented by different equations in different portions of the beam. To illustrate Navier's method of analysis, let us consider a beam loaded at any point C, Fig. 51. Denoting the angle which the tangent at C makes with the horizontal axis by α, he finds the expressions

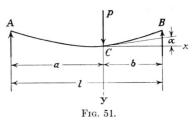

Fig. 51.

$$f_b = \frac{Pa}{l} \frac{b^3}{3EI} + b \tan \alpha \qquad f_a = \frac{Pb}{l} \frac{a^3}{3EI} - a \tan \alpha$$

for the deflections of B and A with respect to the x axis. From the equality of these deflections he concludes that

$$\tan \alpha = \frac{Pab(a-b)}{3EIl}$$

Having α and using the known deflection curve of a cantilever, the deflection curves for the two parts of the beam AB can be written.

He treats the problem of uniform loading on a portion of the beam of Fig. 51 in a similar way. In calculating the maximum stress in this case, he wrongly assumes that the maximum bending moment exists under the center of gravity of the load.

Navier was the first to evolve a general method of analyzing statically indeterminate problems in mechanics of materials. He states that such problems are indeterminate only so long as the bodies are considered as absolutely rigid, but that by taking elasticity into consideration we can always add (to the equations of statics) a number of equations expressing the conditions of deformation, so that there will always be enough relations to allow evaluation of all unknown quantities. Considering, for example, a load P supported by several bars in one plane (Fig. 52), Navier states that if the bars are absolutely rigid the problem is indeterminate. He can take arbitrary values for the forces

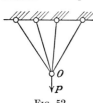

Fig. 52.

in all the bars except two and determine the forces in the last two by using equations of statics. But the problem becomes determinate if the elasticity of the bars is taken into consideration. If u and v are the horizontal and vertical components of the displacement of the point O, he can express the elongations of the bars and the forces in them as functions of u and v. Writing now two equations of statics, he finds u and v and then calculates the forces in all bars.

Examining statically indeterminate problems of bending, Navier begins with the case of a beam built-in at one end and simply supported at the

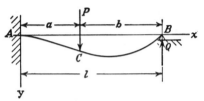

Fig. 53.

other (Fig. 53). Denoting the statically indeterminate reaction at B by Q, he obtains the following equations:

$$EI \frac{d^2y}{dx^2} = P(a - x) - Q(l - x)$$

$$EI \frac{dy}{dx} = P\left(ax - \frac{x^2}{2}\right) - Q\left(lx - \frac{x^2}{2}\right)$$

$$EIy = P\left(\frac{ax^2}{2} - \frac{x^3}{6}\right) - Q\left(\frac{lx^2}{2} - \frac{x^3}{6}\right)$$

for the portion AC of the beam and

$$EI \frac{d^2y}{dx^2} = -Q(l - x)$$

for the portion CB.

Integrating this equation and observing that at C the two parts of the deflection curve have a common tangent and equal ordinates, he finds

$$EI \frac{dy}{dx} = \frac{Pa^2}{2} - Q\left(lx - \frac{x^2}{2}\right)$$

$$EIy = P\left(\frac{a^2x}{2} - \frac{a^3}{6}\right) - Q\left(\frac{lx^2}{2} - \frac{x^3}{6}\right)$$

Since the deflection vanishes for $x = l$, he notes that

$$Q = \frac{P(3a^2l - a^3)}{2l^3}$$

from the last equation. Having this reaction, the equations for both parts of the deflection curve are readily obtained.

In a similar manner Navier analyzes a beam with both ends built-in and a beam on three supports. We see that the methods of arriving at deflection curves by integration, and of calculating statically indeterminate quantities, are completely developed by Navier. But the bending-moment and shearing-force diagrams, so widely used at the present time, are absent from his analysis, and this might well explain why, in some cases, the location of the maximum bending moment is not correctly fixed.

Navier also considers problems in which the bending of prismatical bars is produced by the simultaneous action of axial and lateral forces. After a discussion of the buckling of columns under the action of axial compression, he takes up cases of eccentric compression and tension and also that

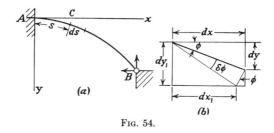

Fig. 54.

of a force acting on the end of the column making a certain angle with the axis. His formulas for calculating maximum bending moment and maximum deflection are more complicated than those for lateral loading alone, and they probably found few applications in his time. Later, however, with the increasing use of slender bars in structures, those formulas became very important and extensive tables were calculated in order to simplify their manipulation.

Navier made an important contribution to the theory of bending of curved bars in his book. Euler had already made the assumption that, when an initially curved bar is bent, the bending moment is proportional to the change in curvature. Navier takes the formula

$$EI\left(\frac{1}{\rho} - \frac{1}{\rho_0}\right) = M$$

and uses it to investigate the bending of a curved bar AB (Fig. 54a) built-in at the end A. Taking an element ds at a point C, he concludes from the above equation that the angle between the two adjacent cross sections changes, due to bending, by the amount $M\,ds/EI$. Thus the cross section C rotates during bending, through the angle

$$\delta\phi = \int^s \frac{M\,ds}{EI}$$

Due to this rotation, the initial projections dx and dy of the element ds will be changed to dx_1 and dy_1 (Fig. 54b), and we have

$$dx_1 - dx = -ds \cdot \delta\phi \cdot \sin\phi = -dy \int_0^s \frac{M\,ds}{EI}$$

$$dy_1 - dy = ds \cdot \delta\phi \cdot \cos\phi = dx \int_0^s \frac{M\,ds}{EI}$$

Integrating these formulas, he obtains the components of the displacement of any point C during bending in the following form:

$$x_1 - x = -\int_0^s dy \int_0^s \frac{M\,ds}{EI}$$

$$y_1 - y = \int_0^s dx \int_0^s \frac{M\,ds}{EI}$$

Using these equations, statically indeterminate problems of bending of curved bars can be solved. Considering, for example, a symmetrical two-hinged arch loaded at the crown (Fig. 55), we have a statically indeter-

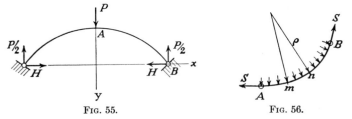

FIG. 55.   FIG. 56.

minate thrust H, whose magnitude can be found from the condition that the horizontal displacement of the hinge B vanishes. Therefore

$$\int_0^s dy \int_0^s \frac{M\,ds}{EI} = 0$$

From this equation Navier calculates the thrust H for parabolic and circular arches. He makes similar calculations for the case where the load is uniformly distributed along the span. All this analysis is done on the assumption that the length of elements, such as are shown in Fig. 54b, remains unchanged. In conclusion Navier shows how the compression produced by the axial force can be taken into account.

The last chapter of the book deals with thin shells, and it too contains some of Navier's original work. He begins with a discussion of the form of equilibrium of a perfectly flexible and inextensible string AB subjected to normal pressure in the plane of the curve (Fig. 56) and concludes (from the conditions of equilibrium of an element mn) that

$$S = \text{const} \quad \text{and} \quad \frac{S}{\rho} = p$$

where S is the tension in the string and ρ is the radius of curvature. The curvature at any point must thus be proportional to the pressure there. Taking an indefinitely long channel (Fig. 57a) which contains liquid and is supported by uniformly distributed forces S, he inquires under what conditions there will be no bending of the shell and shows that this is the case if the curvature at any point m is proportional to the depth y.

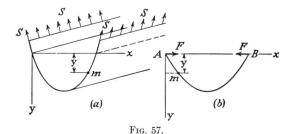

FIG. 57.

The curve satisfying this requirement is produced by bending a slender initially straight strip AB by forces F as shown in Fig. 57b, since the bending moment and the curvature of the deflection curve at any point m are evidently proportional to y.

Considering a thin shell subjected to uniform tension S and normal pressure p, Navier finds the equation

$$S\left(\frac{1}{\rho} + \frac{1}{\rho_1}\right) = p$$

from the conditions for equilibrium. From it, he finds an expression for the tensile stress in a spherical shell of thickness h, namely,

$$\sigma = \frac{S}{h} = \frac{p\rho}{2h}$$

Navier made[1] tests with thin spherical iron shells about 1 ft in diameter and 0.1 in. thick. Subjecting them to internal pressures sufficient to produce rupture, he found that the ultimate strength of the material is approximately the same as it is when obtained from simple tensile tests.

[1] See *Ann. chim. et phys.*, vol. 33, p. 225, Paris, 1826.

Navier's book had additional chapters dealing with retaining walls, arches, plates, and trusses. These will be treated later. We see that the book contains satisfactory solutions of many structural problems although, to make it complete from the present point of view, it would be necessary to add discussions of shearing stresses in the bending of a beam and of the bending of a beam in a plane which does not coincide with that in which the forces act. These two problems, as we shall see, were solved later, after Navier's death.

19. The Experimental Work of French Engineers between 1800 and 1833

The engineers of the eighteenth century were chiefly interested, when testing materials, in ultimate strength and, although they accumulated a considerable amount of data relating to failure, very little heed was taken of the elastic properties of their specimens. We shall now see that the engineers who passed through the École Polytechnique at the beginning of the nineteenth century, usually showed not only practical but also some scientific interest in their experimental work. This was due to their comprehensive training in mathematics, mechanics, and physics.

F. P. C. Dupin[1] (1784–1873) was an outstanding pupil of the new generation and he graduated from the École Polytechnique in 1803. While still in the school, Dupin showed his great capability as a mathematician and published his first paper in geometry. In 1805 he was sent, as a marine engineer, to the Ionian Islands where he was put in charge of the arsenal at Corfu. There he made his important investigations of the bending of wooden beams.[2] Testing beams which are supported at the ends he finds that, up to a certain limit, the deflections are proportional to the load. Beyond this limit the deflections increase at a greater rate and, as he shows, the load deflection relation can be represented by a parabolic curve. Using different kinds of wood, he finds that their resistance to bending increases with their specific weight. Comparing the deflection produced by a concentrated load applied to the middle of a beam with that produced by a uniformly distributed load of the same magnitude, he finds that the deflection in the second case is equal to $\frac{19}{30}$ of that in the first. We see that the ratio found experimentally is in quite close agreement with the value $\frac{5}{8}$, given by the theory.

From experiments with rectangular beams, Dupin finds that deflections are inversely proportional to the width and to the cube of the thickness. He also finds that deflections are proportional to the cube of the span. Considering geometrically similar beams of the same material, he concludes that the curvature at the centers produced by their own weights

---

[1] For Dupin's biography, see J. Bertrand, "Éloges Académiques," p. 221, 1890.

[2] See Expériences sur la flexibilité, la force et l'élasticité des bois, *J. école polytech.*, vol. 10, 1815.

is constant and that their deflections are proportional to the squares of linear dimensions. Investigating the shape of the deflection curve produced by a load applied at the middle of a beam, he finds that the curve can be represented with good accuracy by a hyperbola. From these experiments Dupin draws several conclusions regarding the strength and deflections of the hulls of wooden ships. All these results were obtained before the appearance of Navier's book on strength of materials.

A very extensive series of tests of iron and iron structures was made by A. Duleau,[1] another graduate of the École Polytechnique. In the first part of his paper, Duleau derives the necessary formulas for the bending and buckling of prismatical bars, for the bending of arches, and for the torsion of shafts. In assigning the position of the neutral axis in bending, he incorrectly assumes that the moment of tensile forces about it is equal

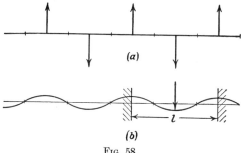

Fig. 58.

to the moment of compressive forces. Since most of his work deals with rectangular and circular beams, this mistake does not affect the validity of his conclusions. At the very beginning, he defines the modulus of elasticity in tension and compression and, assuming that during bending cross sections of a beam remain plane, he derives the differential equation of the deflection curve. He applies this equation to a cantilever and to a beam simply supported at the ends. To arrive at a formula for the deflection of a beam with built-in ends, he considers an infinitely long bar loaded as shown in Fig. 58a. Taking a portion of the resulting wavy deflection curve, as shown in Fig. 58b by the length l, he concludes that building in the ends reduces the deflection at the center to a quarter of the corresponding value for a simply supported beam of the same span.

The results of his experiments in bending of beams show very good agreement with his theory in the case of rectangular beams but, when the cross section has the form of an equilateral triangle, a smaller rigidity is found than that which the theory predicts. This discrepancy can be

[1] A. Duleau, "Essai théorique et expérimental sur la résistance du fer forgé," Paris, 1820.

explained in terms of his mistaken notion regarding the position of the neutral axis.

The next tests to be mentioned deal with axial compression of prismatical iron bars. He used very slender bars for his experiments and took pains to ensure central application of the load. In this way he reached results which show satisfactory agreement with Euler's theory.

Duleau conducted several tests with built-up beams of the type shown in Fig. 59. In calculating the flexural rigidity he uses the quantity $b(h^3 - h_1^3)/12$ for the moment of inertia of the cross section. The experiments show that, to get satisfactory agreement with theory, it is very important to prevent any sliding of the upper portion of the beam with respect to the lower. This can be accomplished by tightening the bolts. The deflections obtained experimentally with these structures were always somewhat larger than those calculated, and the difference became increasingly noticeable when the distance h_1 between the two

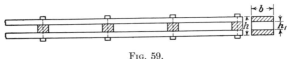

Fig. 59.

parts of the beam was increased. The cause of this inconsistency becomes clear if we observe that, in Duleau's calculations, the influence of shearing force on deflections is not accounted for. With increase of the distance h_1 this influence becomes greater while the total deflection is diminished so that relatively the deflection due to shear becomes more and more important.

Having learned from his experiments that the strength of the beam and its rigidity increase with the distance h_1, Duleau observes that it is advantageous to use beams consisting of two flanges connected by a web. He also recommends the tubular cross sections which were already being used in arches of cast-iron bridges.[1]

Duleau made his next series of tests with thin iron arches which had two hinges. He found that when the arch was loaded at the middle there was no change of curvature at the third points (in the spanwise sense). Assuming that there are hinges in those points, he developed an approximate solution of the problem and this proved satisfactory for arches of the proportions used in his experiments. An acceptable theory of bending of arches was offered later by Navier as we know (see page 77).

The final tests that Duleau carried out dealt with the torsion of prismatical iron bars. Starting with circular shafts and assuming that cross sections remain plane and that radii in those cross sections remain straight

[1] Duleau mentions that Gauthey recommended such tubular arches in 1805.

he works out a formula for the angle of twist which coincides with Coulomb's. He makes the same assumption in calculating the angle of twist of circular tubes. Here again he mentions the advantage of using tubular sections. Regarding the torsion of rectangular bars, Duleau observes that the assumptions made for circular shafts are no longer applicable. At that time it was generally accepted that torsional stresses were proportional to distance from the axis of the bar, but Duleau's experiments proved that this is not so.[1] We shall see later that Cauchy improved this theory and that the problem was finally rigorously solved by Saint-Venant.

We see that Duleau's experiments were always conducted within the bounds of the elastic limit. His material followed Hooke's law and he always tried to verify his theoretical formulas by experimentation. He considered that only this type of experiment can furnish useful information for engineers and criticized such tests as those made by Buffon (see page 55) in which the goal was always to find the ultimate load.

In conclusion, interesting and important results were arrived at by Séquin, Lamé, and Vicat during this period. The first of these famous engineers published the results of his tests of the wire which he used in erecting the first French suspension bridge.[2] Lamé's researches dealt with the mechanical properties of Russian iron,[3] while Vicat advocated tests of long duration as a protection against the phenomenon of creep, which he was the first to notice.[4] Vicat also studied the resistance of various materials to shear[5] and showed by direct test that, in short beams, the effect of shearing force on the strength becomes very important. Since he worked with short beams and used such materials as stone and brick which do not follow Hooke's law, he dealt with cases to which the simple bending theory cannot be applied satisfactorily. Thus his work was of little theoretical value except that it drew attention to the importance of shearing forces in beams.

20. The Theories of Arches and Suspension Bridges between 1800 and 1833

During the period 1800–1833, designers of arches usually followed Coulomb's theory and assumed that if an arch fails, it does so in the manner shown in Fig. 47, *i.e.*, it breaks into four pieces. The chief difficulty

---

[1] Saint-Venant in his historical review (see p. 181) states that Duleau's experimental results, obtained by twisting rectangular bars, surprised Navier.

[2] See his book "Des ponts en fil de fer," 1823; 2d ed., 1826. For a biography of this famous engineer, see Émile Picard, "Éloges et Discours Académiques," Paris, 1931.

[3] The results of his experiments can be found in Navier's book, 2d ed., p. 34.

[4] See *Ann. ponts et chaussées*, 1834 (Note sur l'allongement progressif du fil de fer soumis à diverses tensions).

[5] *Ann. ponts et chaussées*, 1833, p. 200.

of this analysis lay in finding the location of the *cross section of fracture* BC (Fig. 60). The theory supposes that the horizontal thrust H, applied at the topmost point A of the cross section AD, together with the weight P of the portion $ADBC$ of the arch and the load on that portion gives a resultant R passing through the point B. Also, the cross section BC is located such that H has its maximum value. The problem of finding the location of this cross section of fracture was usually solved by a trial-and-error method. For some assumed position of the section, the weight P and the point of its application can be found and then the corresponding value of the thrust H can be calculated from the equations of statics. To find the maximum value of H with sufficient accuracy, this calculation must be repeated several times, and since at that time the work was done analytically, it demanded the expenditures of much time.

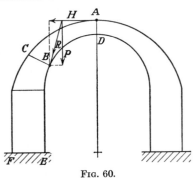

Fig. 60.

To simplify the work, tables were prepared giving the weight and the position of the center of gravity of the portion $BCAD$ of the arch for any location of BC in certain types of arches.[1] When the cross section of fracture and the corresponding value H_{max} have been found, the dimension EF of the lower portion $EFCB$ of the structure must be fixed such that the moment of the weight of the half arch $ADEF$ with respect to the axis F will withstand the action of H_{max} applied at A with a factor of safety.

Some interesting extensions of this theory were given by M. G. Lamé and E. Clapeyron[2] who, at that time, were in the service of the Russian government. Both were professors at the Institute of Engineers of Ways of Communication in St. Petersburg. They were asked to investigate the stability of cylindrical arches and of the dome of the Saint Isaac cathedral, which at that time was being constructed. For a circular arch of constant cross section, they found the position of the cross section of fracture analytically and gave a formula for H_{max}. They also showed that, for symmetrical arches of any shape, the calculation of the position of the cross section of fracture can be greatly simplified if instead of radial cross sections vertical ones are contemplated. They proved that if this were done the required cross section would be found from the condition

[1] Poncelet in his "Examen critique et historique des principales théories ou solutions concernant l'équilibre des voutes," *Compt. rend.*, vol. 35, pp. 494, 531, and 577, 1852, mentions the formulas of Audoy, used for this purpose in the military school of Metz.

[2] *Ann. mines*, vol. 8, 1823.

Strength of Materials between 1800 and 1833

that the tangent to the intrados (or inner surface) at B passes through the intersection point of the forces P and H. A very simple graphic method of locating the cross section BC follows from this fact.

Navier, in the notes he appended to Gauthey's books on bridges (see page 57), follows Coulomb's theory. But, in his "Résumé des Leçons . . . " of 1826, he adds a very important discussion of stresses in arches. Coulomb had already mentioned that the forces H and R (Fig. 60) must act at some distance from points A and B so that there will be sufficient areas for the distribution of stresses. Navier assumes that normal stresses along the cross sections AD and BC are distributed according to a linear law and that they vanish at points D and C where, according to Coulomb's theory, splitting commences when failure occurs. From this it follows that, at points A and B, the stresses are twice as great as those which would be obtained on the assumption of a uniform distribution of stresses (over the cross sections AD and CB). It also follows that the resultant forces H and R must be applied at distances of one-third of the depth of the cross sections from the points A and B. Taking these new lines of action of the forces H and R, Navier finds somewhat larger values for $H_{\max}$ than those arrived at hitherto. He applies these new values in computing stresses and in finding the dimension EF (Fig. 60) of the structure. The new positions of the forces H and R won general acceptance and were adopted in future work in arch analysis. It is interesting to note that the solution of the problem of eccentric compression of bars obtained by Thomas Young (see page 95), and published in his "Natural Philosophy" (1807), was not noticed by Navier who arrived at his conclusion regarding the significance of the third points quite independently.

Navier was the first to solve several problems of the deformation of curved bars; but it did not occur to him that these solutions can be applied in calculating the magnitude of the thrust H in stone arches. The view that arches should be treated as curved elastic bars was expressed, perhaps for the first time, by Poncelet in the article previously mentioned dealing with the history of arch theories. It took a considerable length of time, as we shall see, for this idea to affect the practical design of arches.

The first suspension bridges to be built to withstand the rigors of regular and relatively severe use were erected in the United States of America and in England. Little theory was used in designing them. Barlow, who made some calculations and did some tests for Telford,[1] used the equation of a catenary to determine the tensile force in the cables.[2] But

---

[1] See Barlow, "Treatise on the Strength of Materials," 6th ed., p. 209, 1867.

[2] Davies Gilbert prepared tables to simplify such calculations. See his article, "Tables for Computing All the Circumstances of Strain, Stress, . . . of Suspension Bridges," *Phil. Trans.*, 1826.

the engineers of that time considered it more important to attack problems of strength by means of tests. Telford thought it desirable to test the strength of his cable by submitting it to strains as nearly resembling those in the bridge itself as possible. Figure 61 represents the arrangement which he used for spans of up to 900 ft. It was found that the wire began to stretch rapidly when the force in it was a little larger than a half of its ultimate strength; therefore it was decided that the cable must be designed so that, under the worst possible conditions, the tensile force should not exceed one-third of the ultimate strength.

Further work in the theory of suspension bridges is to be found in Navier's report (see page 73). Navier was very impressed by the bridges he saw in England and praises this type of structure. In his first chapter he gives a historical review of the subject and also describes the new British bridges. The second chapter is devoted to the solutions of various problems concerning stresses and deformations in suspension

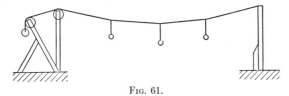

Fig. 61.

bridges. After investigating the deflection produced by a load that is uniformly distributed along the cable, or along the span, Navier examines the deflections produced by a concentrated force and shows that the effect of such a load becomes smaller and smaller as the structure is made larger and heavier. He reaches a similar conclusion when investigating the vibrations produced by the impact of concentrated loads on suspension bridges. On the basis of this analysis, Navier decides that "le succés est d'autant mieux assuré, que l'entreprise est plus grande et semble plus hardie." Subsequent study of suspension bridge construction has revealed that this opinion was a correct one. Large and heavy suspension bridges no longer have the undesirable characteristic of excessive flexibility which was prevalent in earlier short span structures. They can not only support highway traffic but also carry heavy railroad trains. In the third part of his book, Navier gives all the data relating to his own design of a suspension bridge over the Seine in Paris and to the design of an aqueduct.

In the second edition of his report and in a memoir, the French engineer gives a detailed discussion of the factors which finally led to the dismantlement of his suspension bridge before its completion.[1]

[1] The weight of the blocks to which the cables were attached proved insufficient and some sliding was noticed during the process of construction. The story of this

It is interesting to note that the first suspension bridge to be constructed by French engineers was erected in St. Petersburg, Russia (1824–1826), over the Fontanka River.

21. Poncelet (1788–1867)

Poncelet was born in a poor family in Metz.[1] His outstanding work in the grammar school there won him a chance to obtain a scholarship to the *lycée* in his home town. In 1807, he successfully passed the competitive entrance examination of the École Polytechnique, and there he became a pupil of Monge. On graduating (in 1810) he entered the military engineering school in Metz[2] and, after completing its course, joined Napoleon's army in 1812. During the retreat from Moscow in November, 1812, he was taken prisoner and had to live as such for about two years in Saratov on the Volga River. In his confinement, without scientific books or any other contact with Western European culture, but with plenty of time for scientific meditation, Poncelet developed the basic ideas of his new geometry (projective geometry).

FIG. 62. J. V. Poncelet.

When peace was declared, Poncelet returned to France and took a position in the arsenal at Metz. He had sufficient time there to continue his scientific research so that his famous book, "Traité des propriétés projectives des figures," was published in Paris in 1822. At that time, French mathematicians were most interested in applications of mathematical analysis to the solution of physical problems and Poncelet's work in pure geometry did not win proper appreciation. This discouraged the scientist who discontinued his mathematical research and threw all his energy into solving problems of engineering mechanics; there were many important mechanical questions raised in the arsenal. Very soon, Poncelet acquired a reputation in these studies and, in 1825, he was made professor of engineering mechanics at the military school of Metz. In

---

shortcoming in Navier's design is given by Prony. See *Ann. ponts et chaussées*, 1837, p. 1.

[1] For Poncelet's biography, see J. Bertrand, "Mémoires de l'Académie des Sciences," vol. 41, 1879.

[2] The school of Mézières was transferred to Metz in 1794.

1826, his book "Cours de mécanique appliqué aux machines" was published and, in 1829, his book "Introduction a la mécanique industrielle" appeared. Poncelet's work in mechanics gained a quick recognition and he was elected to membership of the Academy of Sciences (1834). He moved to Paris where, for some time, he taught mechanics at the Sorbonne.

During the years 1848–1850 he was commandant of the École Polytechnique; in 1852 he retired with a view to spending the rest of his life revising and reediting his numerous works in geometry and applied mechanics.

While Poncelet's main claims to fame lie in the fields of geometry and dynamics, he also made important contributions to mechanics of materials and they were incorporated in his "Mécanique industrielle." The portion of this book which deals with strength of materials constitutes perhaps the most complete presentation of the knowledge of mechanical properties of structural materials of that time. Poncelet not only gives the results obtained from mechanical tests, but he carefully discusses the practical significance of these results as far as the designing engineer is concerned. To show the results of tensile tests of various materials, he introduces tensile-test diagrams; these have proved very useful in comparing the mechanical properties of various kinds of iron and steel. Using these diagrams, Poncelet shows how the working stress of a material can be selected. He is very conservative in this selection and seldom recommends values greater than one-half of the elastic limit of the material. In this recommendation Poncelet is guided by his study of the dynamical resistance of structural materials.

Thomas Young was the first to show (see page 93) how important the dynamical effect of a load can be. Poncelet, influenced by the design practice of the suspension bridges of his time, goes into a more detailed study of this dynamical action. By using his test diagrams he shows that, up to the elastic limit, an iron bar can absorb only a small amount of kinetic energy and that conditions of impact can easily produce a permanent set. For structural members subjected to impact, he recommends wrought iron, which gives a large elongation in tensile tests and can absorb a greater amount of kinetic energy without failure. Poncelet demonstrates analytically that a suddenly applied load produces twice the stress that the same load would produce if applied gradually. He investigates the longitudinal impact effect on a bar and the longitudinal vibrations produced by such an impact. He also shows that, if a pulsating force acts on a loaded bar, the amplitude of forced vibrations can be built up under conditions of resonance and explains why a detachment of soldiers marching over a suspension bridge may be dangerous. We find an interesting discussion of Savart's experiments on the longitudinal vibration of bars and a demonstration of the fact that large amplitudes and large stresses can be produced by small friction forces acting along the surface.

We find a reference to the important phenomenon of fatigue of metals caused by repeated cycles of stress in Poncelet's "Industrial Mechanics." He states that, under the action of alternating tension and compression, the most perfect spring may fail in fatigue.[1]

Some further investigations in mechanics of materials were made by Poncelet in connection with his lectures at the Sorbonne. These lectures have never been published but exist in manuscript form. Some portions were used by Morin in his book "Résistance des matériaux," paragraphs 213 to 219, Paris, 1853. Saint-Venant, in the third edition of Navier's "Résumé des leçons," refers to Poncelet's unpublished lectures on pages 374, 381, and 512; Dr. Schnuse, the editor of the German translation of Poncelet's "Mécanique appliquée aux machines," added (to this translation) Art. 220 to 270, which contain material from the unpublished work. From this source we learn that Poncelet must be credited with the introduction of the effect of shearing force in the formulas for deflection of beams. If a rectangular cantilever beam of length l, breadth b, and height h carries a uniformly distributed load of intensity q, he gives the formula

$$f = \frac{3}{2} \frac{ql^4}{Ebh^3} \left(1 + \frac{9}{8} \frac{h^2}{l^2}\right)$$

for the maximum deflection and observes that the second term in the bracket, representing the effect of shearing force, is of a practical importance only in comparatively short beams.

In debating the question of selecting safe stresses, Poncelet favors the maximum strain theory and asserts that failure occurs when the maximum strain reaches a definite limit. Thus the condition of rupture in the case of compression of such brittle materials as stone and cast iron rests upon the lateral expansion. The maximum strain theory was consistently used later by Saint-Venant and found extensive use on the Continent, whereas English writers continued to base their design calculations on that of maximum stress.

Poncelet's researches also embraced the theory of structures. In discussing the stability of retaining walls, he offers a graphic method of finding the maximum pressure on the wall.[2] In dealing with the stresses in arches, he was the first to point out that a rational stress analysis can be

---

[1] See p. 317 of the third edition of "Introduction a la Mécanique Industrielle," Paris, 1870.

[2] See his "Mémoire sur la stabilité des revêtements et de leurs fondations" in *Mém. officier génie*, 13, 1840.

developed only by considering an arch as an elastic curved bar (see page 323).

22. Thomas Young (1773-1829)

Thomas Young[1] was born in a Quaker family at Milverton, Somerset.

FIG. 63. Thomas Young.

As a child he showed a remarkable capacity for learning, especially in mastering languages and mathematics. Before he was fourteen years old, he had a knowledge not only of modern languages but also of Latin, Greek, Arabic, Persian, and Hebrew. From 1787 to 1792 he earned his living working for a rich family as a tutor. This position left him sufficient free time to continue his studies, and he worked hard in the fields of philosophy and mathematics. In 1792, Young started to study medicine, first in London and Edinburgh and later at Göttingen University, where he received his doctor's degree in 1796.

In 1797, after his return to England, he was admitted as a Fellow Commoner of Emmanuel College, Cambridge, and spent some time studying there. A man who knew Young well at that time describes him in the following words:[2] "When the Master introduced Young to his tutors, he jocularly said: 'I have brought you a pupil qualified to read lectures to his tutors.' This, however, he did not attempt, and the forbearance was mutual; he was never required to attend the common duties of the college. . . . The views, objects, character, and acquirements of our mathematicians were very different then to what they are now, and Young, who was certainly beforehand with the world, perceived their defects. Certain it is, that he looked down upon the science and would not cultivate the acquaintance of any of our philosophers. . . . He never obtruded his various learning in conversation; but if appealed to on the most difficult subject, he answered in a quick, flippant, decisive way, as if he was speaking of the most easy; and in this mode of talking he differed from all the clever men that I ever saw. His reply never seemed to cost him an effort, and he did not appear to think there was any credit in being able to make it. He did not assert any superiority, or seem to

[1] A very interesting biography of Thomas Young was written by G. Peacock, London, 1855.
[2] See the book of G. Peacock.

Strength of Materials between 1800 and 1833 91

suppose that he possessed it; but spoke as if he took it for granted that we all understood the matter as well as he did. . . . His language was correct, his utterance rapid, and his sentences, though without any affectation, never left unfinished. But his words were not those in familiar use, and the arrangement of his ideas seldom the same as those he conversed with. He was, therefore, worse calculated than any man I ever knew for the communication of knowledge. . . . It was difficult to say how he employed himself; he read little, and though he had access to the College and University libraries, he was seldom seen in them. There were no books piled on his floor, no papers scattered on his table, and his room had all the appearance of belonging to an idle man. . . . He seldom gave an opinion, and never volunteered one. A philosophical fact, a difficult calculation, an ingenious instrument, or a new invention, would engage his attention; but he never spoke of morals, of metaphysics, or of religion."

Very early Young started his original scientific work. As early as 1793, a paper of his on the theory of sight was presented to the Royal Society. While at Cambridge (in 1798) Young became interested in sound. Regarding this work he writes: "I have been studying, not the theory of the winds, but of the air, and I have made observations on harmonics which I believe are new. Several circumstances unknown to the English mathematicians which I thought I had first discovered, I since find to have been discovered and demonstrated by the foreign mathematicians; in fact, Britain is very much behind its neighbours in many branches of the mathematics: were I to apply deeply to them, I would become a disciple of the French and German school; but the field is too wide and too barren for me." The paper "Outlines and Experiments Respecting Sound and Light" was prepared at Cambridge in the summer of 1799 and was read to the Royal Society in January of the following year. In 1801, Young made his famous discovery of the interference of light.

His great knowledge of the physical sciences was recognized when, in 1802, he was elected a member of the Royal Society. The same year he was installed as a professor of natural philosophy by the Royal Institution. This institution was founded in 1799 "for diffusing the knowledge and facilitating the general and speedy introduction of new and useful mechanical inventions and improvements and also for teaching, by regular courses of philosophical lectures and experiments, the applications of the new discoveries in science to the improvements of arts and manufactures and in facilitating the means of procuring the comforts and conveniences of life." As a lecturer at the Royal Institution, Young was a failure; for his presentation was usually too terse and he seemed unable to sense and dwell upon those parts of his subject that were the most calculated to cause difficulty. He resigned his professorship in 1803, but continued to be interested in natural philosophy and prepared his course of lectures for publication. Young's principal contributions to mechanics of materials are to be found in this course.[1]

In the chapter on passive strength and friction (vol. 1, p. 136), the principal types of deformation of prismatical bars are discussed. In considering tension and compression, the notion of modulus of elasticity is introduced for the first time. The definition of this quantity differs from that which we now use to specify Young's modulus. It states: "The modulus of the elasticity of any substance is a column of the same substance, capable of producing a pressure on its base which is to the weight causing a certain degree of compression as the length of the substance is to the diminution of its length." Young also speaks of the "weight of the modulus" and the "height of the modulus" and notes that the height of the modulus for a given material is independent of the cross-sectional area. The weight of the modulus amounts to the product of the quantity which we now call Young's modulus and the cross-sectional area of the bar.[2]

Describing experiments on tension and compression of bars, Young draws the attention of his readers to the fact that longitudinal deformations are always accompanied by some change in the lateral dimensions. Introducing Hooke's law, he observes that it holds only up to a certain limit beyond which a part of the deformation is inelastic and constitutes permanent set.

Regarding shear forces, Young remarks that no direct tests have been made to establish the relation between shearing forces and the deformations that they produce, and states: "It may be inferred, however, from the properties of twisted substances, that the force varies in the simple ratio of the distance of the particles from their natural position, and it must also be simply proportional to the magnitude of the surface to which it is applied." When circular shafts are twisted, Young points out that the applied torque is mainly balanced by shearing stresses which act in the cross-sectional planes and are proportional to distance from the axis of the shaft and to the angle of twist. He also notes that an additional resistance to torque, proportional to the cube of the angle of twist, will be furnished by the longitudinal stresses in the fibers which will be bent

[1] Thomas Young, "A Course of Lectures on Natural Philosophy and the Mechanical Arts," 2 vols., London, 1807. The most interesting part of the material dealing with the deformation of beams was not included in the new edition which was edited by Kelland.

[2] Young had determined the weight of the modulus of steel from the frequency of vibration of a tuning fork and found it equal to 29×10^6 lb per in.[2] See "Natural Philosophy," vol. 2, p. 86.

to helices. Because of this, the outer fibers will be in tension and the inner fibers in compression. Further, the shaft will be shortened during torsion "one fourth as much as the external fibers would be extended if the length remained undiminished."[1]

In discussing bending of cantilevers and beams supported at their two ends, Young gives the principal results regarding deflections and strength, without derivation. His treatment of the lateral buckling of compressed columns includes the following interesting remark: "Considerable irregularities may be observed in all the experiments which have been made on the flexure of columns and rafters exposed to longitudinal forces; and there is no doubt but that some of them were occasioned by the difficulty of applying the force precisely at the extremities of the axis, and others by the accidental inequalities of the substances, of which the fibres must often have been in such directions as to constitute originally rather bent than straight columns."

When considering inelastic deformations, Young makes an important statement: "A permanent alteration of form . . . limits the strength of materials with regard to practical purposes, almost as much as fracture, since in general the force which is capable of producing this effect is sufficient, with a small addition, to increase it till fracture takes place." As we have seen, Navier came to the same conclusion and proposed that working stresses should be kept much lower than the elastic limit of the material.

In conclusion, Young gives an interesting discussion of the fracture of elastic bodies produced by impact. In this case, not the weight of the striking body but the amount of its kinetic energy must be considered. "Supposing the direction of the stroke to be horizontal, so that its effect may not be increased by the force of gravity," he concludes that "if the pressure of a weight of 100 lb (applied statically) broke a given substance, after extending it through the space of an inch, the same weight would break it by striking it with the velocity that would be acquired by the fall of a heavy body from the height of half an inch, and a weight of one pound would break it by falling from a height of 50 in." Young states that, when a prismatical bar is subjected to longitudinal impact, the resilience of the bar "is proportional to its length, since a similar extension of a longer fiber produces a greater elongation." Further on, he finds that "there is, however, a limit beyond which the velocity of a body striking another cannot be increased without overcoming its resilience and breaking it, however small the bulk of the first body may be, and this limit depends on the inertia of the parts of the second body, which must not be

[1] The correct value of the shortening is twice as large as that given by Young. (See "Strength of Materials," vol. 2, p. 300.)

disregarded when they are impelled with a considerable velocity." Denoting the velocity with which a compression wave travels along a bar by V and the velocity of the striking body by v, he concludes that the unit compression produced at the end of the bar at the instant of impact is equal to v/V, and that the limiting value for the velocity v is obtained by equating the ratio v/V to the unit compression at which fracture of the material of the struck bar occurs in static tests.

Considering the effects of impact upon a rectangular beam, Young decides that, for a given maximum bending stress generated by the blow, the quantity of energy accumulated in the beam is proportional to its volume. This is because the maximum force P, produced on the beam by the striking body, and the deflection δ at the point of impact are given by the formulas

$$P = k\frac{bh^2}{l} \qquad \delta = k_1\frac{Pl^3}{bh^3}$$

where b is the width, h the depth, and l the length of the beam; k and k_1 are constants depending on the modulus of the material and on the assumed magnitude of the maximum stress. Substituting into the expression for the strain energy U, we obtain

$$U = \frac{P\delta}{2} = \frac{k^2 k_1}{2}bhl$$

which proves the above statement.

We see that Young contributed much to strength of materials by introducing the notion of a modulus in tension and compression. He was also the pioneer in analyzing stresses brought about by impact and gave a method of calculating them for perfectly elastic materials which follow Hooke's law up to fracture.

Some more complicated problems concerning the bending of bars are discussed in the chapter "Of the Equilibrium and Strength of Elastic Substances" in volume II of "Natural Philosophy." The following remarks regarding this chapter are to be found in the "History of the Theory of Elasticity" by I. Todhunter and K. Pearson: "This is a series of theorems which in some cases suffer under the old mistake as to the position of the neutral surface. . . . The whole section seems to me very obscure like most of the writings of its distinguished author; among his vast attainments in sciences and languages that of expressing himself clearly in the ordinary dialect of mathematicians was unfortunately not included. The formulae of the section were probably mainly new at the time of their appearance, but they were little likely to gain attention in consequence of the unattractive form in which they were presented." It is agreed that this chapter of Young's book is difficult to read; but it contains correct solutions of several important problems, which were new in Young's time. For example, we see the solution of the problem of eccentric tension or compression of rectangular bars for the first time. Assuming that the stress distribution in this case is represented by two triangles, as shown in Fig. 64c, Young determines the position of the neutral axis from the condition that the resultant of these stresses must pass through the point O of application of the external force (Fig. 64b). This gives

$$a = \frac{h^2}{12e}$$

where e is the eccentricity of the force. When $e = h/6$, $a = h/2$. The stress distribution is represented by a triangle, and the maximum stress becomes twice as great as that caused by applying the load centrally. The correct value of the radius of the circle to which the axis of the bar in Fig. 64a will be bent, in the case of very small deflections, is also derived in Young's analysis.

As his next problem Young takes up the bending of a compressed prismatical column which is initially slightly curved. Assuming that the initial curvature may be represented by one-half wave of the sine curve, $\delta_0 \sin(\pi x/l)$, he finds that the deflection at the middle after application of the compressive force P will be

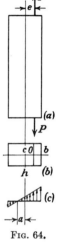

Fig. 64.

$$\delta = \frac{\delta_0}{1 - (Pl^2/EI\pi^2)}$$

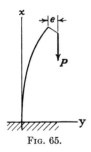

Fig. 65.

From this result he concludes that if $P = EI\pi^2/l^2$, "the deflection becomes infinite, whatever may be the magnitude of δ_0 and the force will overpower the beam, or will at least cause it to bend so much as to derange the operation of the forces concerned." Here we have the first derivation of the column equation allowing for lack of initial straightness. In using Euler's formula for determining the cross-sectional dimensions of a column, Young points out that its application is limited to slender columns and gives certain limiting values of the ratio of the length of the column to its depth. For the smaller values of this ratio, failure of the column is brought about by crushing of the material rather than by buckling.

If a slender column is built-in at one end and free at the other, Thomas Young shows that, when an axial force P acts eccentrically (as in Fig. 65), the deflection y is given by

$$y = \frac{e(1 - \cos px)}{\cos pl}$$

In this formula, e denotes the eccentricity, $p = \sqrt{P/EI}$, and x is measured from the built-in end. In this case then, Young is ahead of Navier.[1]

In his treatment of lateral buckling of columns having variable cross sections, Young shows that a column of constant depth will bend to a circular curve (Fig. 66a) if its width (Fig. 66b) varies along the length as the ordinates y of a circular arch.

Considering the buckling of a column consisting of two triangular prisms (Fig. 67) and assuming that the deflection is such that the radius

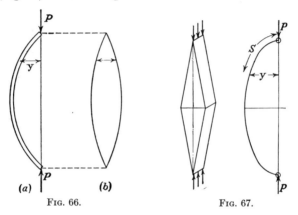

FIG. 66. FIG. 67.

of curvature at the middle is equal to half the over-all length, Young states that the deflection curve is a cycloid. This follows from the equation

$$\frac{Py}{EI} = \frac{1}{\rho} \qquad (a)$$

The moment of inertia I of any cross sections is proportional to s^3 and the ordinates y of the cycloid are proportional to s^2, where s is defined as shown in Fig. 67. Now the left-hand side of Eq. (a) has the dimensions of a number divided by s. Since $\rho = s$ for a cycloid, we can always select such a value for P that Eq. (a) will be satisfied.

When a rectangular beam is cut out of a given circular cylinder, Young finds "that the stiffest beam is that of which the depth is to the breadth as the square root of 3 to 1, and the strongest as the square root of 2 to 1; but the most resilient will be that which has its depth and breadth equal." This follows from the fact that for a given length the stiffness of the beam

---

[1] The solution was given by Navier in a memoir presented to the Academy in 1819. See Saint-Venant's "History," p. 118.

is governed by the magnitude of the cross-sectional moment of inertia, the strength by the section modulus and the resilience by the cross-sectional area. Analogously he solves the problem for thin circular tubes. Young states: "Supposing a tube of evanescent thickness to be expanded into a similar tube of greater diameter, but of equal length, the quantity of matter remaining the same, the strength will be increased in the ratio of the diameter, and the stiffness in the ratio of the square of the diameter, but the resilience will remain unaltered."

From this discussion we see that the chapter on mechanics of materials in the second volume of "Natural Philosophy" contains valid solutions of several important problems of strength of materials, which were completely new in Young's time. This work did not gain much attention from engineers because the author's presentation was always brief and seldom clear. Before leaving the subject of this book, some remarks regarding Young's "Natural Philosophy" will be quoted. They come from the pen of Lord Rayleigh's biographer.[1] In 1892, Lord Rayleigh was appointed as professor at the Royal Institution and his lectures "followed somewhat closely the lines of those which had been given by Thomas Young in the same place nearly a century before, at the birth of the Royal Institution. These were fully written out in Young's Natural Philosophy, published in 1807, and many of the identical pieces of demonstration apparatus figured in that work, which had been preserved in the museum of the Institution, were brought out and used. . . . Rayleigh had studied Young's Lectures and had found them a mine of interesting matter. The pencil marks in his copy show how closely he had gone over the book, and in the lecture he brought to notice some of the good but forgotten work which he had found there and also in other writings of the same author. One of the most striking points was Young's estimate of the size of molecules; (he gave) the molecular diameter as between the two-thousand and the ten-thousand millionth of an inch. This is a wonderful anticipation of modern knowledge, and antedates by more than fifty years the similar estimates of molecular dimensions made by Kelvin. Until Rayleigh drew attention to it, it had fallen completely out of notice, even if we suppose that it had ever attracted notice, and of this there is no evidence."

Young showed his unusual ability not only in solving purely scientific problems but also in attacking practical engineering difficulties. For instance, he presented a report to the Board of Admiralty dealing with the use of *oblique riders*, and concerning other alterations in the construction of ships.[2] He treats the hull of a ship as a beam, assumes a definite

---

[1] "Life of Lord Rayleigh" by his son, fourth Baron Rayleigh, p. 234, 1924.
[2] *Phil. Trans.*, 1814, p. 303.

distribution of weight and a definite shape of waves, and calculates the magnitudes of the shearing forces and bending moments at several cross sections. He also shows how the deflection of a ship can be calculated. He explains that the flexural rigidity of a ship depends, to a great extent, upon the connections between the parts and remarks: "A coach spring, consisting of ten equal plates, would be rendered ten times as strong, if it were united into one mass, and at the same time a hundred times as stiff, bending only one hundredth of an inch with the same weight that would bend it a whole inch in its usual state, although nothing would be gained by the union with respect to the power of resisting a very rapid motion, which I have, on another occasion, ventured to call resilience. Now it appears to be extremely difficult to unite a number of parallel planks so firmly together, by pieces crossing them at right angles, as completely to prevent their sliding in any degree over each other: and a diagonal brace of sufficient strength, even if it did not enable the planks to bear a greater strain without giving way, might still be of advantage, in many cases, by diminishing the degree in which the whole structure would bend before it broke." Basing his arguments upon such considerations, Young favors the adoption of diagonal bracing in the construction of wooden ships. It seems that this report constitutes the first attempt at applying theoretical analysis to the structural design of ships.

After this outline of Young's contributions to the theory of strength of materials, we can perhaps appreciate Lord Rayleigh's[1] remark: "Young . . . from various causes did not succeed in gaining due attention from his contemporaries. Positions which he had already occupied were in more than one instance reconquered by his successors at a great expense of intellectual energy."

23. Strength of Materials in England between 1800 and 1833

During the period 1800–1833 England had no schools equivalent to the École Polytechnique and the École des Ponts et Chaussées in France. The country's engineers did not receive a broad scientific education and the level of its textbooks on strength of materials was much lower than that in France. In describing these textbooks, the English historian of theory of elasticity, Todhunter, states: "Nothing shews more clearly the depth to which English mechanical knowledge had sunk at the commencement of this century."[2] To illustrate the standard of English authors of this time, Todhunter quotes the following criticism of Euler's work from Robison's book "A System of Mechanical Philosophy": "In the old Memoirs of the Academy of Petersburgh for 1778 there is a dissertation by Euler on the subject, but particularly limited to the strain on columns,

---

[1] "Collected Papers," vol. 1, p. 190.
[2] See "History . . . ," vol. 1, p. 88.

in which the bending is taken into account. Mr. Fuss has treated the same subject with relation to carpentry in a subsequent volume. But there is little in these papers besides a dry mathematical disquisition, proceeding on assumptions which (to speak favourably) are extremely gratuitous. The most important consequence of the compression is wholly overlooked, as we shall presently see. Our knowledge of the mechanism of cohesion is as yet far too imperfect to entitle us to a confident application . . . ." and further, "We are thus severe in our observations, because his theory of the strength of columns is one of the strongest evidences of this wanton kind of proceeding and because his followers in the Academy of St. Petersburgh, such as Mr. Fuss, Lexell and others, adopt his conclusions, and merely echo his words." Such a criticism, as Todhunter remarks, is very idle in a writer, who on the very next page places the neutral line on the concave surface of a beam subjected to flexure. . . . It was Robison who perpetuated (in the English textbooks) this error that had already been corrected by Coulomb.

However, although English theoretical work on strength of materials was of such poor quality, British engineers had to solve many important engineering problems. The country was ahead of others in industrial development and the introduction of such materials as cast iron and wrought iron into structural and mechanical engineering presented many new problems and called for investigations of the mechanical properties of these new materials. A considerable amount of experimental work was completed and the results that were compiled found use not only in England but also in France.

A book entitled "An Essay on the Strength and Stress of Timber" was written by Peter Barlow (1776–1862), the first edition of which appeared in 1817. The book was very popular in England and was reedited many times.[1] At the beginning of the book, Barlow gives a history of strength of materials using material from Girard's book (see page 7); but while Girard devotes much space in his book to Euler's study of elastic curves, Barlow is of the opinion that Euler's "instruments of analysis" are "too delicate to operate successfully upon the materials to which they have been applied; so that while they exhibit under the strongest point of view, the immense resources of analysis, and the transcendent talents of their author, they unfortunately furnish but little, very little, useful information." In discussing the theory of bending, Barlow makes the same error as Duleau did (see page 81) and assumes that the neutral line divides the cross section of the beam in such a way that the moment about this line

[1] The sixth edition of the book, prepared by Peter Barlow's two sons, was published in 1867. This edition is of considerable interest in so far as the history of our subject is concerned, for it contains a biography of Peter Barlow and, in an appendix, the work of Willis entitled "Essay on the Effects Produced by Causing Weights to Travel Over Elastic Bars."

of all tensile stresses is equal to the moment of compressive stresses. This error is corrected in the 1837 edition of the book. Barlow attributes the correct determination of the neutral line to Hodgkinson. He seems still to have been ignorant of the fact that Coulomb had made this correction fifty years before.

Of Barlow, Todhunter says: "As a theorist he is another striking example of that want of clear thinking, of scientific accuracy, and of knowledge of the work accomplished abroad, which renders the perusal of the English text-books on practical mechanics published in the first half of this century, such a dispiriting, if not hopeless, task to the historian of theory." It seems that Todhunter is too severe in this criticism, especially in view of the fact that, at that time, engineers of such caliber as Duleau and Navier were making similar mistakes in the determination of the neutral axis.

The book does not add anything to the theory of strength of materials, but it contains descriptions of many experiments that were made in the first half of the nineteenth century and which are of historical interest. Thus we find a report on the tests made on iron wires by Barlow for Telford who was then planning the construction of the Runcorn suspension bridge. The author's experiments with iron rails of different shapes are also very interesting. In them the deflections of rails under rapidly moving loads were measured for the first time and compared with statical results. In the last edition of Barlow's book (1867) the important fatigue tests on plate girders, carried out by Fairbairn, are mentioned. R. Willis's report on the dynamic action of a moving load on a beam is also included. These investigations will be discussed later.

Another English engineer whose books were very popular in the first half of the nineteenth century was Thomas Tredgold (1788–1829). His work "The Elementary Principles of Carpentry" (1820) contains a chapter dealing with strength of materials. There is not much new work to be found in it and all the theoretical results can be seen in Thomas Young's "Lectures" (see page 92). Another book by Tredgold "A Practical Essay on the Strength of Cast Iron" (1822) contains the results of the author's experiments and gives many practical rules for designers of cast-iron structures. It seems that Tredgold was the first to introduce a formula for calculating safe stresses for columns (see page 209). In the period from 1800 to 1833 the use of cast iron in structures rapidly increased and Tredgold's book was widely used by practical engineers. It was translated into French, Italian, and German.

24. Other Notable European Contributions to Strength of Materials

Work on strength of materials was started in Germany as late as the beginning of the nineteenth century, and the first important contribution

Strength of Materials between 1800 and 1833 101

to this science was made by Franz Joseph Gerstner (1756–1832) who graduated from Prague University in 1777. After several years of practical engineering work, Gerstner became an assistant at the astronomy observatory of his old university. In 1789, he became professor of mathematics. While in this position, he continued to be interested in practical applications of science in engineering and planned the organization of an engineering school in Prague. In 1806, "das Bömische technische Institut" was opened in that city and Gerstner was active up to 1832 as a professor of mechanics and as the director. Gerstner's main work in mechanics is incorporated in his three-volume book "Handbuch der Mechanik," Prague, 1831. The third chapter of the first volume of this work deals with strength of materials. Discussing tension and compression, Gerstner gives the results of his experimental work on the tension of pianoforte wires. He finds that the relation existing between the tensile force P and the elongation δ is

$$P = a\delta - b\delta^2 \tag{a}$$

where a and b are two constants. For small elongations, the second term on the right-hand side of Eq. (a) can be neglected and we obtain Hooke's law.

Gerstner also studied the effect of permanent set on the property of tensile-test specimens. He showed that, if a wire is stretched so that a certain permanent set is produced and then unloaded, it will follow Hooke's law up to the load which produced the initial permanent set during a second loading. Only for higher loads than that will deviation from the straight-line law occur. In view of this fact, Gerstner recommends that wires to be used in suspension bridges should be stretched to a certain limit before use.

Gerstner did important work in other branches of engineering mechanics. Of this the best known is his theory of trochoidal waves on water surfaces.[1]

Another German engineer who contributed to our knowledge of engineering mechanics in the early days of the nineteenth century, was Johann Albert Eytelwein (1764–1848). As a boy of fifteen he became a bombardier in the artillery regiment in Berlin. In his spare time he studied mathematics and engineering science and was able, in 1790, to pass the examination for the engineer's degree (architect). Very quickly he gathered a reputation as an engineer with great theoretical knowledge. In 1793 his book "Aufgaben grosstentheils aus der angewandten Mathematik" was published; it contained discussions of many problems of practical importance in structural and mechanical engineer-

[1] See his "Theorie der Wellen," Prague, 1804.

ing. In 1799, he and several other engineers organized the Bauakademie in Berlin and he became the director of that institute and its professor of engineering mechanics.

Eytelwein's course "Handbuch der Mechanik fester Körper" was published in 1800, and the "Handbuch der Statik fester Körper" was published in 1808. In the later handbook, problems of strength of materials and theory of structures are discussed. Regarding this strength of materials section, Todhunter remarks in his "History": "There is nothing original in this section, but its author possesses the advantage over Banks[1] and Gregory[1] of being abreast of the mathematical knowledge of his day." In theory of structures, Eytelwein makes some additions to the theory of arches and to the theory of retaining walls. He also develops a useful method for determining the allowable load on a pile.

After the Napoleonic wars an intensive movement developed for building up the various branches of industry in Germany. This required a considerable improvement in engineering education and, to this end, several schools were built. In 1815, the Polytechnical Institute in Vienna was opened under the directorship of Johan Joseph Prechtl (1778–1854), and the Gewerbeinstitut in Berlin was organized in 1821. These were followed by the founding of polytechnical institutes in Karlsruhe (1825), Munich (1827), Dresden (1828), Hanover (1831), and Stuttgart (1840). These new engineering schools greatly affected the development of German industry and indeed of engineering sciences in general. In no other country did there exist such close contact between industry and engineering education as in Germany, and doubtless this contact contributed much to the industrial progress of that country and to the establishment of its strong position in engineering sciences at the end of the nineteenth century.

British, German, and French progress in the field of strength of materials was followed by a similar development in Sweden. The iron industry is of great importance in that country and the first experimental investigations there dealt with that metal. In the van of this activity, we find the Swedish physicist P. Lagerhjelm.[2] He was acquainted with the theoretical work of Thomas Young, Duleau, and Eytelwein and was guided in his investigations not only by practical but also by theoretical considerations. The tensile-testing machine constructed by Brolling to his design proved very satisfactory; so much so in fact that later a similar

[1] Contemporary English authors of books on engineering mechanics.
[2] Lagerhjelm's work was published in *Jern- Kontorets Ann.*, 1826. A short account of his results was given in *Pogg. Ann. Physik u. Chem.*, Bd. 13, p. 404, 1828. Later Swedish experimenters refer to Lagerhjelm's work. For example, see K. Styffe, who gives a very complete description of Lagerhjelm's machine for tensile tests in *Jern-Kontorets Ann.*, 1866.

machine was constructed for Lamé's experiments in St. Petersburg (see page 83). Lagerhjelm's tests showed that the modulus of elasticity in tension is about the same for all kinds of iron and is independent of such technological processes as rolling, hammering, and the various kinds of heat-treatment but that at the same time the elastic limit and ultimate strength can be changed radically by these processes. He found that the ultimate strength of iron is usually proportional to its elastic limit. He found that the density of iron at the point of rupture is somewhat diminished. Lagerhjelm compared the modulus of elasticity obtained from static tests with that calculated from the observed frequency of lateral vibrations of a bar and found good agreement. He also demonstrated that if two tuning forks give the same note, they continue to do so after one has been hardened.

CHAPTER V

The Beginning of the Mathematical Theory of Elasticity[1]

25. Equations of Equilibrium in the Theory of Elasticity

By the theory of strength of materials, the history of which we have so far traced, solutions which deal with problems of the deflection and stresses in beams were usually obtained on the assumption that cross sections of the beams remain plane during deformation and that the material follows Hooke's law. At the beginning of the nineteenth century, attempts were made to give the mechanics of elastic bodies a more fundamental basis. The idea had existed[2] since Newton's time, that the elastic property of bodies can be explained in terms of some attractive and repulsive forces between their ultimate particles. This notion was expatiated by Boscovich[3] who assumed that between every two ultimate particles and along the line connecting them forces act which are attractive for some distances and repulsive for others. Furthermore, there are distances of equilibrium for which these forces vanish.

Using this theory, with the added requirement that the molecular forces diminish rapidly with increase of the distances between the molecules, Laplace was able to develop his theory of capillarity.[4]

The application of Boscovich's theory in analyzing the deformations of elastic bodies was initiated by Poisson[5] in his investigation of bending of plates. He considers the plate as a system of particles distributed in the middle plane of the plate. However, such a system will resist exten-

---

[1] A more complete discussion of the history of the theory of elasticity can be found in Saint-Venant's notes to the third edition of Navier's book, "Résumé des Leçons . . . ," 1864, in the notes to Saint-Venant's translation of Clebsch's "Théorie de l'élasticité des corps solides," 1883, and in Saint-Venant's two chapters in Moigno's book, "Leçons de mécanique analytique," 1868. See also the article by C. H. Müller and A. Timpe in "Encyklopädie der mathematischen Wissenschaften," vol. 4₄, p. 1, 1906, and H. Burkhardt's "Entwicklungen nach oscillirenden Funktionen," pp. 526–671, Leipzig, 1908.

[2] See Saint-Venant, "Historique Abrégé."

[3] Boscovich, "Philosophiae naturalis," 1st ed., Venice, 1763. Also the new Latin-English edition with a short life of Boscovich, Chicago and London, 1922.

[4] Laplace, *Ann. chim. et phys.*, vol. 12, 1819.

[5] Poisson, *Mém. acad.*, 1812, p. 167.

The Beginning of the Mathematical Theory of Elasticity

sion, but not bending, and thus may represent an ideally flexible membrane, but not a plate.

A further advance was made in the molecular theory of an elastic body by Navier.[1] He assumes that there are two systems of forces ΣF and ΣF_1 acting on the particles of an elastic solid. The forces ΣF balance each other and represent molecular forces acting when external forces are absent. The forces ΣF_1 balance the external forces, such as that due to gravity, for example. They are assumed to be proportional to changes $r_1 - r$ of the distances between the particles and to act along the lines connecting them. If u, v, w are the components of the displacement of a particle $P(x,y,z)$ and $u + \Delta u$, $v + \Delta v$, $w + \Delta w$, the corresponding displacements of an adjacent particle $P_1(x + \Delta x, y + \Delta y, z + \Delta z)$, the change of distance between these particles is

$$r_1 - r = \alpha \Delta u + \beta \Delta v + \gamma \Delta w \qquad (a)$$

where α, β, γ denote the cosines of the angles which the direction r makes with the coordinate axes x, y, z. The corresponding force F_1 is then

$$F_1 = f(r)(\alpha \Delta u + \beta \Delta v + \gamma \Delta w) \qquad (b)$$

where $f(r)$ is a factor which rapidly diminishes with increase of the distance r.

Navier now expands Δu, Δv, Δw in series of the type

$$\Delta u = \frac{\partial u}{\partial x} \Delta x + \frac{\partial u}{\partial y} \Delta y + \frac{\partial u}{\partial z} \Delta z + \frac{1}{2} \frac{\partial^2 u}{\partial x^2} \Delta x^2 + \frac{\partial^2 u}{\partial x \, \partial y} \Delta x \, \Delta y + \cdots$$

and limits himself to terms of the second order. Then, substituting these expressions into Eq. (b) and putting

$$\Delta x = r\alpha \qquad \Delta y = r\beta \qquad \Delta z = r\gamma$$

he obtains the following expression for the x component of the force F_1:

$$\alpha F_1 = rf(r) \left[\alpha^3 \frac{\partial u}{\partial x} + \alpha^2 \beta \left(\frac{\partial u}{\partial y} + \frac{\partial v}{\partial x} \right) + \cdots \right]$$
$$+ r^2 f(r) \left[\frac{\alpha^4}{2} \frac{\partial^2 u}{\partial x^2} + \alpha^3 \beta \frac{\partial^2 u}{\partial x \, \partial y} + \cdots + \frac{\alpha \gamma^3}{2} \frac{\partial^2 w}{\partial z^2} \right] \qquad (c)$$

To compute the total force acting in the x direction on the displaced particle $P(x,y,z)$ a summation of expressions such as (c) must be made for all the particles within the sphere of action of the particle P. Doing this,

---

[1] Navier's paper "Mémoire sur les lois de l'équilibre et du mouvement des corps sólides élastiques" was presented to the Academy, May 14, 1821. It was printed in the *Mém. inst. Natl.*, 1824. An excerpt from it was published in *Bull. soc. philomath. Paris*, 1823, p. 177.

and assuming that the molecular forces are independent of the direction of r, that is, that the body is isotropic, Navier observes that all the terms containing cosines raised to odd powers cancel out. Thus all integrals of the first line of the expression (c) vanish. The evaluation of those integrals of the second line which do not vanish can be reduced to the calculation of one integral, namely,

$$C = \frac{2\pi}{15} \int_0^\infty r^4 f(r) \, dr \qquad (d)$$

Assuming that the value of this integral is known, the projection on the x axis of the resultant force acting on particle P becomes

$$\sum \alpha F_1 = C \left(3 \frac{\partial^2 u}{\partial x^2} + \frac{\partial^2 u}{\partial y^2} + \frac{\partial^2 u}{\partial z^2} + 2 \frac{\partial^2 v}{\partial x \, \partial y} + 2 \frac{\partial^2 w}{\partial x \, \partial z} \right) \qquad (e)$$

In a similar manner the two other projections can be written down.

Introducing the notation

$$\theta = \frac{\partial u}{\partial x} + \frac{\partial v}{\partial y} + \frac{\partial w}{\partial z}$$
$$\nabla = \frac{\partial^2}{\partial x^2} + \frac{\partial^2}{\partial y^2} + \frac{\partial^2}{\partial z^2} \qquad (f)$$

the equations of equilibrium of a particle P can evidently be expressed in the following manner

$$C \left(\nabla u + 2 \frac{\partial \theta}{\partial x} \right) + X = 0$$
$$C \left(\nabla v + 2 \frac{\partial \theta}{\partial y} \right) + Y = 0 \qquad (g)$$
$$C \left(\nabla w + 2 \frac{\partial \theta}{\partial z} \right) + Z = 0$$

where X, Y, Z are the components of the external force acting on the particle $P(x,y,z)$. These relations are Navier's differential equations of equilibrium for isotropic elastic bodies. We see that with the assumptions that he makes, only one constant C is required for determining the elastic properties of the body.[1]

Adding the inertia force of the particle to the external forces X, Y, Z, Navier also obtains the general equations of motion of an elastic body.

In addition to the three equations (g) which have to be satisfied at every point within the body, it is also necessary to establish the require-

[1] Later, Cauchy generalized Navier's derivations and considered anisotropic bodies. He showed that, in the most general case, 15 constants are required to define the elastic properties of such bodies. See "Exercises de mathématique," 3d year, p. 200.

ments at the surface of the body, where the molecular forces must be in equilibrium with the external forces distributed over the boundary. Denoting the three components of this force by $\bar{X}_n$, $\bar{Y}_n$, $\bar{Z}_n$ per unit area at the point with the external normal n and using the principle of virtual displacements, Navier obtains the required boundary conditions in the following form:

$$\bar{X}_n = C \left\{ \left(3 \frac{\partial u}{\partial x} + \frac{\partial v}{\partial y} + \frac{\partial w}{\partial z}\right) \cos \alpha + \left(\frac{\partial u}{\partial y} + \frac{\partial v}{\partial x}\right) \cos \beta \right.$$
$$\left. + \left(\frac{\partial u}{\partial z} + \frac{\partial w}{\partial x}\right) \cos \gamma \right\} \quad (h)$$

. . . .
. . . .

Cauchy later explained the significance of the expressions in the parentheses of Eq. (h) and showed that the components $\bar{X}_n$, $\bar{Y}_n$, $\bar{Z}_n$ of surface forces are linear functions of the six strain components,

$$\frac{\partial u}{\partial x}, \frac{\partial v}{\partial y}, \frac{\partial w}{\partial z}, \left(\frac{\partial u}{\partial y} + \frac{\partial v}{\partial x}\right), \left(\frac{\partial u}{\partial z} + \frac{\partial w}{\partial x}\right), \left(\frac{\partial v}{\partial z} + \frac{\partial w}{\partial y}\right).$$

The next important advance made with the theory of elasticity is due principally to the work of Cauchy. Instead of discussing molecular forces between the individual particles, he introduced the notions of *strain* and *stress* and in this way immensely simplified the derivation of the fundamental equation.

26. Cauchy (1789–1857)

Augustin Cauchy[1] was born in Paris in 1789. His father, a successful lawyer in the service of the government, had to leave Paris at the time of the French Revolution and took refuge in Arcueil, a village near Paris in which the famous scientists Laplace and Berthollet lived at that time. In the time of Napoleon, Laplace's house became a meeting place for the most outstanding scientists of Paris so that young Cauchy grew to know many of them. Among these men was Lagrange, who quickly noticed the boy's unusual mathematical prowess. Cauchy obtained his secondary education in l'École Centrale du Panthéon where he studied classic languages and humanity subjects with great success.

In 1805 he successfully passed the entrance examination to the École Polytechnique and during his period in this school Cauchy showed his great competence in mathematics. After completing this course in 1807

---

[1] For a biography of Cauchy, see "La vie et les travaux du Baron Cauchy," by C. A. Valson, Paris, 1868. See also "Histoire des Sciences Mathématiques et Physiques," vol. 12, p. 144, Paris, 1888.

he elected to enter the École des Ponts et Chaussées for further engineering study and he graduated in 1810. He was placed first in the entrance examination and also in the final one, and his great ability was recognized by his professors. At the age of twenty-one he was already doing important engineering work at the port of Cherbourg. But mathematics attracted Cauchy more than engineering, and he devoted his spare time to mathematical investigations. During the years 1811 and 1812 he presented several important memoirs to the Academy of Sciences, and in 1813 he left Cherbourg and returned to Paris, where he put all his energy into mathematics. In 1816 he became a member of the Academy.

Fig. 68. Augustin Cauchy.

On his return to Paris, Cauchy started to teach at the École Polytechnique and at the Sorbonne. In his lectures on calculus he tried to present the subject in a more rigorous form than had been given before. The originality of this presentation attracted not only students but also professors and scientists from foreign countries to these lectures. The publication of his "Cours d'Analyse de l'École Polytechnique" in 1821 had an important effect upon subsequent trends in mathematics.

About this time, Navier's first work on the theory of elasticity was presented to the Academy of Sciences. Cauchy became interested in that paper and started himself to work in the theory. His results in the early stages of this branch of mechanics were of immense value.[1]

Navier, as we saw in the last article, considered forces acting between the individual molecules of a deformed elastic body in his derivations of the fundamental equations. Cauchy,[2] instead, applies the notion of pressure on a plane (a concept which was familiar to him from hydrodynamics) and assumes that this pressure is no longer normal to the plane upon which it acts in an elastic body. In this way the idea of *stress* was introduced in the theory of elasticity. The total stress on an infinitesimal element of a plane taken within a deformed elastic body is defined as the

[1] The best résumé of Cauchy's work in theory of elasticity was made by Saint-Venant in the two chapters on elasticity, which he wrote for Moigno's "Mécanique Analytique, Statique," 1868.

[2] Cauchy's work, written after his study of Navier's paper (see p. 105), was presented to the Academy of Sciences on Sept. 30, 1822, and an abstract appeared in *Bull. soc. philomath. Paris*, 1823, p. 9.

The Beginning of the Mathematical Theory of Elasticity

resultant of all the actions of the molecules situated on one side of the plane upon the molecules on the other, the directions of which (actions) intersect the element under consideration.[1] By dividing the total stress by the area of the element, the magnitude of stress is obtained.

Considering an elemental tetrahedron (Fig. 69), Cauchy shows that the three components of stress X_n, Y_n, Z_n on an inclined plane abc are obtained from the three equations of equilibrium

$$X_n = \sigma_x l + \tau_{yx} m + \tau_{zx} n$$
$$Y_n = \tau_{xy} l + \sigma_y m + \tau_{zy} n \qquad (a)$$
$$Z_n = \tau_{xz} l + \tau_{yz} m + \sigma_z n$$

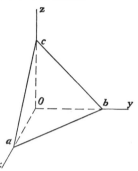

Fig. 69.

where l, m, n are the cosines of the angles which the outward normal to the plane abc makes with the coordinate axes[2] and σ_x, τ_{yx}, . . . are the normal and tangential components of the stress acting at O on the coordinate planes xy, xz, and yz. He also proves that

$$\tau_{yx} = \tau_{xy} \qquad \tau_{zx} = \tau_{xz} \qquad \tau_{zy} = \tau_{yz}$$

which shows that the stress acting on any plane, such as abc, can be specified by the six stress components σ_x, σ_y, σ_z, τ_{xy}, τ_{xz}, τ_{yz}. Resolving the components X_n, Y_n, Z_n [of Eq. (a)] in the direction of the normal to the plane abc, we obtain the normal stress σ_n acting on that plane. Cauchy shows that, if we draw a vector r of length $r = \sqrt{1/|\sigma_n|}$ in every direction n from the origin O (Fig. 69), the ends of these vectors will lie on a surface of the second degree. Cauchy calls the directions of the principal axes of this surface the *principal directions* and the corresponding stresses the *principal stresses*.

Cauchy derives the differential equations of equilibrium for an elemental rectangular parallelepiped; *viz.*,

$$\frac{\partial \sigma_x}{\partial x} + \frac{\partial \tau_{yx}}{\partial y} + \frac{\partial \tau_{zx}}{\partial z} + X = 0$$
$$\frac{\partial \tau_{xy}}{\partial x} + \frac{\partial \sigma_y}{\partial y} + \frac{\partial \tau_{zy}}{\partial z} + Y = 0 \qquad (b)$$
$$\frac{\partial \tau_{xz}}{\partial x} + \frac{\partial \tau_{yz}}{\partial y} + \frac{\partial \sigma_z}{\partial z} + Z = 0$$

---

[1] In this final form the definition of stress is due to Saint-Venant. See *Compt. rend.*, vol. 20, p. 1765, 1845; vol. 21, p. 125. Cauchy used a somewhat different definition in his earlier publications but later he accepted that of Saint-Venant.

[2] For the notations used by various authors in the early stages of the theory of elasticity, see Moigno, "Statics," p. 626. See also, I. Todhunter and K. Pearson, "A History of the Theory of Elasticity," vol. 1, p. 322.

where X, Y, and Z denote the three components of the body force per unit volume at point O. In the case of vibration, inertia forces are added to these body forces.

The French mathematician also examines possible deformations of an elastic body around a point O (Fig. 69) and demonstrates that, when this deformation is small, the unit elongation in any direction and the change of the right angle between any two initially perpendicular directions can be expressed by the six *strain components*:

$$\epsilon_x = \frac{\partial u}{\partial x} \quad \epsilon_y = \frac{\partial v}{\partial y} \quad \epsilon_z = \frac{\partial w}{\partial z} \quad \gamma_{xy} = \frac{\partial u}{\partial y} + \frac{\partial v}{\partial x} \quad \gamma_{yz} = \frac{\partial v}{\partial z} + \frac{\partial w}{\partial y}$$

$$\gamma_{xz} = \frac{\partial w}{\partial x} + \frac{\partial u}{\partial z} \quad (c)$$

These relations give the three strains along the coordinate axes and the distortions of the angles between those axes. If, for every direction, we draw a vector from point O of length $r = \sqrt{1/|\epsilon_r|}$ where ϵ_r is the strain in the direction r, the ends of these vectors will lie on a surface of the second degree. The principal axes of this surface give the *principal directions* of the deformation and the corresponding strains are the *principal strains*.

Cauchy also gives the relations between the six stress components and the six components of strain for an isotropic body. Assuming that the principal directions of deformation coincide with the directions of the principal stresses and that stress components are linear functions of the strain components, he writes the equations

$$\sigma_x = k\epsilon_x + K\theta, \quad \sigma_y = k\epsilon_y + K\theta, \quad \sigma_z = k\epsilon_z + K\theta$$

where k and K are two elastic constants and

$$\theta = \epsilon_x + \epsilon_y + \epsilon_z$$

that is, θ is the unit volume change. For any direction of the coordinate axes, he then obtains

$$\sigma_x = k\epsilon_x + K\theta, \quad \sigma_y = k\epsilon_y + K\theta, \quad \sigma_z = k\epsilon_z + K\theta$$
$$\tau_{xy} = \frac{k}{2}\gamma_{xy}, \quad \tau_{xz} = \frac{k}{2}\gamma_{xz}, \quad \tau_{yz} = \frac{k}{2}\gamma_{yz} \quad (d)$$

The relations (a) to (d) amount to a complete system of equations for solving problems of the elasticity of isotropic bodies. Cauchy himself applies these equations in studying deformations of rectangular bars. He was especially interested in the problem of torsion of rectangular bars and succeeded in finding a satisfactory solution for a bar of narrow rectangular cross section. He shows that cross sections of a twisted bar do not, in general, remain plane but warp during torsion. These conclusions, arrived at by Cauchy, were used later by Saint-Venant, who formulated a more complete theory of torsion of prismatical bars (see page 233).

27. Poisson (1781–1840)

S. D. Poisson was born in the small town of Pithiviers near Paris in a poor family and, until he was fifteen years old, he had no chance to learn more than to read and write. In 1796 he was sent to his uncle at Fontainebleau, and there he was able to visit mathematics classes. His work was so good in that subject that by 1798 he was able to pass the entrance examinations of the École Polytechnique with distinction. Here his excellent progress was noticed by Lagrange (who was giving his course of theory of functions at that time) and also by Laplace. After graduating, in 1800, Poisson remained at the school as an instructor in mathematics and by 1806 he had been put in charge of the course of calculus. His original publications in mathematics made him known as one of the best French workers in that field, and as early as 1812 he became a member of the French Academy. At that time, theoretical physics was undergoing rapid development, and the mathematicians of the day used their skill in solving physical problems theoretically.

Fig. 70. S. D. Poisson.

The theory of elasticity based on the idea of a molecular structure attracted Poisson's interest, and he did much to lay the foundations of that science.

The principal results obtained by Poisson are incorporated in two memoirs,[1] which he published in 1829 and 1831, and in his course of mechanics.[2] Starting his work with a consideration of a system of particles between which molecular forces act, he obtains the three equations of equilibrium and the three conditions at the boundary. These are similar to those given previously by Navier and Cauchy. He proves

---

[1] Mémoire sur l'équilibre et le mouvement des corps élastiques, *Mém. acad.*, vol. 8, 1829, and Mémoire sur les équations générales de l'équilibre et du mouvement des corps élastiques et des fluides, *J. école polytech.* (*Paris*), cahier 20, 1831. Todhunter and Pearson's "History of Elasticity" contains an extensive review of these two memoirs.

[2] "Traité de Mécanique," 2 vols., 2d ed. 1833.

that these equations are not only necessary but that they are also sufficient to ensure the equilibrium of any portion of the body. He succeeds in integrating the equations of motion and shows that if a disturbance is produced in a small portion of a body, it results in two kinds of waves.[1] In the faster moving wave, the motion of each particle is normal to the wave front and is accompanied by volume changes (or dilation); in the other wave the motion of each particle is tangential to the wave front and there is only distortion without volume change during the motion.

Fig. 71. M. V. Ostrogradsky.

In this memoir Poisson refers to the work of Ostrogradsky (see page 142). Applying his general equations to isotropic bodies,[2] Poisson finds that for simple tension of a prismatical bar, the axial elongation ϵ must be accompanied by a lateral contraction of magnitude $\mu\epsilon$ where $\mu = \frac{1}{4}$. The change of volume will then be $\epsilon(1 - 2\mu) = \frac{1}{2}\epsilon$. As a simple example of a three-dimensional problem, the writer discusses the stresses and deformations in a hollow sphere which is subjected to internal or external uniform pressure. He also treats purely radial vibrations of a sphere.

As a two-dimensional solution, Poisson obtains the equation for the lateral deflection of a loaded plate:

$$D\left(\frac{\partial^4 w}{\partial x^4} + 2\frac{\partial^4 w}{\partial x^2 \partial y^2} + \frac{\partial^4 w}{\partial y^4}\right) = q \qquad (a)$$

where the flexural rigidity D has the value corresponding to $\mu = \frac{1}{4}$ and q is the intensity of the load. He discusses boundary conditions. His conditions coincide with those which are now generally accepted for simply supported and for built-in edges. For an edge along which given forces are distributed, he required three conditions (instead of the two that we now use); these are that the shearing force, the twisting moment, and the bending moment (calculated from the molecular forces for any elemental length of the edge) must equilibrate the corresponding quantities furnished by the forces applied at the edge. The reduction of the

[1] See Mémoire sur la propagation du mouvement dans les milieux élastiques, *Mém. acad. . . . France*, vol. 10, pp. 549–605, 1830.

[2] The applications are discussed in the previously mentioned memoir of 1829.

The Beginning of the Mathematical Theory of Elasticity

number of conditions from three to two was accomplished later by Kirchhoff and a physical interpretation of that reduction was given by Lord Kelvin (see page 266). To demonstrate the application of his theory, Poisson considers bending of circular plates when the intensity of the load is a function of radial distance only. To do this, Poisson rewrites Eq. (a) in polar coordinates and gives a complete solution of the problem. Later on he applies this solution to a uniformly distributed load and gives equations for simply supported and clamped edges. He also considers lateral vibration of plates and solves the problem of a vibrating circular plate when the shape of deflections is symmetrical with respect to the center.

Poisson develops equations for the longitudinal, torsional, and lateral vibrations of bars and calculates frequencies of various modes of vibration.

In his book "Traité de mécanique," Poisson does not use the general equations of elasticity and derives others for deflections and vibrations of bars, assuming that, during deformation, their cross sections remain plane. For bending of prismatical bars, he not only uses the second-order equation, which expresses the proportionality of curvature of the elastic line to bending moment, but also the equation

$$B \frac{d^4y}{dx^4} = q \qquad (a)$$

where B is the flexural rigidity and q the intensity of lateral loading. Discussing various applications of Eq. (a), Poisson makes the interesting observation that, if q is a function of x which vanishes at the ends $x = 0$ and $x = l$ of the bar, it can always be represented by a sine series. Then Eq. (a) becomes

$$B \frac{d^4y}{dx^4} = \sum a_m \sin \frac{m\pi x}{l}$$

Integrating this equation and assuming that the deflection is zero at the ends, he obtains the solution

$$By = \frac{l^4}{\pi^4} \sum \frac{a_m}{m^4} \sin \frac{m\pi x}{l} + x(l-x)[Cx + C_1(l-x)]$$

where C and C_1 are constants to be determined so as to satisfy the end conditions. When the ends are simply supported, C and C_1 vanish, and we have

$$y = \frac{l^4}{B\pi^4} \sum \frac{a_m}{m^4} \sin \frac{m\pi x}{l}$$

It seems that here we have the first case in which trigonometric series are used in investigating the deflection curves of bars. In Poisson's time this method did not attract the attention of engineers, but now it has found useful applications.

We see that Poisson did not contribute such fundamental ideas to the theory of elasticity as Navier and Cauchy. However, he did solve many problems of practical importance and because of this we still use his results.

28. G. Lamé (1795–1870) and B. P. E. Clapeyron (1799–1864)

Both of these outstanding engineers graduated from the École Polytechnique in 1818 and from the school of mines in 1820. After completing these courses, they were recommended to the Russian government, as promising young engineers, to help in the work of the new Russian engineering school, the Institute of Engineers of Ways of Communication in St. Petersburg. The new school was later to have a powerful influence upon the development of engineering science in Russia; it was founded in 1809 with the collaboration of such outstanding French engineers as Bétancourt (who was the first director of the school), Bazain, and Potier. Its program and teaching methods were similar to those of the École Polytechnique and of the École des Ponts et Chaussées in Paris. Lamé and Clapeyron had to teach applied mathematics and physics in the school and they had also to help with the design of various important structures in which the Russian government was interested; for example, several suspension bridges were designed and erected[1] in St. Petersburg at that time.

Fig. 72. Augustin de Bétancourt.

To investigate the mechanical properties of the Russian iron used in these bridges, Lamé designed and built (1824) a special testing machine, similar to that of Lagerhjelm in Stockholm.[2] It was a horizontal testing

---

[1] These bridges (constructed 1824–1826) were the first suspension bridges built on the Continent of Europe. See Wiebeking, "Architecture civile," vol. 7, Munich, 1831. See also, G. C. Mehrtens, "Eisenbrückenbau," vol. 1, Leipzig, 1908. Lamé and Clapeyron also helped to design the large suspension bridge over the Neva River (with a single span of 1,020 ft). See *Ann. mines*, vol. XI, p. 265, 1825.

[2] The well-known Woolwich testing machine was built in 1832 on the same principle and also Kirkaldy's machine at the Grove, Southwark Street, London. See C. H. Gibbons, "Materials Testing Machines," Pittsburgh, 1935.

machine which incorporated a hydraulic ram for producing the load and for absorbing the strain. The magnitude of the load was measured by means of weights, as shown in Fig. 74. Lamé noticed, while conducting these tests, that the iron began to stretch rapidly under a load approximately equal to two-thirds of the ultimate strength. The peeling of the oxide scale and the formation of a neck after the yield point was passed were also observed. The results[1] of these tests were used in the design of iron structures in Russia and were mentioned in several books on strength of materials.[2]

In connection with the reconstruction of the cathedral of Saint Isaac in St. Petersburg, Lamé and Clapeyron examined the problem of stability of arches and prepared the memoir to which we have already referred (see page 84).

During their service in St. Petersburg, these two engineers wrote their important memoir "Sur l'équilibre interieur des corps solides homogènes" which was presented to the French Academy of Sciences, reviewed by Poinsot and Navier in 1828, and published in "Mémoires présentés par divers Savants," vol. 4, 1833. This memoir owes its importance to the fact that it contains not only a derivation of the equations of equilibrium (which were already known at that time from the work of Navier and Cauchy) but also some applications of these general equations to the solution of problems of practical interest. In the first section of the memoir they

Fig. 73. Gabriel Lamé.

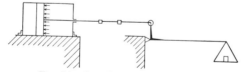

Fig. 74. Lamé's tensile-test machine.

deduce the equations of equilibrium using Navier's notion of the molecular forces (see Art. 25) and show that the same equations are obtained by using the concept of stresses, introduced by Cauchy. In the second

---

[1] The results of these tests, included in Lamé's letter to Baillet, were published in *Ann. mines*, vol. 10, Paris, 1825.

[2] See, for example, Navier, "Résumé des Leçons . . . ," 2d ed., p. 27, 1833.

section the stresses at a point of an elastic body are studied and it is shown that if, for each plane passing through the point, the corresponding stress is represented by a vector drawn from that point, the ends of all such vectors will be on the surface of an ellipsoid. This is the so-called *Lamé stress ellipsoid*. It is also shown how (by using this ellipsoid together with the stress-director quadric) the magnitude and direction of the stress can be obtained for any plane passing through the point.

After this general discussion the authors apply their equations to particular cases. They show how the single elastic constant which enters their equations can be obtained experimentally from tensile tests or from tests involving uniform compression. Then they take up the problem of a hollow circular cylinder and derive formulas for the stresses produced by uniform internal and external pressures. These formulas are used to calculate the necessary thickness of the wall if the magnitudes of the pressures are given. In their analysis they use the maximum stress theory; but they carefully state that each element of the cylinder is in a condition of two-dimensional stress and that the elastic limit obtained from a simple tensile test may be not applicable to this more complicated case. The next problems discussed in this section are those of the simple torsion of circular shafts, a sphere subjected to the forces of gravitation directed toward its center, and a spherical shell subjected to uniform internal or external pressure. In all these cases correct formulas are derived and these have since found many applications.

In the last section, the authors examine more complicated problems. They start with that of an infinite body bounded by a plane upon which given normal forces are distributed. Representing these forces by a Fourier integral, they succeed in obtaining expressions for the components of displacement in the form of quadruple integrals. They apply similar analysis to the case of a body bounded by two infinite parallel planes. Finally, the problem of a circular cylinder of infinite length is taken up. Here, for the first time, cylindrical coordinates are introduced. As an example, the torsion of a cylinder produced by tangential forces distributed over the surface and perpendicular to its axis is considered. It is assumed that the intensity of these forces is a function only of the distance measured along the axis of the cylinder.

Another important aspect of this memoir is that, apart from its interesting conclusions, it contains all the known theoretical results dealing with elastic deformations of isotropic materials. These are presented in a very clear and concise manner. Later, this paper was used by Lamé in the preparation of his famous book "Leçons sur la Théorie Mathématique de l'Élasticité des Corps Solides." This was the first book on theory of elasticity.

Lamé and Clapeyron returned to France in 1831 because, following a revolution in their home country, political relations between France and Russia had deteriorated. There they found that the interest of engineers was attracted by the growth of railroad construction in England and some plans existed for similar work in France. Lamé and Clapeyron helped to outline and construct the line between Paris and Saint-Germain. However, Lamé very soon abandoned this type of work and became professor of physics at the École Polytechnique, in which capacity he remained until 1844. During this period he published his course of physics (1837) and also several important papers on the theory of light.

In 1852, Lamé's book on the theory of elasticity (mentioned above) appeared. The results of the memoir which the author wrote with Clapeyron were included in it, but the form of the equations was somewhat changed since Lamé had come to the conclusion that to determine the elastic properties of an isotropic material, two elastic constants were required. This had been Cauchy's decision (see Art. 26). Lamé introduces problems on vibration and discusses motions of strings, membranes, and bars. The propagation of waves in an elastic medium is also covered. More space is devoted to the use of curvilinear coordinates and to the deformations of spherical shells than was the case in the memoir. Finally, Lamé discusses the application of the theory of elasticity to optics. In this application, he considers the double refraction of light and has to assume that the medium is *aeolotropic*. He changes the elasticity equations accordingly and finds the relations between the components of stress and of strain.

Concluding the review of Lamé's book, in his "History," Todhunter states: "The work of Lamé cannot be too highly commended; it affords an example which occurs but rarely of a philosopher of the highest renown condescending to employ his ability in the construction of an elementary treatise on a subject of which he is an eminent cultivator. The mathematical investigations are clear and convincing, while the general reflections which are given so liberally at the beginning and the end of the Lectures are conspicuous for their elegance of language and depth of thought."

After the publication of his book, Lamé continued to be interested in theory of elasticity and, in 1854, his "Mémoire sur l'équilibre d'élasticité des enveloppes sphériques" appeared in Liouville's *Journal de mathématiques*, vol. XIX. In it the writer gives the complete solution to the problem of the deformation of spherical shells under the action of any surface forces.

In 1859, Lamé's book "Leçons sur les coordonnées curvilignes et leurs diverses applications" was published. This book gives the general theory of curvilinear coordinates and their application to mechanics, to heat, and to the theory of elasticity. The transformation of the equations of elasticity to curvilinear coordinates is given and, as an example, the deformation of spherical shells is treated. In the closing chapters, Lamé discusses the principles upon which the fundamental equations of the theory of elasticity are based. He no longer favors the use of Navier's derivation of the equations (involving molecular forces) and prefers Cauchy's method (in which only the statics of a rigid body is used). Then he takes up Cauchy's assumption that the components of stress are linear functions of the components of strain. With isotropic materials, this assumption leads us to require two elastic constants which can be found from simple tension and simple torsion tests. In this way, all the necessary equations are obtained without making any hypotheses regarding molecular structure and molecular forces.

In 1843 Lamé was elected a member of the French Academy of Sciences, and in 1850 he became a professor at the Sorbonne, where he lectured on various branches of theoretical physics. He was not an outstanding lecturer, but his books were widely read and greatly influenced mathematical physicists for many years. Poor health forced Lamé to discontinue his teaching in 1863, and he died in 1870.

Clapeyron, after his return to France in 1831, was always active in practical work connected with the construction of French railroads. His main occupation was with the application of thermodynamics to locomotive design. Starting in 1844, he gave the course of steam engines at the École des Ponts et Chaussées and he was an excellent teacher. The combination of great theoretical knowledge with broad practical experience which he possessed especially attracted the students to his lectures. While engaged upon the design of a multispan bridge in 1848, Clapeyron evolved a new method of analyzing stresses in continuous beams, which will be discussed later (see page 145).

In his book on elasticity, Lamé describes another contribution of his former companion, which he calls *Clapeyron's theorem*. This theorem states that the sum of the products of external forces applied to a body, and the components of the displacements in the directions of these forces at their points of application, is equal to twice the value of the corresponding strain energy of the body. It appears that this theorem was enunciated by Clapeyron much earlier than the date on which Lamé's book was published, and that it marked the first time that the general expression for the strain energy of a deformed isotropic body had been derived. In 1858 Clapeyron was elected a member of the French Academy of Sciences. He continued his work at the Academy and at the École des Ponts et Chaussées until the time of his death in 1864.

29. The Theory of Plates

The first attacks made upon the problem of deflection of elastic surfaces were those of Euler. In describing the vibration of a perfectly flexible membrane, he regarded it as consisting of two systems of stretched strings perpendicular to each other (Fig. 75). He found[1] the relevant differential equation, *i.e.*,

$$\frac{\partial^2 w}{\partial t^2} = A \frac{\partial^2 w}{\partial x^2} + B \frac{\partial^2 w}{\partial y^2} \qquad (a)$$

where w denotes deflections and A and B are constants. Euler also used this idea in investigating vibrations of bells.

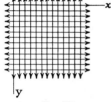

Fig. 75.

Jacques Bernoulli[2] (1759–1789) applied the same notion to the analysis of plates and, in this way, obtained[3] the differential equation

$$D \left(\frac{\partial^4 w}{\partial x^4} + \frac{\partial^4 w}{\partial y^4} \right) = q \qquad (b)$$

where D is the flexural rigidity of the plate and q is the intensity of the lateral load. Bernoulli made it quite clear that his equation was only an approximation and that by taking two systems of beams which are not perpendicular to each other a somewhat different result will be arrived at. He published the work only as a first attempt at solving the problem of bending of plates.

Great interest in the theory of plates was aroused by E. F. F. Chladni's work in acoustics[4] and especially by his experiments with vibrating plates. By covering a plate with fine sand, Chladni was able to show the existence of nodal lines for various modes of its motion and to determine the corresponding frequencies. In 1809 the French Academy invited Chladni to give a demonstration of his experiments, and Napoleon, who attended the meeting, was very impressed by them. At the emperor's suggestion the French Academy proposed, as the subject for a prize essay, the problem of deriving a mathematical theory of plate vibration and of comparing theoretical results with those obtained experimentally.

---

[1] See *Novi Comm. Acad. Petrop.*, vol. X, p. 243, 1767.

[2] Jacques was a nephew of Daniel Bernoulli. Between the years 1786 and 1789 he was a member of the Russian Academy of Sciences.

[3] *Nova Acta*, vol. V, 1789, St. Petersburg.

[4] The first edition of Chladni's "Die Akustik" appeared in Leipzig in 1802. A French translation was published in 1809 and in it a short autobiography is to be found.

In October, 1811, the closing date of the competition, only one candidate appeared, Mlle. Sophie Germain.

Sophie Germain[1] (1776–1831) became interested in mathematics while still quite young and learned Latin in order to read Newton's "Principia." She continued her study under the difficult conditions imposed by the French Revolution, and corresponded with the greatest mathematicians of her time, *viz.*, Lagrange, Gauss, and Legendre. After the battle of Jena, when the French army entered Göttingen, she greatly assisted Gauss by writing letters to General Pernetty, who was in command of the army of occupation, and by collecting money to pay the fine imposed by Pernetty on Gauss. When she heard about the prize that was offered by the Academy, she decided to work on the theory of plates. She was acquainted with Euler's work on elastic curves (see page 32) in which variational calculus was used to derive the differential equation of deflection from the integral representing the strain energy of bending. She decided to proceed in a similar manner and assumed that the strain energy of a plate is given by the integral

$$A \iint \left(\frac{1}{\rho_1} + \frac{1}{\rho_2}\right)^2 ds \qquad (c)$$

where ρ_1 and ρ_2 are the principal radii of curvature of the deflected surface. Sophie Germain made a mistake in calculating the variation of the integral (c), and therefore did not succeed in finding the correct equation. She did not win the prize. However, Lagrange,[2] who was one of the judges, noticed her error and, after making some corrections, he obtained a satisfactory form of the required equation:

$$k\left(\frac{\partial^4 w}{\partial x^4} + 2\frac{\partial^4 w}{\partial x^2 \partial y^2} + \frac{\partial^4 w}{\partial y^4}\right) + \frac{\partial^2 w}{\partial t^2} = 0 \qquad (d)$$

The Academy proposed the subject again, with a new closing date in October, 1813, and Sophie Germain again entered as a candidate for the prize. She now had the correct equation (d), but the judges required a physical justification of the fundamental assumption (c). She was again unsuccessful. But the Academy decided to offer the reward once more. The third time, Sophie Germain won the prize[3] (in 1816), although the judges were not quite satisfied with her work. Her paper does not contain a proper explanation of the basic expression (c) and we must agree with

---

[1] A biography of Sophie Germain is given in her book, "L'état des Sciences et des Lettres," which was published in Paris in 1833, after her death.

[2] See *Ann. chim.*, vol. 39, pp. 149, 207, 1828.

[3] Her work, "Recherches sur la théorie des surfaces élastiques," was published in Paris in 1821.

The Beginning of the Mathematical Theory of Elasticity

Todhunter's statement[1] that "the judges must have been far from severe, as they awarded the prize. . . . "

A further attempt to improve the theory of plates was made by Poisson.[2] To give a physical meaning to Eq. (d), Poisson assumes that the plate consists of particles between which molecular forces (proportional to changes in molecular distances) act. He succeeded in obtaining Eq. (d) from the conditions of equilibrium of this system of particles. But, since he assumes that all particles are distributed in the middle plane of the plate, the constant k in his equation (d) is proportional to the square of the plate thickness and not to the cube as it should be. In the same memoir, Poisson shows that Eq. (d) can be obtained not only from the integral (c) but also from the integral

$$A \iint \left[\left(\frac{1}{\rho_1} + \frac{1}{\rho_2}\right)^2 + m \left(\frac{1}{\rho_1^2} + \frac{1}{\rho_2^2}\right) \right] ds$$

which, with a proper selection of the constants A and m, is the correct expression for the strain energy of bending of an elastic plate. This explains why the correct form (d) of the differential equation for plates was found by Sophie Germain, although the quantity (c) does not give the strain energy of a bent plate.

Navier must be credited with having given the first satisfactory theory of bending of plates. In his paper (which was presented to the Academy on Aug. 14, 1820, and published in 1823),[3] Navier assumes, as Poisson did, that the plate consists of molecules; but he has them distributed through the thickness and assumes that their displacements during bending are parallel to the middle plane of the plate and proportional to the distance from that plane. Thus he finds the correct differential equation for any lateral loading, viz.,

$$D \left(\frac{\partial^4 w}{\partial x^4} + 2 \frac{\partial^4 w}{\partial x^2 \partial y^2} + \frac{\partial^4 w}{\partial y^4}\right) = q \qquad (e)$$

where q is the intensity of the load and D is the flexural rigidity of the plate. Since Navier also makes the assumptions which were described in Art. 25, his results are given in terms of only one elastic constant and his value of D coincides with the value which is now generally accepted when we take Poisson's ratio equal to one quarter. Navier applies his equation (e) to the problem of a simply supported rectangular plate for which he states the correct boundary conditions and gives the cor-

---
[1] See his "History," vol. 1, p. 148.
[2] See his memoir published in the *Mémoires* of the French Academy in 1814.
[3] See *Bull. soc. philomath. Paris*, 1823, p. 92.

rect solution in the form of a double trigonometric series. This he uses for the cases of a uniformly distributed load and a load concentrated at the middle. These solutions were the first satisfactory ones to be given for problems on the bending of plates.

Navier also considers lateral buckling of plates under the action of compressive forces T uniformly distributed along the boundary, and derives the correct differential equation for the buckled surface

$$D \left(\frac{\partial^4 w}{\partial x^4} + 2 \frac{\partial^4 w}{\partial x^2 \partial y^2} + \frac{\partial^4 w}{\partial y^4} \right) + T \left(\frac{\partial^2 w}{\partial x^2} + \frac{\partial^2 w}{\partial y^2} \right) = 0 \qquad (f)$$

He applies this equation to the complicated problem of a rectangular plate which is supported at the four corners. Here, however, he fails to reach an acceptable solution.

CHAPTER VI

Strength of Materials between 1833 and 1867

30. Fairbairn and Hodgkinson

During the first half of the nineteenth century, French engineers, with their superior mathematical background, were occupied with the mathematical theory of elasticity. In this period, English engineers studied strength of materials experimentally. England was then the leading country as regards industrial development and, after the work of James Watt, the machine industry grew rapidly. The construction of a reliable locomotive gave impetus to railroad construction and, whereas in 1827 England produced 690,500 tons of iron, the production was 3,659,000 tons in 1857. At this time of rapid industrial expansion, very little was done toward improving British engineering education and the numerous engineering problems that presented themselves had to be solved largely by self-taught men. These men had no great scientific knowledge and preferred the experimental method of attack. Although this experimental work did not contribute much to the general theory of strength of materials, it was of great use to practical engineers in providing answers to their immediate problems. The results of these tests were widely used not only in England but also on the Continent and were quoted in the engineering literature[1] of France and Germany. The best known of these English engineers are William Fairbairn and Eaton Hodgkinson.

Fig. 76. William Fairbairn.

[1] See, for example, A. Morin's "Résistance des Matériaux," 3d ed., Paris, 1862; G. H. Love, "Des diverses Résistances et outres Propriétés de la fonte, du Fer et de l'Acier . . . ," Paris, 1859; J. Weisbach, "Lehrbuch der ingenieur- und maschinen- Mechanik," 1845–1862 (3 vols.). The last book was translated into English, Swedish, Russian, and Polish.

William Fairbairn (1789–1874)[1] was born in Kelso, Scotland, the son of a poor farmer. After studying at an elementary school, he had to begin helping his father on the farm. However, at the age of fifteen, he started to work as an apprentice in mechanical engineering at Percy Main Colliery near North Shields. Fairbairn was very diligent; after a day's work in the power station he would spend his evenings in studying mathematics and English literature. During the seven years of the apprenticeship he considerably improved his education in this way. In 1811, after taking up his indentures, Fairbairn moved to London where, at that time, the Waterloo Bridge was under construction. The fame of the builder, John Rennie (1761–1821),[2] attracted Fairbairn but he was unable to find work on the bridge and had to continue as a mechanic. He spent his evenings in libraries and continued in this way to make up for the schooling which had been denied him. After two years in London, and short periods spent in the south of England and in Dublin, Fairbairn moved to Manchester in 1814. There he expected to find a better market for his knowledge as a mechanic and at first he worked there in various workshops. But he started his own business in 1817 in partnership with another mechanic, James Lillie.

From the very beginning, he was in constant contact with the mechanical installations of cotton mills. This was new to him but, by careful analysis, he soon discovered the main defects of these mills. He found that they were driven by means of large square shafts which carried wooden drums of considerable diameter and ran at the very low speed of 40 rpm. By increasing the speed, he was able to reduce the diameter of the drums. He also improved the couplings and, in this way, arrived at a more reliable and economical system. This success brought him a great volume of work in the cotton industry, and he became known as an expert in mechanical equipment. In 1824, Fairbairn was asked by the Catrine Cotton Works in Scotland to improve their water power station. Here again he was able to introduce improvements, this time into the design of water wheels, and his work in this field won such recognition that he was soon designing and building hydraulic power stations. In this respect his fame was international and he was invited to Switzerland to remodel and erect water wheels for G. Escher of Zurich. His advice was also asked by manufacturers in Alsace.

About this time the production of malleable iron was considerably improved, and Fairbairn became interested in studying mechanical properties of the new material and its application to various engineering

[1] For Fairbairn's biography, see "The Life of Sir William Fairbairn, Bart." by William Pole, London, 1877.

[2] For John Rennie's biography, see Smiles, "Lives of the Engineers," vol. 2.

structures. He met Eaton Hodgkinson, who was already well known for his work in strength of materials, and proposed that the latter should carry out experiments in his works. A machine for testing materials (Fairbairn's lever) was designed and built (Fig. 77), and most of the research work of Fairbairn and Hodgkinson was made with it. At the request of the British Association for Advancement of Science, they started by studying the mechanical properties of cast iron. While Hodgkinson made experiments in tension and compression, Fairbairn undertook bending tests. In his work, he was especially interested in

Fig. 77. Fairbairn's lever.

time effect and temperature effect. He found that after application of the load the deflection of a bar increased with time and he wanted to find the limiting load below which this was not encountered. His study of the temperature effect showed that the breaking load decreased considerably with increase of temperature.

In the 1830's, Fairbairn became interested in iron ships and made an extensive experimental study of the strength of wrought-iron plates and their riveted joints.[1] He found that the strength of iron plates is about the same in all directions. Using riveted joints of each kind, he showed the higher quality of machine (as opposed to hand) riveting. The use of iron plates in which Fairbairn became most interested was that in iron

[1] The results of this investigation, made in 1838–1839, were published much later. See *Phil. Trans.*, part II, pp. 677–725, 1850.

bridges. The experimental work that he did in connection with the design and construction of such bridges will be discussed later (see page 156).

Fairbairn investigated the resistance to collapse of tubes submitted to external pressure in his work on the design of boilers.[1] His experiments with tubes of various sizes, all having fixed ends, showed that the critical value of the external pressure is given with good accuracy by the formula $p_{cr} = Ch^2/ld$, where h is the thickness of the wall, l the length, and d the diameter of the tube. C is a constant depending on the elastic properties of the material. To get tubes and spherical shells of more perfect form, Fairbairn later used glass. He examined the elastic properties and found the ultimate strength of several kinds of glass and used tubes and globes made of it in his experimental determination of the critical value of the external pressure.[2]

Fairbairn presented the results of his numerous experiments in a series of lectures and later in form of books which became very popular among engineers[3] of that time. In recognition of his scientific work and for his contributions to the designs of the Conway and Britannia tubular bridges, Fairbairn was elected a Fellow of the Royal Society. In 1860, he was awarded a medal by that body.

He was chosen to be one of the jurors of the Paris Exhibition of 1855 and, after its conclusion, he made a report which contains the following interesting remarks: "I firmly believe, from what I have seen, that the French and Germans are in advance of us in theoretical knowledge of the principles of the higher branches of industrial art; and I think this arises from the greater facilities afforded by the institutions of those countries for instruction in chemical and mechanical science . . . . Under the powerful stimulus of self-aggrandisement we have perseveringly advanced the quantity, whilst other nations, less favoured and less bountifully supplied, have been studying with much more care than ourselves the numerous uses to which the material may be applied, and are in many cases in advance of us in quality." This experience caused Fairbairn to become interested in education. He contacted the men who, at that time, were conducting an enquiry into the ignorance of the masses and stated emphatically that England would have to improve its educational system.

Eaton Hodgkinson (1789–1861) was born in Anderton near Northwich,

[1] See *Phil. Trans.*, 1858, pp. 389–413. A very complete discussion of this work is given in Morin's book mentioned before (see p. 123).

[2] See *Phil. Trans.*, 1859, pp. 213–247.

[3] See his "Useful Information for Engineers," London, 1856, and, with the same title, 2d series, London, 1860. See also the book "On the Application of Cast and Wrought Iron to Building Purposes," London, 1856; 2d ed., 1858.

Cheshire, in the family of a farmer. He had only an elementary education, and while still a youngster, he started to work on the farm and to help his widowed mother. In 1811 his family moved to Manchester and there Hodgkinson met the famous scientist John Dalton (1766–1844), who noticed the young man's alertness and started to teach him mathematics. In that way, Hodgkinson had a chance to study the classical work of the Bernoullis, Euler, and Lagrange.

In 1822, he started his work on strength of materials and two years later published his paper "On the Transverse Strain and Strength of Materials."[1] Considering bending of prismatic bars, he observes that, at each cross section, the summation of tensile stresses must be equal to that of the compressive stresses and, assuming that the cross sections remain plane, he concludes: "As the extension of the outer fibre on one side is to the contraction of that on the other, so is the distance of the former from the neutral line to that of the latter." This result was not new, since it had been known from the time of Parent and Coulomb, but the merit of the paper, as stated by Todhunter, lies in the fact that it led practical men in England to place the neutral line in its true position. In his analysis, Hodgkinson does not use Hooke's law, and he assumes the tension to be proportional to some power of the extension. For greater generality, he assumes this power to have a different value for the relation between contraction and compression force. In the second part of his paper, the writer gives the results of his experiments with wooden beams and shows that within the elastic limit Hooke's law applies.

In his second paper,[2] Hodgkinson describes his experiments with the bending of cast-iron beams and shows that, when the load on the beam increases, the neutral line changes its position. He finds that the moduli of elasticity, elastic limits, and ultimate strength for cast iron have different values for tension and compression. With this information, Hodgkinson began an experimental search for the form of a cast-iron beam such that, for a fixed amount of material, it would have the maximum carrying capacity. He showed experimentally that an I beam with equal flanges has not the best form and demonstrated that the cross-sectional area of the flange working in tension must be several times larger than that of the one suffering compression.

The next subject of Hodgkinson's practical work was that of horizontal impact on a beam. The results that he obtained agreed well with those predicted by elementary theory, which is based upon the assumptions

[1] Published in the *Mem. Proc. Manchester Lit. & Phil. Soc.*, vol. IV, 1824.

[2] "Theoretical and Experimental Researches to Ascertain the Strength and Best Forms of Iron Beams," Manchester Literary and Philosophical Society, 1830, pp. 407–544.

that (1) the deflection curve of a beam bent by impact is the same as if it had been bent statically and that (2) the striking body and beam proceed together as one mass after impact. In calculating the common velocity of the striking body and of the beam just after impact, one-half of the mass of the beam was considered.[1]

Hodgkinson published the results of his work at Fairbairn's plant.[2] Figure 78 shows pictures of some crushed compression test specimens of cast iron and, regarding these, Hodgkinson notes that all fractures had one thing in common, *i.e.*, in each case a cone or a wedge was formed which slid off at a nearly constant angle. In the same paper, the results of bending tests on cast-iron beams with unequal flanges are given. They show that the maximum carrying capacity of the beam is obtained when the cross-sectional area of the bottom and the top flanges are in the ratio as 6 or $6\frac{1}{2}$ to 1. This ratio is approximately equal to the ratio of the compressive to the tensile strength of the cast iron.

In 1840, Hodgkinson presented his investigation on buckling of columns to the Royal Society.[3] His purpose was to verify the theoretical formula of Euler. The general arrangement of these experiments and the kinds of test specimens can be seen from Fig. 77. Cylindrical, solid, and hollow specimens with rounded and flat ends were tested. For slender, solid struts, good accord was found with Euler's formula (*i.e.*, the ultimate load was proportional to d^4/l^2, where d is the diameter and l is the length of the strut). It was also found that a strut with flat ends has the same strength as a shorter strut of half its length which has rounded ends.

In the case of shorter struts, considerable deviations from the theoretical formula were found. To approximate the results obtained for solid cylinders with rounded ends and with lengths varying from 121 times the diameter down to 15 times, Hodgkinson had to take the ultimate load as proportional to $d^{3.6}/l^{1.7}$.

In connection with the design of the Britannia and Conway tubular bridges, Hodgkinson, together with Fairbairn, conducted some very important experiments on bending and buckling of thin-walled tubes. These experiments will be discussed later (see p. 156).

Hodgkinson became the professor of the mechanical principles of engineering in University College, London, in 1847, and, during his

[1] Hodgkinson's investigations on impact were published in *British Association Reports* for the years 1833–1835.
[2] This work was published in the *British Association Report* of 1837–1838. Most of the material was reproduced in the book "Experimental Researches on the Strength and Other Properties of Cast Iron," London, 1846. This book was translated in French; see *Ann. ponts et chaussées*, vol. 9, pp. 2–127, 1855.
[3] See *Phil. Trans.*, part II, pp. 385–456, 1840.

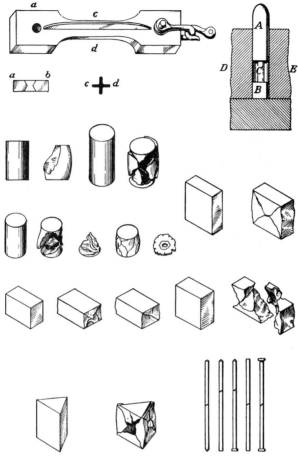

FIG. 78. Crushed compression-test specimens.

tenure of the chair, he participated as a member of the Royal Commission appointed to inquire into the application of iron to railway structures. In 1841, Hodgkinson was awarded a Royal medal for his work on strength of pillars and, in the same year, he was elected a Fellow of the Royal Society.

31. The Growth of German Engineering Schools

Following the Napoleonic wars, the German economic systems had to be rebuilt. It was necessary to promote industry and for this purpose several engineering schools were founded closely following the pattern of the French École Polytechnique. It was decided to build engineering education around a central core of preparation in the fundamental sciences of mathematics, physics, and chemistry. Engineers were to receive a general education comparable with that which was necessary in other professions and which was obtainable in the universities. The new engineering schools were given the status of universities and they were to train engineers who would not only be able to solve technical problems but who would be equipped to carry on the development of engineering sciences. While these fundamental ideas were taken from the École Polytechnique, there was some notable difference. For whereas the French establishment was only a preparatory school for young men who were later to undergo their engineering training in one of the specializing schools (such as the École des Ponts at Chaussées, the École des Mines, or the military academy), the German schools were to cover the complete course of study. During their first two years of study, the students were to receive instruction in the sciences. The second two were to be devoted to one of the branches of engineering. This new system permitted a better regulation of the amount, and type, of theory to be given.

There was another important difference. The French schools trained engineers mainly for governmental service. But the German schools, from the very start, worked in close liaison with private enterprise and helped immensely in the growth of industry. There was also a great administrative difference between the schools. The École Polytechnique had a military regime, while the new schools had the status of German universities and were conducted according to the principle of academic freedom. They were to be autonomous institutions in which the professors' council elected the president of the school and the deans of the various divisions. The students were to be allowed considerable freedom in planning their work and in the way they spent their time.

This new kind of engineering education proved very successful, and German engineering schools very soon became an important factor in the advancement of industry and the engineering sciences. The social status of a professor has always been a very high one in Germany and, as a result, the engineering schools were able to attract the best engineers to teach and to engage in scientific work.

As far as engineering mechanics was concerned, German science was at first greatly influenced by the French books of Navier, Poisson, Poncelet, and others. But before long German engineers began to strike out on their own; for the abstract presentation of mechanics which was so popular in the École Polytechnique did not satisfy them. Thus in this period (1833–1867) under consideration, more practical books in engineering mechanics began to appear, and we shall see the way in which they affected the development of strength of materials.

Strength of Materials between 1833 and 1867

Let us turn our attention to the activity of the most prominent German professors of mechanics of this era. J. Weisbach wrote books on "Mechanics of Machinery and Engineering" which enjoyed a high reputation not only in Europe but also in America, where an English translation was published.[1]

Julius Weisbach (1806–1871) graduated from the famous School of Mines (Bergakademie) in Freiberg in 1826 and later, in order to improve his knowledge of the fundamental sciences, he studied for two years (1827–1829) at the University of Göttingen and for one year at the Vienna Polytechnical Institute. After this study, Weisbach rejected an opportunity to work as a mining engineer and stayed at Freiberg, where he preferred to earn his living by giving private lessons in mathematics. In 1833, he was called upon by the Academy of Mines of Freiberg to teach applied mathematics, and from 1836 to the end of his life he was a professor of mechanics and machine design at that school. The principal achievements of Weisbach concerned hydraulics and in his work he followed Poncelet's methods and succeeded in obtaining important practical advances by combining experimental results with very elementary theoretical analysis.

FIG. 79. Julius Weisbach.

Problems of strength of materials are ably dealt with in his book on mechanics and his original work in this field was centered around the design of machine parts which are subjected to the action of combined stresses.[2] He uses the maximum strain theory as the basis for his selection of safe dimensions for machine parts. This method of analysis was suggested by Poncelet (see p. 89) and followed by Saint-Venant.

J. Weisbach was very interested in methods of teaching engineering mechanics and he organized a laboratory in which students had to verify the principles of statics, dynamics, and strength of materials experimentally.[3] His students carried out experiments on bending of solid and

---

[1] Edited by W. R. Johnson, Philadelphia, 1848.
[2] Z. Ing., vol. 1, pp. 252–265, 1848.
[3] Weisbach published a description of these problems in the journal *Civiling.*, vol. 14, pp. 339–370, 1868.

built-up beams, models of trusses, torsion of shafts, and combined torsion and bending. For these tests wooden models were used and they were of such proportions that small forces would produce deformations which were large enough to permit easy measurement. It appears that this was the first time that students had had to work experimentally in strength of materials.

F. Redtenbacher (1809–1863) was another outstanding German professor of this period, who advocated the introduction of scientific analysis into machine design. He graduated in 1829 from the Vienna Polytechnical Institute, and since he was an outstanding student, he was invited to take up the position of instructor in engineering mechanics at the Institute. In 1833 he became a teacher of mathematics at the Zurich technical school.[1] Here he supplemented his teaching activities with the work of a designer at the Escher and Wyss Manufacturing Company so that he acquired practical experience in machine design. The Polytechnical Institute at Karlsruhe offered him the professorship in applied mechanics and machine design in 1841, and he worked there as a professor (and later as rector of the school) to the end of his life. Combining his wide knowledge of mechanics with his practical experience, he completely reorganized the teaching of machine design and introduced theoretical analysis into the solution of design problems. He became a leader in the growth of engineering education in Germany. His publications were used by many mechanical engineers[2] and his methods were adopted in industry. He was interested in the application of strength of materials to the determination of safe dimensions of machine parts and, in his books "Principien der Mechanik und des Machinenbaues" (1852) and "Der Machinenbau" (1862–1865), we find solutions of such problems as the analysis of stresses in hooks, leaf springs, helical springs, elliptic rings of chains, etc. At that time, little was known regarding the

FIG. 80. F. Redtenbacher.

[1] In 1855, this school was reorganized into the Eidgenössische Technische Hochschule in Zürich by General Dufour, a former pupil of the École Polytechnique in Paris.

[2] His book "Resultate für den Machinenbau," 1848, was translated into French.

Strength of Materials between 1833 and 1867

analysis of such machine parts, and Redtenbacher was a pioneer in that subject.[1]

After Redtenbacher's death, F. Grashof was elected to the chair of applied mechanics at the Karlsruhe Polytechnical Institute. Grashof[2] (1826–1893) obtained his engineering education at the Gewerbeinstitut in Berlin. After his graduation, he spent $2\frac{1}{2}$ years as a seaman in a sailing ship which visited India, Australia, and Africa. Upon returning home (in 1851) he started to prepare himself for the teaching profession and began lecturing in applied mathematics in 1854 at the Gewerbeinstitut.

FIG. 81. F. Grashof.

In 1856, he helped to launch the Verein deutscher Ingenieure and became the editor of its journal. In this journal he published a number of articles dealing with various problems of applied mechanics. This activity brought him a measure of fame and led to his election by the Karlsruhe Polytechnical Institute to replace Redtenbacher. Here Grashof continued to teach in the various branches of engineering mechanics.

He was much interested in strength of materials and, in 1866, published his "Theorie der Elasticität und Festigkeit." In this book he does not confine himself to elementary strength of materials, but introduces the fundamental equations of the theory of elasticity. These he applies to the theory of bending and torsion of prismatical bars and to the theory of plates. In discussing bending of bars, he finds solutions for some shapes of cross section which were not considered by Saint-Venant, who was the originator of the rigorous theory of bending of prismatical bars (see page 237). In giving formulas for the design of machine parts, Grashof takes as his criterion of strength the maximum strain theory and extends Weisbach's investigation on combined stresses.[3] In several places Grashof develops his own methods of solution when dealing with particular problems and he finds some new results. For instance, in treating

---

[1] These solutions of Redtenbacher are quoted in V. Contamin's book "Résistance Appliquée," Paris, 1878.

[2] See "Franz Grashof, ein Führer der deutschen Ingenieure" by Wentzcke. See also R. Plank, Z. Ver. deut. Ing., vol. 70, p. 933, 1926.

[3] See Z. Ver. deut. Ing., vol. 3, pp. 183–195, 1859.

bending of bars with hinged ends, he examines the effects of the longitudinal tensile force which would be produced during bending if the hinges were immovable and shows how the corresponding tensile stresses can be computed. These calculations indicate that in the bending of slender bars this longitudinal force becomes important and its effect on the deflection and on the stresses cannot be neglected.

Treating cylindrical shells which are subjected to internal pressure, Grashof does not merely apply Lamé's formula but discusses the local bending stresses which are produced when the edges of the shell are rigidly connected with the end plates. In this analysis he uses the differential equation governing the deflection of longitudinal strips which are cut out from the shell by two radial sections.[1] Grashof gives complete solutions for several cases of symmetrically loaded circular plates. He also considers uniformly loaded rectangular plates and presents some approximate solutions for several cases.

In presenting his work, Grashof prefers the use of analytical methods and seldom uses figures to illustrate his meaning. He usually begins with the discussion of a problem in its most general form and only later, when a general solution has been found, does he introduce the simplifications afforded in specific cases. This manner of presentation makes reading difficult; therefore the book was not popular with practical engineers and was too difficult for the majority of engineering students. However, the more persevering students could gain a very thorough knowledge of the theory of strength of materials from it. To this day, Grashof's book has not lost its interest, as it was the first attempt to introduce the theory of elasticity into a presentation of strength of materials for engineers.

Fairbairn's experiments dealing with the collapse of circular tubes when submitted to uniform external pressure (see page 126) led Grashof to a very interesting theoretical investigation on buckling of thin circular tubes of great length.[2] Assuming that the cross section of an initially circular tube becomes elliptical upon buckling, Grashof finds that the critical value of pressure is given by formula

$$p_{cr} = \frac{2Eh^3}{d^3}$$

where h is the thickness of the tube and d is its diameter. In his analysis, Grashof considers an elemental ring and disregards the stresses acting between the adjacent rings of which the tube is composed. This explains why this formula contains the modulus E, instead of $E/(1 - \mu^2)$.

[1] It seems that local bending at the ends of circular tubes and at stiffening rings was investigated first by H. Scheffler; see "Organ für Eisenbahnwesen," 1859. See also E. Winkler, *Civiling.*, vol. 6, 1860.

[2] See Grashof's paper in *Z. Ver. deut. Ing.*, vol. 3, p. 234, 1859.

Strength of Materials between 1833 and 1867 135

German experimental work on the mechanical properties of structural materials progressed during the middle third of the nineteenth century. A. K. von Burg, at the Vienna Polytechnical Institute carried out tests on the strength of steel plates.[1] K. Karmarsch, at the Hannover Polytechnicum, studied the properties of metal wires of various diameters.[2] W. Lüders examined the network of orthogonal systems of curves which appear on the surface of soft cast-steel specimens during bending and in other cases where materials undergo considerable straining. He showed that these curves become more prominent if etched with a weak solution of nitric acid.[3]

In 1852 L. Werder designed and built (at Nuremberg) the 100-ton testing machine for checking the tension members of the bridges that were designed by W. Pauli. This machine proved very accurate and was suitable for testing large structural members. Subsequently most European laboratories installed copies of this machine, and there is no doubt that a large proportion of the research work in mechanics of materials conducted during the second half of the nineteenth century was completed on Werder's machines.

32. Saint-Venant's Contributions to the Theory of Bending of Beams.

The main work of Saint-Venant concerns the mathematical theory of elasticity, and it will be discussed later. But he also contributed much to elementary strength of materials (especially to the theory of bending of bars).[4] He was the first to examine the accuracy of the fundamental assumptions regarding bending, *viz.*, (1) that cross sections of a beam remain plane during the deformation and (2) that the longitudinal fibers of a beam do not press upon each other during bending and are in a state of simple tension or compression. He demonstrates that these two assumptions are rigorously fulfilled only in uniform bending when the beam is subjected to two equal and opposite couples applied at the ends. Considering the pure bending of a rectangular beam (Fig. 82a), he shows that the changes in length of fibers, and the corresponding lateral deformations, not only satisfy the above conditions but also the condition of continuity of deformation. He also shows that the initially rectangular cross section changes its shape as shown in Fig. 82b, that is, due to lateral contraction of the fibers on the convex side and expansion on the concave

---

[1] See *Sitz. Akad. Wiss. Wien Math.-Naturer. Klasse*, vol. 35, pp. 452–474, 1859.
[2] *Polytech. Zentr.*, 1859, Leipzig.
[3] *Dinglers Polytech. J.*, Stuttgart, 1860.
[4] Most of Saint-Venant's work in the elementary theory of strength of materials is collected in his notes to the third edition of Navier's "Résumé des Leçons . . . ," Paris, 1864, and also in his lithographed course "Leçons de mécanique appliquée faites par intérim par M. de St.-Venant," 1837–1838.

side, the initially straight line ab becomes slightly bent and the corresponding radius of curvature is ρ/μ, where μ is Poisson's ratio and ρ is the radius of curvature of the axis of the bent bar. Due to this lateral deformation, the distances of the neutral fibers a and b from the upper and lower surfaces of the bar are also slightly altered. The upper and the lower surfaces will be bent to *anticlastic* surfaces. This was the first time that the distortion of the shape of the cross section of a bent bar had been investigated.[1]

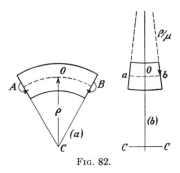

FIG. 82.

Taking a cantilever loaded at the free end (Fig. 83), Saint-Venant shows that shearing stresses act in the planes of cross section such as ab and a_1b_1 and that, due to the presence of these stresses, the cross sections do not remain plane during bending, but suffer warping, as shown in the figure. Since this warping is the same for any two cross sections, it produces no changes in the lengths of fibers and thus will not affect the bending stresses which are calculated on the assumption that the cross sections remain plane during bending.

Navier always assumed, in his book, that the neutral axis is perpendicular to the plane in which the bending forces act. Persy[2] was the first to show that this is legitimate only if the plane xy, in which the bending

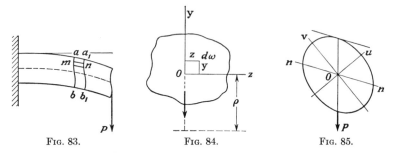

FIG. 83. FIG. 84. FIG. 85.

forces act, intersects cross sections of the beam along one of the principal axes of inertia (Fig. 84). Only then does the moment of internal forces with respect to the y axis $\left(\text{equal to } \dfrac{E}{\rho} \int_\Omega zy\,d\omega\right)$ vanish and the moment of

---

[1] A determination of Poisson's ratio by an optical measurement of the ratio of the two principal curvatures of the anticlastic surfaces was made by Cornu; see *Comp. rend.*, vol. 64, p. 333, 1869.

[2] See Navier, "Résumé des Leçons . . . ," 3d ed., p. 53.

the same forces with respect to z axis balance the external moment. Saint-Venant showed how the direction of the neutral axis can be determined if the plane in which the bending forces act does not pass through the principal axis of the cross sections. If Fig. 85 represents the ellipse of inertia of the cross section of the beam with principal axes ou and ov and if oP is the plane in which the forces lie, then he shows that the neutral axis nn is parallel to the tangent drawn at the points of intersection of the ellipse with the plane OP.

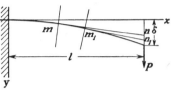

FIG. 86.

Saint-Venant shows[1] that the deflection of a cantilever can be calculated in an elementary way without integration of the appropriate differential equation and introduces the device which now is called the *area-moment method*. The deflection nn_1, corresponding to the curvature of an element mm_1 of the deflection curve (Fig. 86), is

$$\frac{\overline{mm_1}}{\rho}\,\overline{mn} \approx \frac{dx}{\rho}(l - x)$$

Observing now that the curvature at any cross section is equal to

$$\frac{M}{EI} = \frac{P(l - x)}{EI}$$

he obtains the expression

$$\delta = \frac{P}{EI}\int_0^l (l - x)^2\, dx = \frac{Pl^3}{3EI}$$

for the deflection. The required integral can be calculated as the statical moment of the triangle representing the bending-moment diagram. In a similar way he finds the deflection produced by a uniformly distributed load.

Saint-Venant examines large deflections of the cantilever such that the curvature $1/\rho$ cannot be replaced by the approximate value d^2y/dx^2. He gives his solution in the form of a series by means of which the deflection δ can be calculated with any desired degree of accuracy.[2]

All the results discussed above were obtained assuming Hooke's law to prevail. Now Saint-Venant takes up the bending of beams the material of which does not follow this law.[3] He makes a very complete study of the problem by assuming that cross sections of the beam remain plane

[1] Navier, "Résumé des Leçons . . . ," 3d ed., p. 72.
[2] Navier, "Résumé des Leçons . . . ," 3d ed., p. 73.
[3] See Navier, "Résumé des Leçons . . . ," 3d ed., p. 175.

during bending and that, instead of following a linear law, stresses and strains are related by the equations

$$\sigma = A\left[1 - \left(1 - \frac{y}{a}\right)^m\right] \quad \text{for tension}$$

$$\sigma_1 = B\left[1 - \left(1 - \frac{y}{a_1}\right)^{m_1}\right] \quad \text{for compression} \quad (a)$$

where A, B, a, a_1, m, m_1 are certain constants and y is the absolute value of the distance of a point from the neutral axis. Limiting the problem to that of a rectangular beam of cross section bh, the distances y_1 and y_2 of the most remote fibers from the neutral axis may be found from the two equations

$$y_1 + y_2 = h \qquad \int_0^{y_1} \sigma \, dy = \int_0^{y_2} \sigma_1 \, dy \qquad (b)$$

Having the position of the neutral axis, he can calculate the moment of internal forces by using Eq. (a).

Saint-Venant considers the special case where $y_1 = a$ (which means that $A = \sigma_{\max}$). He further assumes that the two curves represented by Eqs. (a) have a common tangent at $y = 0$, so that the modulus of the material is the same for tension and compression if the stresses are small.

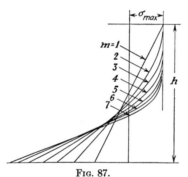

Fig. 87.

If, in addition, the two curves (a) are identical (that is, $m = m_1, a = a_1, A = B$), Saint-Venant finds that the moment of internal forces is given by the equation $M = \alpha b h^2 \sigma_{\max}/6$, where α is a factor whose value is dependent upon the magnitude of the quantity m. The values of this factor are given in the table below.

m	1	2	3	4	5	6	7	8
α	1	$\frac{5}{4}$	$\frac{27}{20}$	$\frac{7}{5}$	$\frac{10}{7}$	$\frac{81}{56}$	$\frac{35}{24}$	$\frac{22}{15}$ = 1.467

We see that when m increases, α also increases and approaches the value $\frac{3}{2}$, which is obtained when the tensile and compressive stresses are distributed uniformly.

When $m_1 = 1$, the material follows Hooke's law in compression. The corresponding stress distributions for various values of m are shown in Fig. 87. In the table below, values of the ratio y_1/y_2, defining the position of the neutral axis, are given. Values of the ratio of the maximum compressive stress to the maximum tensile stress and values of the factor α in the formula

$$M = \frac{\alpha \sigma_{\max} b h^2}{6}$$

are also shown.

m	1	2	3	4	5	6	7	∞
$\|\sigma_{\min}\|/\sigma_{\max}$	1	1.633	2.121	2.530	2.887	3.207	3.500	∞
$y_1:y_2$	1	1.225	1.414	1.581	1.732	1.871	2.000	∞
α	1	1.418	1.654	1.810	1.922	2.007	2.074	3

For $m = 6$, we have approximately $M = \sigma_{\max} b h^2/3$ which coincides with Mariotte's theory (see page 22) and with the values of M_{ult} obtained experimentally by Hodgkinson for rectangular cast-iron beams. When $m = \infty$, we arrive at Galileo's theory by which the tensile stresses are uniformly distributed over the cross section and the compressive stresses give a resultant which is tangential to the concave surface of the beam. This analysis for the bending of rectangular beams which do not follow Hooke's law can be used for calculating the ultimate bending moment for beams which do not possess rectangular cross sections. If, for example, experiments with rectangular cast-iron beams show that an accurate value of M_{ult} is predicted by taking $m = 6$ and $m_1 = 1$, we can use these values to calculate M_{ult} for other shapes of cross section, proceeding in the same manner as for rectangular beams.

Fig. 88.

In an elementary discussion of pure bending of beams (see Fig. 82), Saint-Venant formulates the principle which now carries his name. He states that the stress distribution found for this case conforms to the rigorous solution only when the external forces applied at the ends are distributed over the end cross sections in the same manner as they are distributed over intermediate cross sections. But (he states[1]) the solution obtained will be accurate enough for any other distribution of the forces at the ends, provided that the resultant force and the resultant couple of the applied forces always remain unchanged. He refers to some experiments which he carried out with rubber bars and says that those experiments showed that if a system of self-equilibrating forces is distributed on a small portion of the surface of a prism, a substantial deformation will be produced only in the vicinity of these forces. Figure

[1] See Navier, "Résumé des Leçons . . . ," 3d ed., p. 40.

88 shows one of Saint-Venant's examples. The two equal and opposite forces which act on the rubber bar produce only a local deformation at the end, and the remainder is practically unaffected. *Saint-Venant's principle* is often used by engineers in analyzing stresses in structures. We shall discuss this notion again when Saint-Venant's work on torsion and bending of prismatical bars is considered (see page 233).

The French elastician also investigated the bending of curved bars and introduced additional terms, representing displacements due to the stretching of the axis of the bar and due to shear, into Navier's formulas (see page 78). As examples, he discusses the deformation of a circular ring which is suspended in a vertical plane under the action of gravity forces and that of a circular ring placed vertically upon a horizontal plane and loaded on the top.

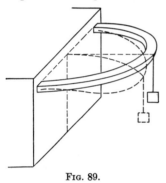

FIG. 89.

Saint-Venant points out that, in describing the deformation of a bar having double curvature, it is not sufficient to specify the shape of the center line, since deformations and stresses can be produced while the center line remains undeflected. Imagine (he says) an elastic circular wire of double curvature, which is placed in a channel of the same shape and which has the same cross section as the wire. It will be possible to rotate the wire in the channel. Due to this rotation, the longer fibers will be shortened and the shorter ones stretched although the shape of the center line will not be changed. From this it follows that the bending and torsional stresses in the wire cannot be completely defined by the deformation of the center line and that it is necessary to account for the angle of rotation of cross sections with respect to the center line of the wire. Saint-Venant does not confine his attention to a general discussion of this subject but applies his theory to particular problems. He gives a solution for a bar with a circular axis which is submitted to forces perpendicular to the plane of the axis (Fig. 89) and also derives an equation for the extension of a helical spring when the forces applied at the ends act along the axis of the helix.[1]

When allowable stresses in structures have to be fixed, Saint-Venant always assumes that the elastic limit of a material is reached when the

---

[1] These results were given in Saint-Venant's earlier memoirs which were collected and reprinted under the title "Mémoires sur la Résistance des Solides suivis de deux notes sur la flexion des pièces à double courbure," Paris, 1844.

distances between the molecules are increased, by the deformation, beyond a certain value which is peculiar to any given material. Thus, his formulas for calculating safe dimensions of structures are derived from maximum strain considerations. For instance, to cover the combined bending and torsion of shafts, he establishes a formula for the maximum strain; this has seen service at the hands of engineers ever since its appearance.

Saint-Venant was the first to show that pure shear is produced by tension in one direction and an equal compression perpendicular to it. Taking Poisson's ratio equal to $\frac{1}{4}$, he concludes that the working stress in shear must be equal to eight-tenths of that for simple tension.

33. Jourawski's Analysis of Shearing Stresses in Beams

Coulomb drew attention to the shearing stresses in a cantilever and mentioned that they become important only in short beams. Thomas Young in his "Lectures on Natural Philosophy"[1] pointed out that the ability to resist shear distinguishes rigid bodies from liquids. Vicat (see page 83) mentioned many cases where this resistance is of primary importance and criticized the theory of bending of beams on the ground that it does not consider shearing stresses. While the first edition of Navier's book (1826) contains no reference to this subject, the second (1833) has an article on the bending of short beams. In it, an average value is taken for the shearing stresses that are distributed over the built-in end of a cantilever and an unsatisfactory method is given for combining it with the longitudinal bending stresses. The rigorous solution of the problem of shearing stresses in beams was given by Saint-Venant in his famous paper, "Mémoire sur la flexion."[2] But it covered only a few of the simplest shapes of cross section, and therefore, for more complicated cases, engineers have been forced to use (to this day) an approximate elementary solution presented by D. J. Jourawski[3] (1821–1891).

Jourawski graduated in 1842 from the Institute of Engineers of Ways of Communication in St. Petersburg which, we recall, was organized by French engineers (see page 114). By the time Jourawski attended the Institute (1838–1842), there were no more French professors there and

[1] See "Lectures on Natural Philosophy," vol. 1, p. 135.
[2] See *J. Liouville*, 1856.
[3] The theory of shearing stresses in rectangular beams was developed by Jourawski in connection with the design of wooden bridges for the St. Petersburg–Moscow railroad in 1844–1850. It was presented to the Russian Academy of Sciences together with other investigations relating to the design of bridges of Howe's system in 1854 (see page 186). Jourawski was awarded Demidoff's prize for the work by the academy.

the teaching was in Russian hands. Mathematics was taught by M. V. Ostrogradsky who was both a well-known mathematician and an outstanding professor.[1] In his lectures, he frequently went far beyond the limits of the required program, and there is no doubt that Jourawski had the opportunity of obtaining a very good mathematical training. He studied the mechanical properties of materials under the direction of A. T. Kupffer (see page 220).

Jourawski's career, after his graduation, was closely bound up with the development of railroad construction in Russia. The first railroads were laid down in that country in 1838. These were two short lines between St. Petersburg and Zarskoje Selo and between St. Petersburg and Peterhoff. In 1842, work on the railroad between St. Petersburg and Moscow

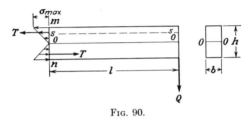

Fig. 90.

was begun, and immediately after graduation Jourawski was assigned to this important project. His capability was soon recognized, and in 1844 he was charged with designing and erecting one of the most important structures on the line, the bridge over the river Werebia (9 spans, 180 ft long, 170 ft above the water surface). In making this bridge, Jourawski had often to use wooden beams of great depth and also built-up wooden beams. The material was very weak in shear along the fibers, and Jourawski correctly concluded that shearing stresses in such beams were important and could not be disregarded. The existing literature had no information regarding the calculation of these stresses, and Jourawski had to solve the problem by himself.

Starting with the simplest case of a rectangular cantilever loaded at the free end (Fig. 90) and considering the neutral plane OO, Jourawski concludes that the normal stresses that are distributed over the cross section mn at the built-in end have a tendency to produce shear in the plane OO.

[1] M. V. Ostrogradsky (1801–1861) was born in a village near Poltava in south Russia. After graduating from the University of Charkov, he went to Paris, where he became a pupil of Cauchy, Poisson, and Fourier. He is best known for his work in variational calculus. In Todhunter's "History of the Calculus of Variations," we find an entire chapter devoted to Ostrogradsky's work. He also worked in the theory of elasticity, and his investigation of the motion of waves in an elastic body is discussed in Todhunter and Pearson, "History of the Theory of Elasticity," vol. I.

The magnitude of the shearing force T is

$$T = \frac{\sigma_{max} bh}{4} = \frac{3Ql}{2h}$$

and the corresponding shear stress, which is uniformly distributed over the neutral plane OO, is

$$\tau = \frac{T}{lb} = \frac{3}{2}\frac{Q}{bh}$$

In a similar way, this writer calculates shearing stresses which act in any plane ss parallel to the plane OO. When the load is uniformly distributed along the length of the cantilever, Jourawski shows that the shearing stresses are no longer uniformly distributed along the neutral plane but increase with the distance from the free end.

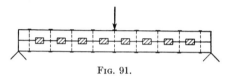

FIG. 91.

With this solution for solid beams, Jourawski turns to built-up wooden beams, shown in Fig. 91, and shows how the forces which act on each individual key can be calculated. Further, he demonstrates that the necessary dimensions of the keys may be computed if the mechanical properties of the materials of the keys and of the beam are known. This Russian engineer also applies his method to the analysis of built-up iron beams and shows how the proper distances between rivets can be calculated if the allowable shearing force per rivet is known.
He analyzes beams of tubular cross section (Fig. 92), and on the strength of this discusses the rivet distribution in the Conway and Britannia tubular bridges (see page 162). He shows that the number of rivets employed could have been considerably reduced by allowing for the fact that the transverse force acting on the tube diminishes as we

FIG. 92.

pass from the ends to the center of a span so that the rivet distance in the middle can be increased without detriment to the strength of the tube.

That portion of Jourawski's work which deals with the analysis of shearing stresses in beams was translated into French.[1] Saint-Venant praised Jourawski's approximate method and, in his notes to the third edition of Navier's book, page 390, he adapts it to a rectangular beam whose depth is much larger than its width. The method was incorpo-

[1] See *Ann. ponts et chaussées*, vol. 12, p. 328, 1856.

rated in textbooks on strength of materials,[1] and since its introduction it has been used by engineers and has proved especially useful in the study of thin-walled structures where shearing stresses are of primary importance and where rigorous solutions of the problems have not been found.

34. Continuous Beams

Navier was the first to grapple with the statically indeterminate problem which we encounter in analyzing continuous beams.[2] In his book "Résumé des Leçons . . . ," he considers a beam on three supports and takes the reaction at one of them as the statically indeterminate quantity. When there are more than three supports, selection of the reactions as the unknown quantities becomes inconvenient, since we obtain as many equations as there are intermediate supports and each contains all the unknowns. Study of the particular case of equal spans and of a uniform load covering the entire length of the beam, or of equal concentrated loads applied at the center of each span, reveals that the problem can be simplified and a linear relation found between the three consecutive reactions. Using this relation, the reactions can be calculated without much difficulty for any number of spans.[3]

Further progress in the analysis of continuous beams was made by Clapeyron. He uses expressions for the angles which the tangents to the deflection curve at the supports make with the initially straight axis of the beam. In the case of a uniformly loaded simple beam of a length l with moments M and M' applied at the ends, these angles are

$$\alpha = \frac{ql^3}{24EI} + \frac{1}{6EI}(2M + M')$$
$$\alpha' = -\frac{ql^3}{24EI} - \frac{1}{6EI}(M + 2M')$$
(a)

For a continuous beam having n spans, he writes $2n$ equations of this kind having $4n$ unknowns (α,α',M,M'). Observing now that, at each intermediate support the two adjacent spans have a common tangent and a common value of the bending moment, he gets $2n - 2$ additional equations. Assuming that the moments vanish at the ends of the beam, he finds that the number of equations is equal to the number of unknown

---

[1] See, for example, Belanger, "Théorie de la Résistance et de la Flexion plane des solides," Paris, 1858; 2d ed., 1862. Bresse "Cours de Mécanique Appliquée," 2d ed., part 1, p. 209, Paris, 1866; E. Collignon, "Cours de Mécanique Appliquée aux Constructions," 2d ed., p. 198, Paris, 1877.

[2] See *Bull. soc. philomath. Paris*, 1825.

[3] This method was developed by G. Rebhann. See his book "Theorie der Holz- und Eisen-Constructionen," Vienna, 1856. In this book bending-moment diagrams appear for the first time and are used for simple and continuous beams.

quantities, all of which can be found readily. This method of analysis was developed in connection with the reconstruction of the d'Asnières bridge near Paris (1849), and it was in use for several years before Clapeyron described it in a paper which he presented to the Academy of Sciences[1] in 1857. It appears in some early books on theory of structures.[2]

The *three moments equation*, in its present form, was published for the first time by the engineer Bertot.[3] It is easy to see, however, that the transformation made by Bertot in converting Clapeyron's equations (*a*) to the three moments equation was a comparatively simple one and we agree with Professor Bresse who states: "M. Clapeyron est, selon moi, le véritable auteur de la découverte, comme en ayant produit l'idée mère, ..."[4] Thus the name Clapeyron's equation which is often given to the three moments equation is justified, although it appeared for the first time in Bertot's paper. In this work, Bertot refers to Clapeyron's idea but he does not derive the theory, giving only the method of solving a system of these equations. When the ends of a continuous beam (of $n-1$ spans) are simply supported, he puts $M_0 = M_n = 0$ and thus obtains the system of equations:

$$2(l_1 + l_2)M_1 + l_2M_2 = \tfrac{1}{4}(w_1l_1^3 + w_2l_2^3)$$
$$l_2M_1 + 2(l_2 + l_3)M_2 + l_3M_3 = \tfrac{1}{4}(w_2l_2^3 + w_3l_3^3) \qquad (b)$$
$$\dots\dots\dots\dots\dots\dots\dots\dots\dots\dots\dots\dots\dots$$
$$\dots\dots\dots\dots\dots\dots\dots\dots\dots\dots\dots\dots\dots$$

for the case of uniform loads w per unit run on each span. Taking a value of M_1 (a_1, say), he calculates the value of M_2 from the first of equations (*b*). Substituting this into the second, he finds M_3 and, proceeding in this way, he finally arrives at some value b_1 for the last moment M_n, from the last equation. The correct result requires that M_n must vanish. Repeating his calculations by taking another value (say, a_2) for M_1, he obtains another value b_2 for M_n. Since there exists a linear relationship between the values of a and b and we have the values b_1 and b_2, corresponding to a_1 and a_2, he can find the value of a for which b vanishes. That value of a is the correct value of M_1. From it the remaining moments are readily calculated. This is the method proposed by Bertot for solving the system of three moments equations.

Clapeyron, in his paper (mentioned above), gives the three moments equation in the same form as Bertot but he does not refer to the latter's

[1] See *Comp. rend.*, vol. 45, p. 1076.

[2] For example, these equations appear in the book by L. Molinos and C. Pronnier, "Traité théorique et pratique de la construction des ponts métalliques," Paris, 1857. See also F. Laissle and A. Schübler, "Der Bau der Brückenträger," Stuttgart, 1857.

[3] See "*Mém. soc. Ing. civils France*," vol. 8, p. 278, 1855.

[4] See *Ann. ponts et chaussées*, vol. XX, p. 405, 1860.

work. Clapeyron then describes his own method for solving these equations. In conclusion, he gives some interesting information regarding the Britannia tubular bridge, which amounted to a continuous beam on five supports. He states that the calculations made by Molinos and Pronnier gave the following values for the maximum stress: (1) at the middle of the first span, 4,270 lb per in.[2]; (2) at the first pillar, 12,800 lb per in.[2]; (3) at the middle of the second span, 7,820 lb per in.[2]; and (4) at the central pillar, 12,200 lb per in.[2] From this he concludes that "Ce magnifique ouvrage laisse donc quelque chose à désirer en ce qui concerne la distribution des épaisseurs de la tôle, qui paraissent relativement trop faibles sur les points d'appui."[1] We shall see later (see page 160) that, in selecting the cross-sectional dimensions of that bridge, experimental data obtained from a simply supported model was used and the bending moments at the supports were equalized with the moments at the centers of the spans by using a special method of construction.

Clapeyron and Bertot always took all the supports of their continuous beams at the same level. If this condition is not fulfilled, some additional moments appear at the supports. These were investigated by the German engineers Köpcke,[2] H. Scheffler,[3] and F. Grashoff.[4] The three moments equation, with additional terms to allow for the vertical positions of the supports, appears for the first time in a paper by Otto Mohr.[5] Further work in the theory of continuous beams was done by M. Bresse and E. Winkler, and it will be discussed in the next two articles.

35. Bresse (1822–1883)

Jacques Antoine Charles Bresse was born in Vienne (province Isère). After graduating from the École Polytechnique (1843), he entered the École des Ponts et Chaussées where he received his engineering education. Soon after leaving the second school, he returned following his election to an instructorship in applied mechanics (in 1848). He assisted Professor Belanger until, in 1853, he succeeded the latter. He taught applied mechanics at the École des Ponts et Chaussées to the end of his life and enjoyed a fine reputation in strength of materials and theory of structures. In 1854, Bresse published the book, "Recherches analytiques sur la flexion et la résistance des pièces courbés," which deals with curved bars and their application in the theory of structures. The year 1859 saw the appearance of the first two volumes of his course in strength of materials

---

[1] "This magnificent structure could be improved by some changes in the thickness of the steel plates, which seem to be relatively too weak at the supports."
[2] *Z. Architek. u. Ing. Ver. Hannover*, 1856.
[3] "Theorie der Gewölbe, Futtermauern und eisernen Brücken," Brunswick, 1857.
[4] *Z. Ver. deut. Ing.*, 1859.
[5] *Z. Architek. u. Ing. Ver. Hannover*, 1860.

and hydraulics. The third volume of the course was a very complete study of continuous beams and it was published in 1865. Bresse was awarded the Poncelet prize by the French Academy in 1874 for his work in applied mechanics and was elected a member of the French Academy in 1880.

Bresse's chief accomplishment in engineering science was his theory of curved bars with its application to the design of arches. In the first chapter of his book, he discusses the eccentric compression of a prismatical bar. The particular case of a rectangular bar loaded in a plane of symmetry had already been discussed by Thomas Young (see page 95). Bresse treats the problem in a general form and shows that, by using the central ellipse of inertia of the cross section of the bar (Fig. 93), the direction of the neutral axis can be fixed easily for any position of the load. If the point of application O of the force moves along a line Cm, the neutral axis always remains parallel to the tangent pq to the ellipse at n, the point of intersection of the ellipse with the line Cm. The neutral axis approaches the centroid C if point O moves away from the centroid and it passes through C when the distance OC becomes infinitely large, *i.e.*, under conditions of pure bending. If a and b are the coordinates of point O, the equation of the neutral axis is

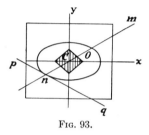

Fig. 93.

$$\frac{ax}{r_y{}^2} + \frac{by}{r_x{}^2} + 1 = 0 \qquad (a)$$

where r_x and r_y denote the radii of inertia of the cross section with respect to the principal axes xy. Having this equation, that portion of the cross section can be determined within which the point O must remain if the neutral axis is not to intersect the cross section of the bar and so that the stresses over the cross section shall all be of the same sign. In this way the concept of the *core of the cross section* was introduced by Bresse.[1]

In studying the deformation of a curved bar in the plane of its curvature, Bresse not only considers the change of curvature, which had been investigated previously by Navier (see page 77), but also the stretching of the axis of the bar. To illustrate Bresse's method for the calculation of deflections of curved bars, let us assume that a cross section a of the bar

---

[1] Todhunter mentions in his "History . . . " that he was in possession of a lithographed course of lectures which were delivered at the École des Ponts et Chaussées in 1842–1843 and which contained a discussion of eccentric compression similar to that of Bresse's. He attributes this course to Bresse, but at that time Bresse was a student at the École Polytechnique and could not have been the author.

is built-in (Fig. 94) and denote the axial tensile force and the bending moment at any cross section of the bar by N and M; then the stretching of an infinitesimal element mn of length ds is $N\,ds/AE$ and the rotation of the cross section n with respect to that at m is $M\,ds/EI$. Due to this rotation, any point c on the axis of the bar will describe an infinitesimal arc $\overline{cc_1}$ equal to $\overline{nc}\cdot M\,ds/EI$. Observing that the infinitesimal triangle cc_1d is similar to the triangle cen, we find that the horizontal displacement $\overline{cd}$ of the point c due to the change of curvature of an element mn of the axis of the bar is

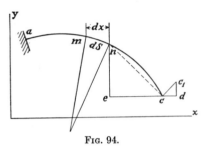

Fig. 94.

$$\overline{cd} = \overline{cc_1}\frac{\overline{cd}}{\overline{cc_1}} = \overline{cc_1}\frac{\overline{ne}}{\overline{nc}} = \frac{M\,ds\cdot\overline{ne}}{EI} = \frac{M\,ds}{EI}(y_n - y_c) \qquad (b)$$

The horizontal displacement of the same point due to stretching of the element mn will be equal to $N\,dx/AE$. To find the total horizontal displacement u of the point c, we have only to make a summation of the displacements resulting from the deformations of all the elements of the bar starting at a and finishing at c. This gives

$$u = \int_a^c \frac{N\,dx}{AE} + \int_a^c \frac{M\,ds}{EI}(y - y_c) \qquad (c)$$

In a similar manner we obtain the displacement v in the direction of the y axis

$$v = \int_a^c \frac{N\,dy}{AE} - \int_a^c \frac{M\,ds}{EI}(x - x_c) \qquad (d)$$

For the angle of rotation α of the cross section c, we have

$$\alpha = \int_a^c \frac{M\,ds}{EI} \qquad (e)$$

Using Eqs. (c) to (e), the components u and v of the displacement and the rotation α can be calculated for any cross section c of the bar, provided that the external forces acting on the bar are known.

These equations can be used for the calculation of statically indeterminate quantities. Considering, for example, a two-hinged arch abc (Fig. 95) and taking the horizontal reaction H as the statically indeterminate quantity, we can determine this quantity from the condition that the horizontal displacement of the point c must be zero. Using Eq. (c),[1] we can separate the action of the force H and put

$$N = N_1 - H\frac{dx}{ds} \quad M = M_1 - Hy$$

Then Eq. (c) gives

$$\int_a^c \frac{N_1\,dx}{AE} - H\int_a^c \frac{dx^2}{AE\,ds} + \int_a^c \frac{M_1\,ds}{EI}y - H\int_a^c \frac{y^2\,ds}{EI} = 0$$

and we find

$$H = \frac{\int_a^c \frac{N_1\,dx}{AE} + \int_a^c \frac{M_1 y}{EI}\,ds}{\int_a^c \frac{dx^2}{AE\,ds} + \int_a^c \frac{y^2\,ds}{EI}} \quad (f)$$

When an arch has built-in ends, we have a problem with three statically indeterminate quantities. The necessary three equations can readily be obtained from Eqs. (c) to (e) if we observe that, for the built-in cross section (c), the two components u and v of the displacement and the angle of rotation α must all vanish. Bresse also shows that expansion due to temperature changes can easily be taken into consideration, for, in the example shown in Fig. 95, it is only necessary to add the quantity

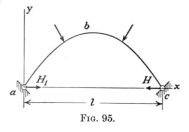

Fig. 95.

$\epsilon t l$ to the numerator in formula (f), where ϵ is the coefficient of thermal expansion, t is the rise of temperature, and l is the span of the arch. Bresse not only gives the general solution of the problem of arches but presents a detailed discussion of various particular cases of loading. Here he gives a very important discussion of the *principle of superposition* and shows that, for small deformations which follow Hooke's law, the displacements are linear functions of the external loads and can be obtained by summing the displacements produced by individual loads. Thus for vertical loads it is sufficient to investigate the effect of a single vertical force. Then the stresses and deflections which are produced by a system of vertical loads may be obtained by summation. For symmetrical arches, a further simplification can be achieved if we note that the thrust will not be changed if the load P is moved from point a (Fig. 96a) to

---

[1] This equation can be used here, although the cross section a may have some rotation. This follows from the fact that any small rotation of the arch about the hinge a only produces a vertical displacement of the point c.

the symmetrically located point a_1. This means that, in calculating the statically indeterminate quantity H, we can consider the symmetrical case (Fig. 96b) instead of the unsymmetrical (actual) one of Fig. 96a, and H is obtained by dividing the thrust found for the symmetrical case by 2. A similar simplification can be effected in the case of an inclined force acting on the arch.

Bresse now applies these theoretical considerations to specific problems on symmetrical circular two-hinged arches of constant cross section. He gives numerical tables for such arches of various proportions, by the use of which the thrust may readily be obtained for a single load, for a load uniformly distributed along the axis of the arch or along the horizontal projection of this axis. A table is also given for facilitating the calculation of the thrust produced by a rise of temperature. All these carefully prepared tables are still of some practical value.

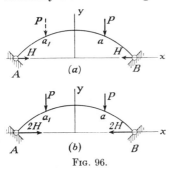

Fig. 96.

In this discussion of Bresse's book on curved bars, we have seen how the author was not content with getting certain theoretical results but wanted to make them useful for engineers. To this end, he did not hesitate to undertake lengthy calculations of tables.

Bresse's course of applied mechanics is composed of three volumes.[1] Of these only the first and third deal with problems of strength of materials. Bresse makes no attempt to incorporate any results of the mathematical theory of elasticity into the elementary theory of strength of materials. In all cases of deformation of bars, it is assumed that cross sections remain plane during deformation. Under this assumption, eccentric tension and compression are discussed, using the central ellipse of inertia as explained before (see page 147). He shows how the problem can be treated if the modulus of the material varies over the area of the cross section. The hypothesis that cross sections remain plane is also used in the theory of torsion, and Bresse tries to justify it by saying that in practical applications the cross sections of shafts are either circles or regular polygons; hence, warping of cross sections can be disregarded. In the theory of bending, Jourawski's analysis of shearing stresses is given. In the chapters dealing with curved bars and arches, the content of Bresse's (previously discussed) book is used.

In covering the analysis of stresses in cylindrical boilers, an interesting

[1] Bresse, "Cours de mécanique appliquée," 1st ed., 1859; 2d ed., 1866; 3d ed., 1880. In our discussion we refer to the second edition.

problem is discussed which concerns the bending stresses in rings which have small initial ellipticity and which are submitted to uniform internal pressure. Professor Bresse shows that if ϵ_0 is the initial eccentricity of the ellipse, then the eccentricity after application of the internal pressure p is

$$\epsilon^2 = \frac{\epsilon_0{}^2}{1 + (pd^3/2Eh^3)}$$

where h is the thickness of the wall and d is the diameter of the boiler. Where the pressure is applied externally, the same theory gives

$$\epsilon^2 = \frac{\epsilon_0{}^2}{1 - (pd^3/2Eh^3)}$$

The critical value of p is obtained from the condition that

$$1 - \frac{pd^3}{2Eh^3} = 0$$

or

$$p_{cr} = 2E \frac{h^3}{d^3}$$

This result, as we have seen (see page 134), was obtained by Grashof.

Bresse takes up problems of longitudinal and lateral vibrations of prismatical bars in his course. Dealing with lateral vibration, he was the first to take the rotatory inertia of the elements of the bar into consideration. He also considers the dynamical deflection of a simply supported beam under the action of a moving load. This last problem will be discussed later (see page 174).

As mentioned before, the third volume[1] of the course contains a very complete study of continuous beams. In the first chapter, the general problem of continuous beams is considered and whereas Clapeyron and Bertot required that all spans be equal and that a uniform load be distributed along the entire length of the beam, Bresse removes these limitations. Further, he allows the supports not to be at the same level, and thus obtains the three moments equation in its general form. Bresse divides the applied forces into two parts: (1) the uniformly distributed dead load such as the weight of the beam and (2) a moving (live) load which may cover only part of the beam. The moments produced by the permanent load at the supports are found by solving the three moments equations. In dealing with the moving load, the main question is to find its most unfavorable position for each cross section of the beam. In attacking this problem,[2] Bresse examines the case of a single concentrated

[1] "Cours de mécanique appliquée," part 3, Paris, 1865.
[2] E. Winkler was the first to investigate this problem. See his paper in *Civiling.*, vol. 8, pp. 135–182, 1862.

force and finds the two fixed points in each unloaded span, which are the inflection points when the load is to the right or to the left. Having these points, the bending-moment diagram for any position of the single concentrated load can be drawn and, from it, the most unfavorable position of the moving load can be found. We see that in this analysis, Bresse came very close to the concept of *influence lines*.[1] However, he did not introduce them and he did not use a graphic method of selecting the most unfavorable load distribution. He preferred analytical approach and prepared numerous tables, for beams with various numbers of spans, in order to simplify the work of designing engineers.

36. E. Winkler (1835–1888)

E. Winkler was born in 1835 near Torgau, Saxony, and went to school at the Torgau Gymnasium. When his father died, he was forced to interrupt his study and to work for some time as an apprentice to a bricklayer. However, he was able to overcome his difficulties and to finish his high school career. After that he entered the Dresden Polytechnicum, where he took up structural engineering. At this technical school he shone in the theoretical analysis of engineering problems. Soon after graduating, he published his important paper on the theory of curved bars.[2] In 1860 he began working as an instructor in strength of materials at the Dresden Polytechnicum, and in 1863 he became a lecturer at the same school in bridge engineering. He obtained his doctor's degree from Leipzig University in 1860 for his theory of retaining walls, and in 1862 his important work on continuous beams was published (see page 151). He was not only an outstanding engineer but also a good teacher, and in 1865 he was elected to the chair of bridge and railroad engineering in the polytechnical institute of Prague. There he continued his scientific work and published (in 1867) his book on strength of materials,[3] in which he incorporated his own solutions of a number of important engineering problems. 1868 saw Winkler's election to a professorship at the Vienna Polytechnicum and here he started writing the books on railroad and bridge engineering which subsequently played important roles in those branches of engineering science, not only in Germany and Austria but also abroad.

In 1877 Berlin started reorganizing its Bau-Akademie, with a view to bringing it up to the level of other German polytechnical institutes, and

[1] The method of influence lines was introduced by E. Winkler, *Mitte. Architek. u. Ing. Ver. Böhmen*, 1868, p. 6; and by O. Mohr, *Z. Architek. u. Ing. Ver. Hannover*, 1868, p. 19.

[2] E. Winkler, Formänderung und Festigkeit gekrümmter Körper, insbesondere der Ringe, *Civiling.*, vol. 4, pp. 232–246, 1858.

[3] E. Winkler, "Die Lehre von der Elasticität und Festigkeit," Prague, 1867.

Strength of Materials between 1833 and 1867   153

Winkler was invited to help and to take charge of the theory of structures and bridge engineering. It was here that he became interested in experimental stress analysis. He used rubber models to investigate the stresses in riveted joints, investigated the pressure distribution of sand on retaining walls and wind pressure on various kinds of lattice trusses, and did some work on stress analysis in arches. For this last purpose, an experimental arch was built in the basement of the school.

Winkler's main contribution to strength of materials was his theory of bending of curved bars. Navier and Bresse, when dealing with this type of bar, calculated deflections and stresses from formulas derived for the prismatical case. This method of analysis is adequate if the cross-sectional dimensions of the bar are small in comparison with the radius of curvature of the axis. But in hooks, rings, chains, links, etc., this condition is not fulfilled, and formulas derived for straight bars are not accurate enough to be reliable. Toward a more satisfactory theory, Winkler retains the hypothesis that the cross sections remain plane during bending, but he allows for the fact that, due to the initial curvature, the longitudinal fibers between two adjacent cross sections have unequal lengths and thus the stresses in them are no longer proportional to their distances from the neutral axis. Also this axis does not pass through the centroid of the cross section.

In the bending of a curved bar in the plane of its initial curvature, we denote the length of an element of the axis by ds and the initial angle between these cross sections by $d\phi$. Let Δds be the elongation, $\epsilon_0 = (\Delta ds/ds)$ the longitudinal strain of the axis, and $\Delta d\phi$ the change in the angle between the two adjacent cross sections during bending. Then the elongation of a fiber at a distance v from the centroidal axis of the cross section is

$$\Delta ds_v = \Delta ds + v\, \Delta d\phi$$

and its longitudinal strain is

$$\epsilon_v = \frac{\Delta ds + v\, \Delta d\phi}{ds + v\, d\phi} = \left(\epsilon_0 + \frac{v\, \Delta d\phi}{ds}\right)\frac{r}{r+v}$$

where r is the radius of curvature of the axis of the bar. Under the assumption that the longitudinal fibers do not press upon each other during bending, we find the stress

$$\sigma = E\epsilon_v = E\left(\epsilon_0 + \frac{v\, \Delta d\phi}{ds}\right)\frac{r}{r+v} \qquad (a)$$

and arrive at the following expressions for the axial force N and the bending moment M:

$$N = \int_A \sigma\, dA = Er\epsilon_0 \int_A \frac{dA}{r+v} + Er\frac{\Delta d\phi}{ds} \int_A \frac{v\, dA}{r+v} \qquad (b)$$

$$M = \int_A \sigma v\, dA = Er\epsilon_0 \int_A \frac{v\, dA}{r+v} + Er\frac{\Delta d\phi}{ds} \int_A \frac{v^2\, dA}{r+v} \qquad (c)$$

Introducing the notation

$$r \int_A \frac{v^2\, dA}{r+v} = \Theta \qquad (d)$$

we have

$$r \int_A \frac{dA}{r+v} = \int_A dA - \frac{1}{r}\int_A v\, dA + \frac{1}{r}\int_A \frac{v^2\, dA}{r+v} = A + \frac{1}{r^2}\Theta$$

$$r \int_A \frac{v\, dA}{r+v} = \int_A v\, dA - \int_A \frac{v^2\, dA}{r+v} = -\frac{1}{r}\Theta$$

From these relations Winkler obtains

$$\epsilon_0 = \frac{N}{AE} + \frac{M}{AEr} \qquad (e)$$

$$\frac{\Delta d\phi}{ds} = \frac{M}{E\Theta} + \frac{M}{Aer^2} + \frac{N}{AEr} \qquad (f)$$

$$\sigma = \frac{N}{A} + \frac{M}{Ar} + \frac{Mrv}{\Theta(r+v)} \qquad (g)$$

from Eqs. (b), (c), and (a). In any statically determinate case, we can readily find N and M for a particular cross section. The quantity Θ is calculated from formula (d) for a known shape of the cross section and the stresses found from Eq. (g). In a statically indeterminate case, Eqs. (e) and (f) can be used in the calculation of the indeterminate quantities.

Winkler applies this general theory to the analysis of stresses in hooks, rings of various cross sections, and chain links. He shows that when the cross-sectional dimensions of a curved bar are not small in comparison with the radius r, the elementary bending formula of straight beams loses its value and the new theory must be used.

In his next important article,[1] which deals with deformations which are symmetrical about an axis, Winkler discusses a cylindrical tube under uniform internal and external pressures and derives Lamé's formula. In choosing the proper thickness of the wall of the tube, Winkler uses the maximum strain theory and arrives at a new formula which is somewhat different from that of Lamé. Winkler also considers the conditions at the ends of the tube and discusses spherical and flat ends. For both of these cases, he gives equations for stresses and shows that the cylindrical

[1] See *Civiling.*, vol. 6, pp. 325–362 and pp. 427–462, 1860.

tube will be subjected to some local bending at the ends. Regarding this bending, he introduces some corrections to the theory previously developed by Sheffler (see page 134). In the conclusion, Winkler derives relations governing stresses in rotating disks and uses them in analyzing the stresses in flywheels.

Winkler's paper on continuous beams was published in 1862.[1] At that time he was interested in bridge engineering, and most of the paper deals with questions concerning the most unfavorable loading. Tables are included for a beam with four spans.[2]

In the book on strength of materials, which appeared in 1867, Winkler does not keep to an elementary presentation of the subject but gives the general equations of theory of elasticity and their application to the theory of bending of beams. He makes a comparison of the usual approximate solutions with the results given by the theory of elasticity and shows that the elementary equations are accurate enough for practical purposes. In deriving formulas for safe dimensions of structures, Winkler follows Saint-Venant and always uses the maximum strain theory as his guide. The chapter on bending of beams contains a very complete discussion of continuous beams. Concerning the lateral buckling of axially compressed bars, the author gives several solutions for cases of bars having variable cross sections. The "beam-column" problem (of the combined action of an axial force and lateral loading) is discussed in much detail, and some useful formulas are derived for calculating maximum deflections and maximum bending moments. The bending of a beam on an elastic foundation is discussed for the first time and the applicability of the theory to the analysis of stresses in railroad tracks is indicated. The chapter which deals with curved bars contains the general theory (which we have already discussed) and its applications to arch analysis. Regarding two-hinged arches, Winkler describes the cases which had already been treated by Bresse, but in the case of a hingeless arch, a number of new results are given. Tables are calculated for various loading conditions of circular and parabolic arches of constant cross section in order to facilitate stress analysis.

Winkler's book is written in a somewhat terse style and it is not easy to read. However, it is possibly the most complete book on strength of materials written in the German language and is still of use to engineers. In later books in this subject, we shall see a tendency to separate strength of materials from the mathematical theory of elasticity and to present it in a more elementary form than that employed by Winkler.

[1] See *Civiling.*, vol. 8, pp. 135–182, 1862.
[2] A further development of the theory of continuous beams is given in Winkler's "Theorie der Brücken," Vienna, 1872.

CHAPTER VII

Strength of Materials in the Evolution of Railway Engineering

37. *Tubular Bridges*

The construction of the first railroads greatly affected the development of strength of materials by presenting a series of new problems (especially in bridge engineering) which had to be solved. The materials used for building bridges were then stone and cast iron. It was known that the latter was very suitable when working in compression as in arch bridges, but that cast-iron girders were unreliable, since the metal fails by fatigue under the action of the varying stresses which are produced by heavy moving loads. Cast-iron girders reinforced by iron tensile members were tried without much success. It became necessary to introduce a more reliable material and, after 1840, the use of malleable iron in bridge construction won quick acceptance. The use of iron-plate girders of I section became very common in bridges of small spans, but for larger structures it was realized that other kinds of construction were needed for carrying railroad trains. Suspension bridges having long spans already existed, but their great pliability under heavy moving loads made them unsuited to railroad traffic.

More rigid structures were needed and, as a possible solution tubular bridges were tried in England in the construction of the London–Chester–Holyhead railroad. In the course of this work, the engineers encountered a very difficult problem in crossing the River Conway (Fig. 97) and the Menai Straits. It was required that the bridges should create no hindrance to shipping so that the arch bridges, which were initially proposed, were inadmissible. The chief engineer, Robert Stephenson,[1] suggested building the bridges in the form of large tubes of sufficient size that the trains could pass through them. Stephenson asked (1845) for the assistance of Fairbairn (see page 123), who had gained great experience with malleable iron-plate structures in connection with his work in shipbuild-

---

[1] Robert Stephenson, son of George Stephenson, the "Father of Railways," gained a reputation as a railway engineer. He was the second president of the Institute of Mechanical Engineers and became a Fellow of the Royal Society in 1849. He was the president of the Institution of Civil Engineers in 1856 and 1857.

ing, to help him develop this novel idea and to aid in the construction of these unusually large bridges (the longest span was 460 ft). Stephenson's original notion was not adopted completely, for he spoke about a tube supported by chains. Only after Fairbairn's preliminary tests was it decided to discard the chains and to design the tubes so that they would be strong enough to support the heaviest trains. Fairbairn's first experiments with tubular iron beams of various cross sections[1] showed that they failed, not on the convex side, as was usually the case with cast-iron beams, but on the concave side where the material works in compression. It was clearly noted that the failure occurred because the comparatively

Fig. 97. Conway bridge.

thin walls of the tubes became unstable and buckled. Fairbairn states: "Some curious and interesting phenomena presented themselves in the experiments—many of them are anomalous to our preconceived notions of the strength of materials, and totally different to anything yet exhibited in any previous research. It has invariably been observed, that in almost every experiment the tubes gave evidence of weakness in their powers of resistance on the top side, to the forces tending to crush them." Here we have the first experiments with thin-walled structures which fail through instability.

[1] For a complete description of these experiments and of the final design, see W. Fairbairn, "An Account of the Construction of the Britannia and Conway Tubular Bridges," London, 1849. See also E. Clark, "The Britannia and Conway Tubular Bridges," 2 vols., London, 1850. The latter work contains the results of the experimental work of Hodgkinson (which also appeared in the *Reports of the Commissioners* . . . .) and W. Pole's study of continuous beams.

Fairbairn asked his friend Hodgkinson (see page 126), a man of greater theoretical knowledge, to examine these results. Hodgkinson found that it is not sufficient to know the ultimate strength of the material when calculating the carrying capacity of tubular beams by means of applying the usual bending stress formula. He states: "It appeared evident to me, however, that any conclusions deduced from received principles, with respect to the strength of thin tubes, could only be approximations; for these tubes usually give way by the top or compressed side becoming wrinkled, and unable to offer resistance, long before the parts subjected to tension are strained to the utmost they would bear. To ascertain how far this defect, which had not been contemplated in the theory, would affect the truth of computations on the strength of the tubes . . . ," Hodgkinson proposes that a number of fundamental tests be made. Some of these tests were made by him and will be discussed later. Fairbairn could not act upon this suggestion due to the lack of time and he was forced to make a decision regarding the required cross-sectional dimensions of the bridge on the basis of the bending tests that he carried out on tubular beams of various shapes.

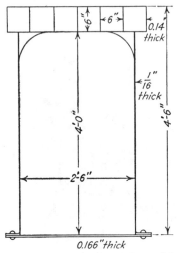

FIG. 98. Cross section of the model tested by Fairbairn.

From practical considerations, a rectangular cross section was selected and, to ensure equal strength in tension and compression, more material was to be used on the upper side.

It was decided to make tests with a large model having a span of 75 ft and the cross section shown in Fig. 98. To strengthen the upper side, a cellular structure was employed and the ratio of the cross-sectional area at the top compared with that at the bottom was taken as 5:3, with the idea of subsequently increasing the strength of the tension side should the experiments indicate the necessity of so doing. The experiments showed that the cellular structure contributed much to the stability of the beam, and the first failure occurred at the bottom side. This side was then reinforced in stages, and it was found that each succeeding arrangement exhibited a higher resistance. Equality of strength in tension and compression was achieved when the ratio of the area of compression to that of tension was equal to 12:10 so that the cellular structure considerably reduced the disparity between the resistances of material in tension and compression. These same experiments also showed that the sides of the tube were not sufficiently stable, and to eliminate wave formation in the side plates, vertical stiffeners were attached.

After these tests were completed, the final dimensions of the cross section of the tubular bridges were chosen on the assumption that the carry-

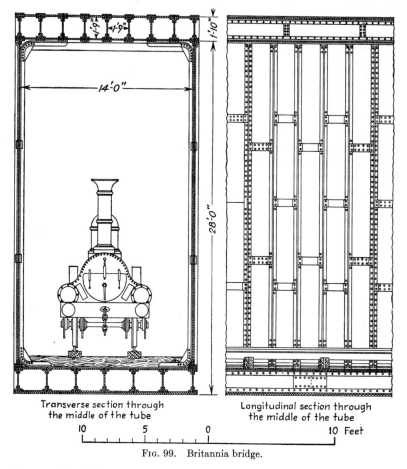

Fig. 99. Britannia bridge.

ing capacity of the tube increases as the square of linear dimensions and its weight increases as the cube. Figure 99 shows the cross section and a side view of the Britannia Bridge in its ultimate form. When the design of the Conway tube was completed, Hodgkinson was asked to calculate the maximum load which the bridge would be able to carry and the deflection which this would bring about. Although the problem was a simple one, requiring merely a calculation of the maximum stress and maximum deflection of a simply supported and uniformly loaded beam of constant cross section, this kind of analysis was beyond the power of Fairbairn and of the resident engineer E. Clark. It required assistance from "the mathematician" (Hodgkinson). This analysis was considered so important that, with all its details, it was included in Clark's book and in the *Report of the Commissioners*. . . . We also find it reproduced in Weale's "Theory, Practice, and Architecture of Bridges." Hodgkinson found that, with the dimensions taken in the design and allowing a maximum compressive stress of 8 tons per in.^{2}, the load concentrated at the middle could be as much as $W = 1{,}130.09$ tons when the corresponding deflection would be $\delta = 10.33$ in. For a uniformly distributed load, W could be doubled and the deflection multiplied by 5/4. The modulus of elasticity in this calculation was taken as 24×10^6 lb per in.^{2} After the completion of the bridge, the deflections produced by a load applied at the center were made and a deflection of 0.01104 in. per ton was found. This is about 20 per cent higher than that which the calculations predicted.

The construction of the Britannia and Conway Bridges constituted a great advance in our knowledge of strength of engineering structures. Not only was the general strength of tubular bridges established by the model tests, but the strength of iron plates and various types of riveted joints were investigated.[1] Also, the effects of lateral wind pressure and of nonuniform heating by sunlight were studied. Fairbairn took out a patent for tubular bridges and several bridges of this type were built later.[2]

The erection of these tubular bridges attracted the interest of foreign engineers, and we find descriptions of this great work in many books and papers dealing with strength of materials and theory of structures. Clapeyron's criticism has been mentioned before (page 146). He indicated that the maximum stresses at the pillars of the Britannia Bridge were much greater than at the middle of the spans. It seems that Clapeyron did not take the procedure used in the assembly of the Britannia Bridge into consideration, for, from Clark's book (vol. 2, p. 766), we see that the engineers clearly understood the effects of uniformly loading a continuous beam. It is stated there that "we have to observe that the object in view was not to place the tubes in their place in precisely the same conditions as though they had been originally constructed in one length and then

---

[1] As a result of these experiments it was concluded that the friction produced between the plates of a joint by the rivets is sufficient to resist shearing forces and that the deflection is the result of elastic deformation alone.

[2] He also introduced tubular sections into the construction of cranes, and this proved very successful.

deposited on the towers, for in such a continuous beam (supposed of uniform section) the strain would have been much greater over the towers than at the center of each span, whereas the section is nearly the same; the object, therefore, was to equalise the strain." To accomplish this equalization, a special procedure was followed in the process of assembly. The large tube CB (Fig. 100) was placed in position first. In making the joint B, the adjacent tube AB was tilted to some angle as shown, under which condition the riveted joint was made. Now, when the portion AB was put into the horizontal position, a bending moment at the support B was induced, the magnitude of which depended upon the amount of tilt. A similar procedure was adopted in assembling the joints C and D. In

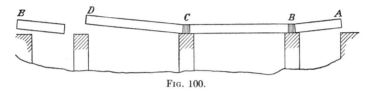

Fig. 100.

selecting the proper angles of tilt, the engineers were assisted by W. Pole, who investigated the problem of bending of continuous beams by using Navier's method.

An extensive criticism of the design of the Britannia Bridge was made by Jourawski.[1] He begins his discussion with a consideration of latticed trusses and correctly concludes that the buckling of the sides of a tubular bridge is caused by compressive stresses acting in the sides at an angle of 45 deg to the horizontal. He recommends placing the stiffeners in the direction of maximum compressive stresses. To prove his point, he made some very interesting experiments with models which were made of thick paper reinforced by cardboard stiffeners. In discussing his selection of such materials, he adds an interesting general discussion of English experiments. He does not consider it satisfactory to judge the strength of a construction by comparison with the magnitude of the ultimate load, since, as the load approaches the ultimate value, the stress conditions of the members of the structure may be completely different from those which occur under normal working conditions. He suggests the testing of models under loads corresponding to the service conditions of structures and suggests using materials which have small moduli of elasticity for the models so that deformations within the elastic limit will be large enough to be measured easily. Using his paper models, Jourawski measures the deformation in the plane of the web near the neutral axis and shows that

[1] The French translation of this work appeared in *Ann. ponts et chaussées*, vol. 20, p. 113, 1860.

the largest compressive strain there is in the 45-deg direction (with respect to the vertical). He was also able to study the direction of waves which are formed during buckling of the sides. Comparing the effectiveness of the stiffeners, he found that the model with inclined stiffeners could carry 70 per cent more load than that with vertical stiffeners. At the same time, the cross-sectional area of the inclined stiffeners was taken as half that of the vertical stiffeners.

As mentioned before (page 143), Jourawski also criticized the rivet distances chosen for tubular bridges. Using his theory of shearing stresses in beams, he had no difficulty in selecting rivet distances properly for each particular case.

An important investigation on buckling of thin-walled tubes under compression was made by Hodgkinson[1] in connection with the great tubular bridges. He shows that the strength of rectangular tubes decreases if the ratio of the thickness of the tube to the width of its sides diminishes. He also notes the advantage of having a (circular) cylindrical form of tube. This work amounts to the first experimental study of buckling of compressed plates and of thin-walled tubes.

For bridges of shorter spans, thin-walled plate girders were introduced. Brunel's plate girders became popular since tests on them gave very favorable results.[2] Some experimentation with built-up I beams was carried on in Belgium by Houbotte.[3] Testing these built-up I beams (with unstiffened webs) by applying a concentrated load at the middle, this engineer found that in all cases failure occurred due to local buckling of the web near the point of application of the load.

38. Early Investigations on Fatigue of Metals

The fact that the subjection of a metal bar to many cycles of stresses can produce fracture, by much smaller forces than would be required for static failure, has been recognized by practical engineers for a very long time. Poncelet, in his lectures for the workers of Metz, speaks of the *fatigue* of metals under the repeated action of tension and compression.[4]

---

[1] The results of this investigation were published as Appendix AA to the *Report of the Commissioners Appointed to Inquire into the Application of Iron to Railway Structures*, London, 1849. See also E. Clark, "The Britannia and Conway Tubular Bridges," London, 1850.

[2] A description of some of these bridges is given in Fairbairn's book "The Application of Cast and Wrought Iron to Building Purposes," 2d ed., p. 255, London, 1857–1858. See also E. Clark's book in which the experiments made by Brunel are described.

[3] *Ann. Travaux Publique de Belgique*, 1856–1857.

[4] See his "Mécanique industrielle," 3d ed., p. 317, 1870. Since this edition is simply a reprint of the second edition (1839), it is possible that Poncelet was the first to discuss the property by which materials resist repeated cycles of stress and that he introduced the term "fatigue."

Morin in his book[1] discusses the very interesting reports of the two engineers in charge of mail coaches on the French highways. The engineers recommended a careful inspection of the coach axles after 70,000 km of service because experience showed that after such a period thin cracks were likely to appear at points where the cross-sectional dimensions of the axles suffered sharp changes and especially at sharp reentrant corners. They also give interesting pictures of the gradual onset of these cracks and discuss their brittle character. However, they do not agree with the (then usually accepted) theory that the cycles of stresses give rise to some recrystallization of the iron. Static tests of axles which had seen much service revealed no change in the internal structure of the metal.

With the expansion of railroad construction, the problem of fatigue failures of locomotive axles assumed great importance. Perhaps the first English paper in this field was presented by W. J. Macquorn Rankine.[2] The unexpected fracture of apparently good axles after running for several years was usually explained on the hypothesis that the fibrous texture of malleable iron gradually assumes a crystallized structure. Rankine shows that a gradual deterioration takes place without loss of the fibrous texture. He says: "The fractures appear to have commenced with a smooth, regularly-formed, minute fissure, extending all round the neck of the journal and penetrating on an average to a depth of half an inch. They would appear to have gradually penetrated from the surface towards the centre, in such a manner that the broken end of the journal was convex, and necessarily the body of the axle was concave, until the thickness of sound iron in the centre became insufficient to support the shocks to which it was exposed. A portion of a fiber which is within the mass of the body of the axle will have less elasticity than that in the journal, and it is probable that the fibres give way at the shoulder on account of their elastic play being suddenly arrested at that point. It is therefore proposed, in manufacturing axles, to form the journals with a large curve in the shoulder, before going to the lathe, so that the fibre shall be continuous throughout."

During the years 1849–1850 the same problem was discussed at several meetings of the Institution of Mech. Engineers. James E. McConnell presented a paper[3] "On Railway Axles" in which he states: "Our experience would seem to prove that, even with the greatest care in manufacturing, these axles are subject to a rapid deterioration, owing to the vibration and jar which operates with increased severity, on account of their peculiar form. So certain and regular is the fracture at the corner of the

---

[1] A. Morin, "Résistance des Matériaux," 1st ed., 1853.
[2] See *Proc. Inst. Civil Engs.* (*London*), vol. 2, p. 105, 1843.
[3] See *Proc. Inst. Mech. Engrs.* (*London*), 1847–1849 (the meeting of Oct. 24, 1849).

crank from this cause, that we can almost predict in some classes of engines the number of miles that can be run before signs of fracture are visible. . . . The question of deterioration of axles arising from various causes, which I have enumerated, is a very important one to all railway companies; that some change in the nature of the iron does take place is a well-established fact, and the investigation of this is most deserving of careful attention. . . . I believe it will be found that the change from the fibrous to the crystalline character is dependent upon a variety of circumstances. I have collected a few specimens of fractured axles from different points, which clearly establish the view I have stated.

"It is impossible to embrace in the present paper an exposition of all the facts on this branch of the subject; but so valuable is a clear understanding of the nature of the deterioration of axles, that I am now registering each axle as it goes from the workshops, and will endeavor to have such returns of their performances and appearances at different periods as will enable me to judge respecting their treatment. When it is considered that on the railways of Great Britain there are about 200,000 axles employed, the advantage of having the best proportions, the best qualities, and the best treatment for such an important and vital element of the rolling stock must be universally acknowledged." McConnell gives an important piece of advice: "All my experience has proved the desirableness of maintaining . . . [journals of axles] as free as possible from sharp abrupt corners and sudden alteration in diameter or sectional strength."

The subsequent discussion of the paper centered largely about the question of whether the structure of iron changes under the action of cycles of tension and compression or not. No general agreement was reached, and the chairman, Robert Stephenson, concluded the meeting with the remark: "I am only desirous to put the members on their guard against being satisfied with less than incontestable evidence as to a molecular change in iron, for the subject is one of serious importance, and the breaking of an axle has on one occasion rendered it questionable whether or not the engineer and superintendent would have a verdict of manslaughter returned against them. The investigation hence requires the greatest caution; and in the present case there is no evidence to show that the axle was fibrous beforehand, but crystalline when it broke. I therefore wish the members of the Institution, connected as you are with the manufacture of iron, to pause before you arrive at the conclusion that iron is a substance liable to crystalline or to a molecular change from vibration. . . . "

The discussion was continued in the volume covering the meetings of 1850. A very interesting suggestion was made at one of these meetings by P. R. Hodge:[1] "To arrive at any true results as to the structure of iron it would be necessary to call in the aid of the microscope, to examine the fibrous and crystalline structure." This counsel was followed up by R. Stephenson who reported[2] that he had (since the last meeting) examined a piece of iron called "crystalline," and a piece of iron called "fibrous" under a powerful microscope and that it would probably surprise the members to know that no real difference could be perceived.

About the same time as that of the above discussion on fatigue at the meetings of the Institution of Mechanical Engineers in London, some interesting work in the subject was done by the commission appointed in 1848 to inquire into the application of iron to railway structures. As part of the commission's work Capt. Henry James and Capt. Galton undertook

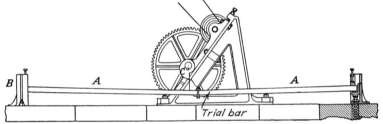

FIG. 101. James and Galton's fatigue-testing machine.

an experimental investigation (at Portsmouth) of the strength of iron bars when subjected to many cycles of loading.[3] A rotating eccentric was used to deflect the bar and then to release it suddenly (see Fig. 101), the frequency of the deflections being from four to seven per minute. Of the bars subjected to these tests, "three bore 10,000 depressions equal to what would be produced by one-third of the statical breaking-weight, without having their strength to resist statical pressure apparently at all impaired; one broke with 51,538 such depressions, and one bore 100,000 without any apparent diminution of strength; whilst three bars, subjected by the same cam to a deflection equal to what would be produced by half the statical breaking-weight, broke with 490,617 and 900 depressions, respectively. It must therefore be concluded that iron bars will scarcely bear the reiterated application of one-third their breaking-weight without injury."

[1] *Proc. Inst. Mech. Engrs.*, (*London*), Jan. 23, 1850.
[2] *Proc. Inst. Mech. Engrs.* (*London*), April, 1850.
[3] The results of this investigation were published in Appendix B5 to the *Report of the Commission*, p. 259. They are also discussed in Fairbairn's paper, *Phil. Trans.*, vol. 154, 1864.

Fairbairn was very much interested in the strength of tubular bridges as affected by the repeated loading caused by the weight of passing trains. He wanted to find the maximum stress which could be applied for an indefinitely large number of times without injury to the material. Thus he would be able to compute a safe working stress. He states[1] that, in the design of the great tubular bridges, the dimensions were chosen in such a way that the ultimate load for the bridge "should be six times the heaviest load that could ever be laid upon them after deducting half the weight of the tube. This was considered a fair margin of strength; but subsequent considerations such as generally attend a new principle of construction with an untried material induced an increase of strength, and, instead of the ultimate strength being six times, it was increased to eight times the weight of the greatest load." After this Fairbairn dis-

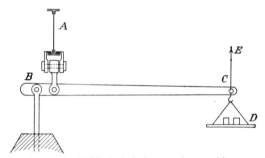

Fig. 102. Fairbairn's fatigue-testing machine.

cusses the requirement of the Board of Trade "that all future bridges for railway traffic shall not exceed a strain of 5 tons per in.[2]" He observes (correctly) that in the case of tubular bridges the plate in compression may buckle at comparatively low stresses and that a cellular construction must be used.

In order to find a safe value of working stress for bridges, Fairbairn decided to carry out tests in which the strain which bridges suffer by the passage of heavy railway trains was initiated as closely as possible. In the tests, an I beam A (Fig. 102) composed of iron plates riveted to angle irons, 22 ft long and 16 in. deep, was used. The initial deflection was produced by the load D applied at the end C of the lever BC. To get cycles of stress, vertical motion of the end C was produced by the rod CE, which was attached to a uniformly rotating eccentric. In this way, from seven to eight cycles of stress per minute were inflicted upon the beam. The experiments showed "that wrought-iron girders, when loaded to the extent of a tensile strain of 7 tons per in.[2], are not safe, if that strain is

[1] *Phil. Trans.*, vol. 154, 1864.

subjected to alternate changes of taking off the load and laying it on again, provided a certain amount of vibration is produced by that process; and what is important to notice is, that from 300,000 to 400,000 changes of this description are sufficient to ensure fracture. It must, however, be born in mind that the beam from which these conclusions are derived had sustained upwards of 3,000,000 changes with nearly 5 tons tensile strain on the square inch and it must be admitted from the experiments thus recorded that 5 tons per in.² of tensile strain on the bottom of girders, as fixed by the Board of Trade, appears to be an ample standard of strength."

We see that, by the middle of the nineteenth century, engineers knew of the ill effects of fluctuating stresses on the strength of metals and that, by making a few tests, they came to the conclusion that a load equal to one-third of the ultimate can be considered as safe when fluctuating. A much more complete insight of this phenomenon of "fatigue" was gained later by Wöhler.

39. The Work of Wöhler

A. Wöhler (1819–1914) was born in the family of a schoolmaster in the province of Hannover and he received his engineering education at the Hannover Polytechnical Institute. Being an outstanding student, he won a scholarship after his graduation, which enabled him to get a prac-

FIG. 103. A. Wöhler.

tical training, both at the Borsig locomotive works in Berlin and in the construction of the Berlin–Anhalter and Berlin–Hannover railways. In 1843 he was sent to Belgium to study locomotive production. After his return he was put in charge of the railway machine shop in Hannover. In 1847, Wöhler was given charge of the rolling stock and of the machine shop of the Niederschlesisch–Märkische railway and this caused him to remain in Frankfurt on the Oder for the next twenty-three years. There, he had to solve many problems concerning the mechanical properties of materials and started his famous investigations upon the fatigue strength of metals. His position enabled him to put the results of his experiments to practical use.

The uniformity of the properties of materials is of vital importance in practical applications, and in an effort to get the uniformity, he drew up exacting specifications for the materials to be used on railroads. He also fostered the organization in Germany of a chain of material-testing laboratories and helped to bring about uniformity in the mechanical testing of metals. His influence in this direction was very great in Germany, and the testing machines which he designed and built for his work were the best in his time. As a mark of their historical importance, they are preserved in the Deutschen Museum in Munich.

Wöhler's main work on fatigue of metals was carried out with a view to remedying the occurrence of fractures of railway axles on the Niederschlesisch–Märkischen railway.[1] To find the maximum forces acting on an axle in service, the device diagramed in Fig. 104 was used. The part mnb was rigidly attached to the axle pq and the pointer bc was connected to the rim of the wheel by a hinged bar ab. Owing to bending of the axle, indicated by the dotted line, the hinge a moved and the end c of the pointer bc made a scratch on the zinc plate mn. In service, the size of the scratches varied considerably and became especially large under the action of the impact forces produced by irregularities in the railroad track. By measuring the peak values of the scratches, the greatest deformations of the axle were found. To find the corresponding forces, static bending tests were made by applying known horizontal forces to the axle at the points a and a_1 and measuring the corresponding change in the distance aa_1. Having this information, the maximum bending stresses in service

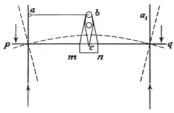

FIG. 104. The device for recording deflections of axles in service.

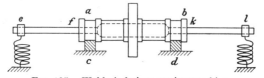

FIG. 105. Wöhler's fatigue-testing machine.

could be readily calculated. In a similar manner, the peak torsional stresses were also determined by recording the maximum angle of twist suffered by the axle in service.

Knowing the magnitudes of the maximum forces acting, Wöhler started his investigations of the strength of axles when submitted to constant reversal of the stresses. A special machine was constructed for the work (shown in Fig. 105). A cylinder ab was made to rotate in the bearings c

---

[1] Z. Bauwesen, vol. 8, pp. 641–652, 1858; vol. 10, pp. 583–616, 1860; vol. 16, pp. 67–84, 1866; vol. 20, pp. 73–106, 1870.

and d at about 15 rpm. The two axles ef and kl were fixed in the rotor and were bent by the forces of the springs which were transmitted through the ring bearings e and l. During each revolution, the stresses in the fibers passed through one cycle. The stresses could be obtained by proper adjustment of the springs. Axles of various materials and diameters were tested with this machine, and certain results regarding the relative strengths were obtained. But for a fundamental study of the fatigue strength of steel, a large number of experiments were required, and Wöhler decided to use, instead of actual axles (which had diameters of from $3\frac{3}{4}$ in. to 5 in.), smaller specimens which could be machined from circular bars of $1\frac{1}{2}$ in. in diameter. With the adoption of these smaller specimens, came the possibility of increasing the speed of the testing machine to about 40,000 revolutions per day. Thus many millions of cycles of stress could be applied to the specimens. Using several identical specimens with different forces applied at the ends and comparing the numbers of cycles required to produce fracture, Wöhler was able to draw some conclusions regarding the fatigue strength of his material. He did not use the so-called *Wöhler's curve* in this analysis but he speaks about *limiting stress* (*bruchgrenze*).

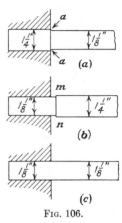

FIG. 106.

Wöhler recognized that, to eliminate the effect of sharp corners (Fig. 106a), the fillets a should be introduced. Tests were made with the specimens shown in Figs. 106b and 106c and it was found that specimens of uniform diameter exhibited the greater fatigue strength. It was also found that, in the case of the specimens of Fig. 106b, the fracture always occurred at the section mn where the sharp change in diameter is situated, so that by adding material to a shaft (corresponding to the increase in diameter at mn) we may make it weaker. Wöhler explains this in terms of irregularity of the stress distribution at the section mn and states that this weakening effect (of sharp corners) is of importance not only in axles, but that it must be considered in the design of other kinds of machine parts. Further experiments showed him that this weakening effect depends on the kind of material and that the reduction in strength produced by sharp corners varies from 25 to 33 per cent. Again it was found

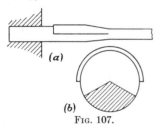

FIG. 107.

that, if the sharp change in the cross section extends only around a part of the circumference, as shown in Figs. 107a and 107b, the fatigue crack always begins at the sharp corner. The shaded portion of the cross section in Fig. 107b shows the coarse portion of the crack which can be seen after fracture.

The experiments done with the machine of Fig. 105 involved complete reversal of the stresses. To obtain other kinds of cycles of stress, Wöhler designed and built another type of fatigue-testing machine in which rectangular specimens with simply supported ends were used. The varying deflections of the middle of the specimens were produced by an eccentric so that the stress in the most remote fiber from it (on the tension side) could be made to vary between its maximum value and zero or between the maximum value and some smaller tensile stress. Comparing the results obtained with this machine with those found before with complete reversal of stresses, he found that the fracture always began at the tension side and that the magnitude of the maximum stress which the material could stand depended very much upon the difference between the maximum and the minimum of stress (*i.e.*, upon the range of stress). For a particular kind of axle steel, Wöhler gives the following limiting values of maximum and minimum stresses which the axle steel could withstand when subjected to a large number of cycles:

Maximum stress, lb per in.²	30,000	50,000	70,000	80,000	90,000
Minimum stress, lb per in.²	−30,000	0	25,000	40,000	60,000
Range of stress, lb per in.²	60,000	50,000	45,000	40,000	30,000

All the experiments showed that, for a given maximum stress, the number of cycles required to produce fracture diminished as the range of stresses was increased. From these tests, Wöhler concludes that in bridges of large span and in railroad-car springs (where the greater part of the maximum stress is produced by dead load), a much higher allowable stress can be used than in axles or piston rods, where we have complete reversal of the stresses.

All these conclusions were made on the strength of bending tests, and Wöhler decided to verify them by taking recourse to direct stress, for which purpose a special tensile-test machine was designed. In the new machine, the specimens were subjected to the action of tensile-stress cycles, and the results obtained with it showed satisfactory agreement with the previous ones given by bending tests. This led Wöhler to recommend that the same magnitude of working stress be used for similar stress cycles irrespective of the way in which these cycles were produced.

The next step in this investigation of the fatigue strength of metals was a study of cycles of combined stress. Here, Wöhler assumes that the strength depends upon the cycles of maximum strain (following the maximum strain theory) and uses, in his calculations of strain, a Poisson's ratio of ¼. He now applies his general considerations to torsion, for which the assumed theory states that the fatigue limit for complete reversals of stress is equal to 0.80 of that found for tension/compression. To verify this, Wöhler built a special machine in which circular bars were submitted to cycles of torsional stress. The experiments done with it, using solid circular specimens, upheld the theory. Thus Wöhler recommends a stress equal to 0.80 of that used in tension/compression for a working stress in shear. He also noticed that the cracks in torsional tests

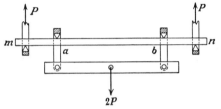

FIG. 108. Wöhler's apparatus for static bending.

appeared in a direction at 45 deg to the axis of the cylinder and were produced by the maximum tensile stress.

Since fatigue tests require much time and expense, Wöhler naturally tried to find some correlation between the fatigue strength of a material and its other mechanical properties which could be determined by static tests. It seems that he was especially interested in the elastic limit of the materials which he tested in fatigue. The determination of this limit from tensile tests requires accurate measurement of very small elongations, and suitable instruments for that purpose were not available then. For this reason, Wöhler decided to use bending tests for his measurements of the elastic limit, although he clearly states that this method is not an accurate one since the limiting stress will be reached first only by the outermost fibers and the beginning of yielding becomes noticeable only after a considerable portion of the material has passed through the elastic limit. To make the measurement as accurate as possible, Wöhler used the system shown in Fig. 108 in his specially built machine. The portion ab of the specimen mn undergoes pure bending, and from accurate measurement of its deflections the modulus of elasticity E and the load at which a permanent set began were determined.

In connection with these static tests, Wöhler became interested in the plastic deformation of bars which are initially loaded beyond the elastic limit and in the residual stresses which are produced by such a deformation. He discusses these stresses in rectangular bars, specimens which he loaded beyond the elastic limit and then unloaded (Fig. 108). Assuming that the material follows Hooke's law in the process of unloading, he shows that the residual stresses are distributed as shown by the shaded areas of Fig. 109a. He obtains these areas by superposing the linear dis-

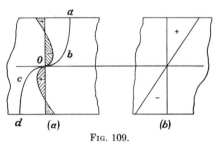

Fig. 109.

tribution of stresses, set up in the process of unloading (Fig. 109b), on the initial stress distribution, represented by the curve *abocd* (Fig. 109a). From this consideration, Wöhler concludes that the relevant range of stresses in fatigue tests is not affected by the initial plastic deformation. This is the first time that we find discussion of the residual stresses produced by plastic bending.

Fig. 110.

Wöhler's consideration of the distortion of the rectangular cross section during plastic bending of the bar is of interest. Assuming that the density of the material remains constant during plastic deformation, he arrives at the conclusion that if the fibers suffer a unit elongation ϵ, then the unit lateral contraction must be of amount $\epsilon/(2 + \epsilon)$. Corresponding to a unit contraction ϵ_1 there is a unit lateral expansion $\epsilon_1/(2 - \epsilon_1)$. As a result of these lateral deformations, the distortion of the rectangular bar when plastically bent must be such as shown in Fig. 110. This agrees with experiments.

Wöhler concludes his work with an interesting discussion of working stresses. He recommends a safety factor of 2 for bars under constant central tensile stress so that the working stress is half of the ultimate strength of the material. In the case of cycles of stress, he again recommends a factor of safety of 2, which means that the working stress would be equal to one-half of the endurance limit. In both instances, it is assumed that the most unfavorable loading conditions are taken into consideration. At the joints of a structure, the working stresses should be much smaller to compensate for any irregularities in stress distribution. He calls for a careful investigation of these irregularities.

Strength of Materials in the Evolution of Railway Engineering

We see that these experiments were of a fundamental nature, and it can justly be said that scientific investigation in the field of fatigue of materials began with Wöhler's work. For each kind of test, Wöhler designed and built all the necessary machines and measuring instruments. In designing these machines, he imposed stringent requirements upon the accuracy with which forces and deformations were to be measured so that his machines represent an important advance in the technique of testing structural materials.

40. Moving Loads

In studying the strength of bridges, the problem of stresses and deflections of a beam under the action of a moving load presents itself. We shall now trace the progress that was made with this question. Appendix

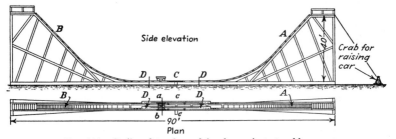

FIG. 111. Railroad track used for dynamic tests of beams.

B of the *Report of the Commissioners* . . . (see page 162) describes some very interesting investigations made by Prof. R. Willis, Capt. H. James, and Lt. D. Galton.[1] At the middle of the nineteenth century there was no agreement among engineers over the effect of a moving load on a beam. While some assumed that a load moving with high speed acts like a suddenly applied load and may produce deflections larger than those corresponding to static action, others argued that at very high speeds there was insufficient time for the load to drop through the distance of the expected dynamical deflection.

It was decided to make tests which were to be as realistic as possible. A railroad track, supported by a special scaffold, was constructed at the Portsmouth dockyard (Fig. 111), and a carriage having two axles was used as the moving load. The height of the top of the track above the horizontal portion was about 40 ft so that the trolley could attain velocities up to 30 mph. The trial cast-iron bars C (Fig. 111) were 9 ft long and had rectangular cross sections of three different sizes; *viz.* 1 by 2 in.,

---

[1] The Willis report describing the experiments can be found also in the appendix of an edition of Barlow's book "A Treatise on the Strength of Timber," London, 1851.

1 by 3 in., and 4 by 1½ in. In the static experiments, deflection of the bars under the weight of the carriage was measured with gradually increasing loads, and their ultimate static strength was determined. In dynamic tests, the carriage, with its minimum load, would be drawn up to the point of the inclined track corresponding to the selected speed and suddenly released. The experiment was repeated several times with gradually increased loads and the dynamical deflections were measured. Finally, the ultimate load for the selected speed would be determined.

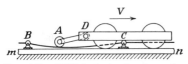

FIG. 112. Willis's device for dynamic tests of beams.

In this way, it was shown that dynamical deflections with a given load are larger than those produced statically by the same load and that the bars broke under a moving load which was considerably smaller than the ultimate statical load. By repeating the tests with higher speeds, it was found that the ill effects increased with increase of the speed. Dynamical deflection two and sometimes three times larger than statical were obtained at higher speeds.

Experiments made on actual bridges did not show the effects of speed in such a marked way. To find an explanation for this discrepancy, Willis repeated the experiments at Cambridge in a simplified form and also developed an analytical theory. In his arrangement, the trial bar BC (Fig. 112) was subjected to the action of a roller A, which was hinged to a carriage moving along rails. A third rail was placed between the two supporting the carriage and the bar BC formed part of it. By recording the relative movement of the frame AD with respect to the carriage rolling along the rigid track mn, the path of the roller A was obtained.

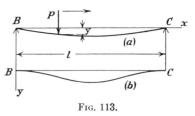

FIG. 113.

In developing his theory, Willis neglects the mass of the trial bar BC as being small in comparison with the mass of the roller A and uses the expression for the statical deflection y under the force P (Fig. 113), which is

$$y = \frac{Px^2(l-x)^2}{3lEI} \qquad (a)$$

where l denotes the length and EI is the flexural rigidity of the bar. Using this expression, we see that, when the force P moves along the bar, its point of application describes a curve with horizontal tangents at the ends (Fig. 113b). To adapt Eq. (a) to the case of a moving roller A

(Fig. 112), Willis substitutes the pressure of the roller on the bar for the force P. This pressure consists of two parts, viz., the gravity force W of the roller and its inertia force $-W\ddot{y}/g$ corresponding to the vertical movement. Thus he obtains

$$y = \frac{W}{3lEI} x^2(l-x)^2 \left(1 - \frac{\ddot{y}}{g}\right) \qquad (b)$$

Observing now that

$$\frac{dy}{dt} = \frac{dy}{dx}\frac{dx}{dt} = v\frac{dy}{dx} \quad \text{and} \quad \frac{d^2y}{dt^2} = v^2\frac{d^2y}{dx^2}$$

we obtain Willis's equation

$$y = \frac{W}{3lEI} x^2(l-x)^2 \left(1 - \frac{v^2}{g}\frac{d^2y}{dx^2}\right) \qquad (c)$$

Since $-d^2y/dx^2$ is the curvature of the line along which the point of contact of the roller A moves, we see from Eq. (c) that the dynamical action of the roller A is covered if we add the centrifugal force $-(Wv^2/g)(d^2y/dx^2)$ to the gravity force W. To calculate this centrifugal force, we must integrate Eq. (c) and find the equation of the roller's path. But Eq. (c) is a complicated one, and Willis suggested that an approximate solution of it could be obtained by assuming that the dynamical effect is small and that it would introduce little error if, on the right-hand side, the expression

$$y = \frac{Wx^2(l-x)^2}{3lEI} \qquad (d)$$

were substituted. This presents the trajectory of the roller calculated on the basis of the statical deflection [see Eq. (a)] and it has the symmetrical shape shown in Fig. 113b. Near the supports, the centrifugal force acts *upward* and the roller pressure on the bar is somewhat smaller than the gravity force W. At the middle, the centrifugal force is directed *downward* and is added to the gravity force. The maximum pressure on the bar will be

$$W - \frac{W}{g}v^2\left(\frac{d^2y}{dx^2}\right)_{x=\frac{l}{2}} = W\left(1 + \frac{16\delta_{st}v^2}{gl^2}\right) \qquad (e)$$

where

$$\delta_{st} = \frac{Wl^3}{48EI}$$

Assuming, for example, that $v = 32.2$ ft per sec, $l = 32.2$ ft, $g = 32.2$ ft per sec^2, and $\delta_{st} = 0.001l$, we find, from Eq. (e), that the maximum pressure exceeds the gravity force by only 1.6 per cent. In the Portsmouth experiments, the span l was about a quarter as large as this and δ_{st}/l was very often more than ten times greater. In such a case the dynamical effect is about 64 per cent of the gravity force (for a velocity of 32.2 ft per sec) and becomes more pronounced with increase of the velocity v. In this way, Willis was able to explain that the very large dynamical effect obtained in the Portsmouth dockyard was due principally to the great flexibility of the bars under test.

Willis's experiments also showed that the path of the roller was not a symmetrical curve as shown in Fig. 113b. During motion of the roller from B to C (Fig. 113), the maximum deflection occurs nearer the support C than the figure shows and the deviation from symmetry becomes larger and larger as the quantity

$$\frac{1}{\beta} = \frac{16\delta_{st}v^2}{gl^2}$$

increases. The approximate solution (e) is not sufficient to cover cases where $1/\beta$ is not small and the complete solution of Eq. (c) is required.

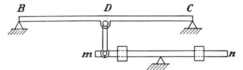

FIG. 114. Willis's inertial balance.

This solution was found by G. G. Stokes,[1] and for large values of β it shows good agreement with Willis's approximation. For smaller values of β, it gives nonsymmetrical trajectories and shows that the cross section of maximum stresses becomes displaced from the middle of the span toward the support C (Fig. 113). This last result exhibited good agreement with experiments, which always produced fracture in the portion of the bar between the center and the support C. This solution further shows that complete agreement between the theoretical paths and the curves obtained experimentally cannot be achieved so long as the mass of the beam is neglected in the derivation of the equation.

To get some idea of the effect of the mass of the test bar on the shape of the trajectory of the moving roller, Willis used an ingenious device which he called an *inertial balance*. The lever mn carrying two weights is hinged to the center of the trial bar BC (Fig. 114). The weights are in equilibrium so that no static force is transmitted to the bar at D. When the roller moves along the bar BC, the inertial balance is brought into motion and an inertia force acts at D. This action is similar to that of a

[1] See *Trans. Cambridge Phil. Soc.*, vol. 8, p. 707, 1849.

mass attached at D. If the deflection curve of the bar can be taken as half a sinusoid, a mass attached at D is dynamically equivalent to the doubled mass uniformly distributed along the span. Thus, by introducing the inertial balance, the influence of the mass of the trial bar for various speeds of the roller can be investigated.

The same device also made it possible to employ a ratio of the mass of the roller to the mass of the bar such as could be expected in an actual bridge. Arranging the experiment in such a way that this ratio of masses and the quantity β are the same for the model as for the actual bridge, we can obtain the dynamical deflections of the bridge from the deflection curves of the model.[1]

Stokes, who worked with Willis at Cambridge, also succeeded in getting an approximate solution for another extreme case, *viz.*, that in which the mass of the bridge is considered, the mass of the moving load is neglected, and it is assumed that a constant force moves along the beam. Considering only the fundamental mode of vibration, Stokes shows that the magnitude of the dynamical deflection depends upon the ratio of the period of the beam's fundamental mode of vibration to the time that it takes the moving force to cross the span.

The results obtained by Willis and Stokes were criticized by Homersham Cox.[2] From energy considerations, he concludes that the dynamical deflection cannot exceed the doubled statical deflection. However, in his analysis he overlooks the fact that not only the energy corresponding to the gravity force must be considered but also the kinetic energy of the moving roller. The reaction of the beam, which is normal to the deflection curve, gives a component in the direction of motion along the first portion of the trajectory and in the opposite direction along the second. Due to the unsymmetrical shape of the trajectory the net effect of this reaction is some reduction in the speed of the roller. The lost kinetic energy of the roller is transformed into potential energy of bending which makes it possible to have deflections that are greater than twice the statical deflections of the beam.

It can be seen that Willis and Stokes were able to explain why the Portsmouth experiments showed such large dynamical effects. They also showed that in an actual bridge the dynamical effect will be comparatively small. In this way the practical problem which confronted the commissioners was solved although a complete mathematical solution was not found. Since that time, many engineers have tried[3] to improve our knowledge concerning the dynamical action of a moving load on the

[1] This was shown by Stokes.
[2] *Civil Engrs, Archits. J.*, vol. 11, pp. 258–264, London, 1848.
[3] See Phillips, *Ann. mines*, vol. 7, pp. 467–506, 1855. Also see Renaudot, *Ann. ponts et chaussées*, 4th séries, vol. 1, pp. 145–204, 1861.

deflection of a beam, but little progress was made toward the solution in the nineteenth century.

At the same time as the dynamical effects of moving loads on bridges were being studied in England, some work was done in this direction in Germany. The deflections of cast-iron bridges on the Baden railroad were measured for various speeds of locomotives. For this purpose, an instrument was used in which a plunger worked in a reservoir of mercury and forced the fluid along a capillary tube, thus magnifying the deflection. The experiments showed some increase of the deflection due to the speed of the locomotives, but the effect was small for speeds up to 60 ft per sec.[1]

41. Impact

The commission that was appointed in 1848 to enquire into the application of iron to railway structures was also concerned with the effects of impact on beams, and in its report we find the experimental results which have already been discussed (page 127). Further progress toward solution of this problem was made by Homersham Cox,[2] who took up the problem of a central horizontal impact on a simply supported beam of uniform cross section. In this work, he gives some justification for an assumption (due to Hodgkinson) that at the instant of impact the impinging ball loses as much of its velocity as it would had it struck another free ball having half the mass of the beam.

Cox assumes that the deflection curve of the beam during impact does not differ much from the elastic curve produced by statical pressure at its center and uses the equation

$$y = \frac{f(3l^2x - 4x^3)}{l^3} \quad (a)$$

where l is the span, f is the deflection at the center, and x is measured from the end of the beam. Denoting the velocity which the beam acquires at its center at the instant of impact by v, he assumes that the velocity of the beam at any other cross section will be given by

$$\dot{y} = \frac{v(3l^2x - 4x^3)}{l^3} \quad (b)$$

The kinetic energy of the beam is then

$$T = \int_0^l \frac{q}{2g} (\dot{y})^2 dx = \frac{17}{35} ql \frac{v^2}{2g}$$

[1] See the paper "Die gusseisernen Brücken der badischen Eisenbahn," by Max Becker in *Ingenieur*, vol. 1, 1848, Freiberg.

[2] *Trans. Cambridge Phil. Soc.*, vol. 9, pp. 73–78. (This paper was read on Dec. 10, 1849.)

where ql denotes the weight of the beam. It is seen that the kinetic energy is the same as if we had a mass equal to $\frac{17}{35}$ of the mass of the beam concentrated at the middle of the span. Using this "reduced" mass and denoting the velocity of the impinging ball (of weight W) by v_0, Cox finds the common velocity v from the equation of conservation of momentum. This gives

$$v = v_0 \frac{W}{W + \frac{17}{35}ql} \qquad (c)$$

To determine the maximum deflection f of the beam, he assumed that, after acquiring the common velocity v, the ball moves with the beam until the total kinetic energy of the system is transformed into potential energy of bending. This gives the equation

$$\frac{24EIf^2}{l^3} = \left(W + \frac{17}{35}ql\right)\frac{v^2}{2g} = \frac{Wv_0^2}{2g}\frac{1}{1 + \frac{17}{35}(ql/W)} \qquad (d)$$

We see that the maximum deflection f is proportional to the velocity v_0 of the impinging ball. This agreed satisfactorily with Hodgkinson's experiments. In calculating the strain energy in Eq. (d), Hooke's law was assumed to prevail. In order to improve the agreement between his theory and Hodgkinson's experimental results, Cox assumes that, for the cast iron used in those experiments, the stress is related to strain by the equation

$$\sigma_x = \frac{\alpha \epsilon_x}{1 + \beta \epsilon_x}$$

where α and β are two constants. By proper selection of these constants, Cox obtained the good agreement that he sought.

Further improvement of the theory of impact was provided by Saint-Venant.[1] He considers a system consisting of a prismatical bar at the center of which is attached a mass W/g. He assumes that, at the instant of impact, a velocity v_0 is given to this mass while the rest of the bar remains stationary. He considers the different modes of vibration which ensue and calculates the maximum deflection at the middle. This calculation shows that, if only the first (most important) terms of the series representing the maximum deflection are retained, the result coincides closely with that given by the approximate equation (d). An examination of the second derivative d^2y/dx^2, representing the curvature of the deflected form, showed that the curve may differ considerably from that assumed in Eq. (a) and that the difference becomes more marked with increase of the ratio ql/W. Calculated values found by Saint-Venant for

[1] *Bull. soc. philomath. Paris*, 1853–1854; *Compt. rend.*, vol. 45, p. 204, 1857.

the ratio of the maximum stress σ_{max} as obtained from his series to the stress σ_{max}' found by using the curve (a) are given in the table below.

ql/W	$\frac{1}{2}$	1	2
$\sigma_{max}/\sigma_{max}'$	1.18	1.23	1.49

The discrepancy between the σ_{max} and σ_{max}' increases further with increase of the ratio ql/W.

It should be noted that Saint-Venant's investigation cannot be considered as yielding a complete solution of the problem of transverse impact. The local deformation at the point where the impinging ball strikes the beam is not considered and the assumption that the ball after the instant of impact remains in contact with the beam until the maximum deflection is reached is not borne out in practice. The ball may rebound before the maximum deflection is reached, in which case Saint-Venant's solution cannot be applied.[1]

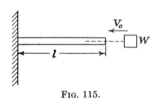

Fig. 115.

Saint-Venant also discusses longitudinal impact of bars. Considering a prismatical bar fixed at one end and subjected to a longitudinal impact at the other (Fig. 115), he assumes that during impact a uniform compression of the bar will be set up. To allow for the mass of the bar, he finds that we must assume that one-third of that mass is concentrated at the free end. The impinging body, which is considered as being absolutely rigid, assumes a velocity v which it has in common with the end of the bar after the instant of impact where

$$v = \frac{v_0}{1 + \frac{1}{3}(ql/W)}$$

The maximum value (f) of the compression may be found from the equation

$$\frac{AEf^2}{2l} = \frac{Wv_0^2}{2g} \frac{1}{1 + \frac{1}{3}(ql/W)} \qquad (e)$$

To get a more satisfactory solution, Saint-Venant considers a system consisting of a bar fixed at one end and having a mass rigidly attached to the other. Free vibrations of such a system had already been studied by Navier.[2] To adapt his teacher's solution to the case of impact, Saint-

---

[1] Saint-Venant assumes that his solution is exact. In his "Historique Abrégé . . . " (see p. 238) he states: "Pour calculer sa résistance, il faut, sous peine de commettre de graves erreurs qui inspireraient une fausse sécurité, se livrer au long calcul de la formule nouvelle et *exacte*."

[2] *Bull. soc. philomath.*, Paris, 1823, pp. 73–76; 1824, pp. 178–181.

Venant assumes that, while the bar is stationary at the instant of impact, the mass attached to its end suddenly takes a velocity v_0. Now using Navier's solution, he represents the motion of the end of the bar in the form of a series and, by means of summation, he finds the maximum deflection. This agrees closely with the approximate solution obtained from Eq. (e). Using the series representing longitudinal vibrations of the bar after the instant of impact, Saint-Venant also computes the maximum longitudinal strain and finds that it may differ greatly from the uniform strain assumed in the approximate equation (e). The results of Saint-Venant's calculations are given in the table below. They show that the

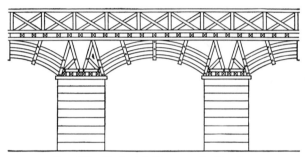

FIG. 116. Trajan's bridge over the Danube.

ratio of the maximum stress σ_{max} (calculated with the series) to the stress σ_{max}' [obtained by using Eq. (e)] increases as the ratio ql/W increases.

ql/W	$\frac{1}{4}$	$\frac{1}{2}$	1	2	4
$\sigma_{max}:\sigma_{max}'$	1.48	1.59	1.84	2.67	3.47

It should be noted once again that the possibility of rebounding of the impinging body was not considered by Saint-Venant in this solution, as he assumes that the body becomes rigidly attached to the end of the bar. Later on[1] Saint-Venant returned to this problem and found that little headway can be made toward a solution by using Navier's trigonometric series and that it is necessary to find a solution in finite terms. This solution was finally obtained by Boussinesq and also by Sébert and Hugoniot. The work of these scientists will be discussed later (page 239).

42. The Early Stages in the Theory of Trusses

The first uses to which trusses were put were in the construction of wooden bridges and roofs. Trusses were used by the Romans. Figure

---

[1] See Saint-Venant's "note finale du 60" of the book "Théorie de l'élasticité des corps solides de Clebsch, traduite par MM. Barré de Saint-Venant et Flamant, avec des Notes étendues de M. de Saint-Venant," Paris, 1883.

116 shows the wooden superstructure of the famous bridge over the Danube, which was erected at the order of Emperor Trajan, as it appears in the bas-reliefs of the Trajan column in Rome.[1] At the time of the Renaissance, Italian architects became interested in wooden trusses, and the famous architect Palladio was perhaps the first to construct such structures of any great span. Figure 117 shows a wooden bridge with a span of about 100 ft constructed by Palladio between Trent and Bassano. Some other Palladio bridge designs are shown in Figs. 118 and 119.

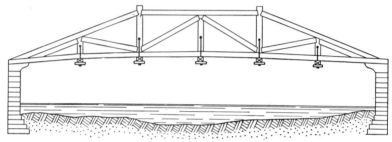

FIG. 117. Palladio's bridge near Trent.

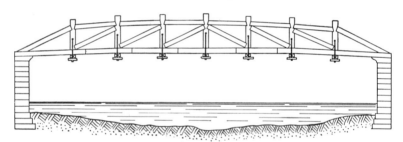

FIG. 118. Palladio's truss bridge.

Further use of wooden bridges was made in Switzerland. Figure 120 depicts one of the two spans of the famous bridge over the Rhine at Schaffhausen; it was constructed in 1757 by the gifted carpenter Jean-Ulrich Grubenmann. The two spans were 171 ft and 193 ft in length and proved strong enough to carry carriages up to 25 tons in weight with safety. In 1778, Grubenmann and his brother built a longer bridge over

[1] For the early history of bridges, see H. Gautier, "Traité des Ponts," Paris, 1765. See also "Oeuvres de M. Gauthey," published by M. Navier, Paris, 1809–1813. Interesting information on wooden bridges can be found in the book by Gaspar Walter, "Brückenbau oder Anweisung wie allerley Arten von Brücken sowohl von Holz als Steinen nach den besten Regeln der Zimmerkunst dauerhaft anzulegen sind," Augsburg, 1766.

the Limmat near Wettingen with a span of about 390 ft. Both of these famous bridges were burnt by the French during the war of 1799.[1] Many bridges of this kind, but with smaller spans, were later built in Switzerland[2] and Germany. Subsequent development of wooden bridges involved a change over to arch construction because of the greater rigidity that it admits.[3]

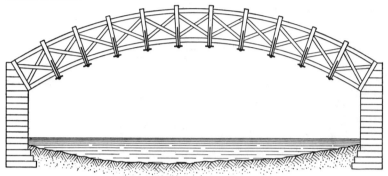

Fig. 119. Palladio's arch bridge.

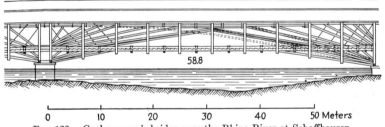

Fig. 120. Grubenmann's bridge over the Rhine River at Schaffhausen.

With the beginning of railroad construction, the proper design and construction of bridges became vital. In Western Europe, where the first railways were built in densely populated areas, the bridges had to be of a permanent character and stone arches and cast-iron girders and arches came into vogue.[4] Conditions in the United States and Russia were

[1] A detailed description of these bridges is given by Chrétien de Mechel, "Plans, coupes et élévations des trois ponts de bois les plus remarquables de la Suisse."
[2] Regarding Swiss wooden bridges, see the book by Dr. J. Brunner, "Der Bau von Brücken aus Holz in der Schweiz," Zürich, 1925.
[3] These bridges were developed by Gauthey in France and especially by Wiebeking in Bavaria.
[4] Many types of early railway bridges are described by J. Weale, "The Theory, Practice, and Architecture of Bridges," London, 1853. See also William Humber, "A Practical Treatise of Cast and Wrought Iron Bridges and Girders," London, 1857.

quite different, for, with sparse populations and long distances, economic considerations required extreme economy in initial expenditure. All construction was of a provisional nature; therefore wood was widely used in bridgebuilding and various systems of trusses, similar to those used by Palladio, were devised. The systems due to Long[1] and that of Town are shown in Figs. 118 and 121, while Howe's truss is seen in Fig. 122; all these were frequently used.

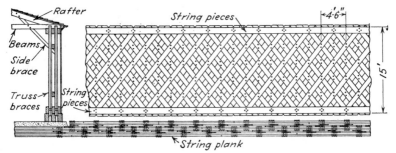

Fig. 121. Town's bridge.

Fig. 122. Howe's truss.

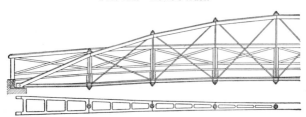

Fig. 123. Whipple's bridge.

The first all-metal trusses were built in the United States in 1840.[2] Figure 123 shows one of these that was built by S. Whipple; it had cast-iron upper members and wrought-iron lower chords and diagonals. Figure 124 depicts the bridge constructed by Whipple (1852–1853) for the Rensselaer–Saratoga railway. It is similar to the wooden bridges that

[1] Long's truss is similar to Palladio's truss shown in Fig. 118.
[2] See Mehrtens, "Eisenbrückenbau," Leipzig. The first volume of this work is given over principally to the history of bridges.

were built by Howe and it also has compression members of cast iron and tension members of wrought iron.

Whipple was the first to publish a book[1] containing useful information regarding the analysis of trusses. He not only considers the uniformly distributed dead load but also discusses moving loads for which he finds the most unfavorable position as far as any particular member is concerned. In analyzing a truss such as that shown in Fig. 125, he assumes that the diagonals can work only in compression and in this way obtains a statically determinate system. Starting at one of the supports, he finds the forces in the members by applying the parallelogram of forces to the

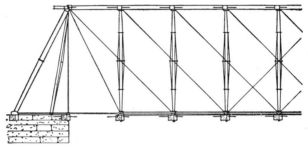

Fig. 124. Whipple's bridge on Rensselaer-Saratoga railway.

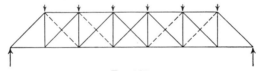

Fig. 125.

consecutive joints of the system. In this way Whipple clearly outlined perfectly valid methods, both analytical and graphic, for solving determinate trusses. In the graphic method he drew a force polygon around each joint on his line diagram of the truss. It appears that Whipple's work went unheeded by practical engineers, for in the preface to H. Haupt's book[2] (1851) it is stated: "When his attention was first directed to the subject of properly proportioning the parts of bridges, by being called upon in the discharge of professional duties to superintend their construction, he found it impossible to procure satisfactory information, either from engineers and builders, or from books." His own book is, moreover, much less clear and satisfactory than that of Whipple, which was published five years earlier.

[1] S. Whipple, C. E., "An Essay on Bridge Building," Utica, N.Y., 1847.
[2] Herman Haupt, "General Theory of Bridge Construction," New York, 1851 and 1869.

The first metal trusses in England were built in 1845. They were *lattice trusses*, similar to wooden Town trusses (Fig. 121). In 1846, the triangular Warren system was introduced (Fig. 126), and the method of analyzing them was known as early as 1850. Fairbairn, in his book "On the Application of Cast and Wrought Iron" (2d ed., p. 202, 1857) states that in 1850 he obtained such an analysis from W. B. Blood, and then gives correct formulas for calculating stresses in the members, not only for uniformly distributed dead load, but also for the most unfavorable distribution of moving load. In the book by W. Humber (see page 183), interesting experimental data is given on the stresses in Warren trusses (p. 56). The experiments were made by Blood and Doyn on a model that was 154 in. long and 12.12 in. deep. The forces were measured by a special dynamometer which was placed in the model in place of the member whose stress was to be determined. The test results were in very satisfactory agreement with the theoretical results.

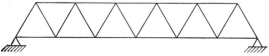

FIG. 126. The triangular Warren system.

French engineers were also interested in the analysis of trusses. As early as 1848, M. Michon was discussing the methods of calculating stresses in roof trusses of large span in his lectures at the military school of Metz.[1] A railway bridge truss of the Palladio type (see Fig. 117) having a span of 5 m, was tested at the Conservatoire des Arts et Métiers by Morin.[2] The stresses measured by dynamometers agreed well with the results of calculation. Further progress in analyzing Howe trusses was made by D. I. Jourawski (page 141). For the Werebia Bridge, which he designed and built, he selected Howe's type, for it had already seen several years' use on American railroads. However, at that time (1844) there was no theory of truss analysis in existence, and Jourawski had not only to design the bridge but also to develop a method of calculating stresses in the members. Jourawski succeeded in this work and offered a general method of analyzing trusses with parallel chords.

Considering a truss such as shown in Fig. 127, Jourawski observes that the kind of joints in Howe's system is such that the diagonals can work only in compression, and in the case of uniformly distributed load only the diagonals shown by full lines will work. In this way, a statically determinate system is obtained. In analyzing it, Jourawski concludes from symmetry that the load will be equally distributed between the two

[1] See "Historique abrégé . . . ," p. 211. Michon's methods are incorporated in Morin's book "Résistance des Matériaux," 3d ed., vol. 2, p. 141, 1862.
[2] Morin, "Résistance des Matériaux," 1st ed., Arts. 324–326, 1853.

supports, and we can assume that the loads to the left of the center of the span and half of the centrally applied load are transmitted to the left-hand support and the rest of the loads to the right-hand support. Beginning now from the upper central joint 0, Jourawski concludes that the load $\frac{1}{2}P$, which has to be transmitted to the left-hand support, will produce in the diagonal 0-1 a compressive force $P/(2 \cos \alpha)$. This force, transmitted to joint 1 will produce a tensile force $P/2$ in the vertical bolt 1-2 and a horizontal force $(P/2) \tan \alpha$ acting along the lower chord. Considering now the joint 2, we see that in addition to the tensile force $P/2$ of the bolt the load P will act vertically down. These two forces together will produce, eventually, a compressive force $\frac{3}{2}P/(\cos \alpha)$ in the diagonal 2-3 and the horizontal force $\frac{3}{2}P \tan \alpha$ will act along the upper chord. Continuing

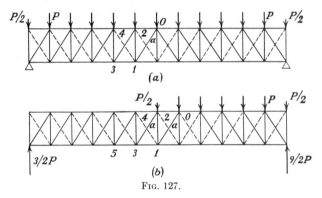

Fig. 127.

in the same manner with the joints 3, 4, . . . , Jourawski finds for the tensile forces in the remaining bolts the values $\frac{3}{2}P$, $\frac{5}{2}P$, . . . and for the compressive forces in the diagonals $\frac{5}{2}(P/\cos \alpha)$, $\frac{7}{2}(P/\cos \alpha)$, ... It is seen that the forces in the bolts and diagonals increase from the center of the span to the supports. At the same time, the forces in the chords are greatest at the center of the span and diminish toward the supports.

In nonsymmetrical loading, shown in Fig. 127b, Jourawski begins by finding the diagonals which will work. Knowing the reactions, he concludes that the first two loads from the left will be transmitted to the left-hand support and the working diagonals will be those indicated in the figure by full lines. Starting now from joint 0, and proceeding as before, the forces in all members can be readily calculated. Having this method, Jourawski found the most unfavorable distribution of load for each member of the bridge and calculated the corresponding maximum forces which must be used in selecting the proper cross-sectional dimensions of the members. Jourawski made a model of the bridge in which the vertical bolts were made of strings. The pitch of the strings, when the model was loaded, gave him the magnitudes of the tensile forces produced.

The actual problem of determining forces in members of Howe's truss is more complicated, since, by screwing the bolts, considerable initial stresses are usually produced in the system which cannot be disregarded in stress analysis. Jourawski goes into an investigation of initial stresses, considering first one isolated panel (Fig. 128a), and shows that by tightening the vertical bolts, compression of diagonals and tension of chord members are produced. But, he states, these stresses should not be simply added to stresses produced by external loads.[1] To prove this, he assumes that the bolt ab is fixed and considers the action of a vertical load Q along the bolt cd. It is easy to see that due to this action of the force Q, the

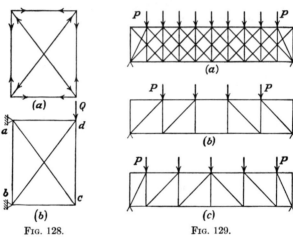

FIG. 128.   FIG. 129.

initial compression in the diagonal ac will be diminished while that in the diagonal bd will increase. The best effect of the initial tightening of the bolts in any panel of the bridge is obtained when, under the action of the maximum shearing force Q in that panel, there remains in such a diagonal as ac (in Fig. 128b) only the small compressive force necessary to keep the diagonal in its place. In such a case the total shearing force Q will be transmitted by the second diagonal and the values of the maximum forces in bolts and diagonals will be of the same magnitude as that obtained before from the analysis illustrated in Fig. 127.

In his further discussion, Jourawski also considers more complicated systems, such as that shown in Fig. 129a, and he proposes to calculate forces in the members by superposition of the forces which can easily be found for the two simple trusses shown in Figs. 129b and 129c.[2]

[1] Such superposition was practiced by some engineers long after Jourawski's investigation. See, for example, G. Rebhann's book, "Theorie der Holz und Eisen-Constructionen," p. 517, Vienna, 1856.

[2] The spans are assumed to be equal.

In conclusion, Jourawski discusses trusses on three supports and gives a method of calculating stresses for a uniformly loaded truss. To apply the previously used method of analysis, Jourawski needs to determine the position of the cross section mn, defining the load which is distributed along the portion $a0$ of the truss and which is transmitted to the left-hand support A. If the position of that cross section is found, the directions of those diagonals which are working become known, and he proceeds with the analysis as before, starting with the upper joint 0. Now Jourawski determines the position of the cross section mn by using two assumptions: (1) that this cross section coincides with one of the bolts (bolt 0-0 in Fig. 130) and (2) that this bolt remains vertical during deflection of the truss. To find this cross section, he applies a trial-and-error method. Let bolt 0-0 coincide with the required cross section; then the working diagonals are known and the forces at the joints acting along the

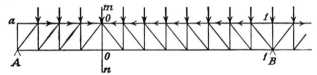

Fig. 130.

upper chord will be determined as before. Considering now the portion 0-1 of the upper chord as a bar fixed at the ends 0 and 1, he can readily determine the force transmitted to the fixed end 0. If the location of the bolt 0-0 was correctly selected, this force must be in equilibrium with the forces acting along the portion 0-a of the upper chord. In this way, Jourawski solves his statically indeterminate problem.

Publication of these investigations was started after the construction of the bridge was finished. In 1850, Jourawski published his method of analyzing trusses.[1] In 1854, he presented his work on bridges of the Howe system in complete form to the Russian Academy of Sciences and obtained Demidov's prize for it.[2]

Further progress in analyzing trusses was due to two German engineers, J. W. Schwedler[3] and A. Ritter.[4] Schwedler uses in his analysis the notions of *bending moment* and *shearing force* and establishes the relation

$$V = \frac{dM}{dx} \tag{a}$$

---

[1] See *Zhurnal Glavnago Upravlenia Putej Soobchenia i Publichnih Rabot*, 1850.

[2] This work was published in 1856–1857 in St. Petersburg. The portion dealing with calculation of shearing forces in beams was translated and published in French; see *Ann. ponts et chaussées*, vol. 12, p. 328, 1856.

[3] J. W. Schwedler, Theorie der Brückenbalkensysteme, *Z. Bauwesen*, vol. 1, 1851, Berlin.

[4] A. Ritter, "Elementare Theorie und Berechnung eiserner Dach und Brücken-Constructionen," 1862.

which later saw extensive use in truss analysis. Using this equation, Schwedler shows that the cross section at which bending moment reaches its maximum value is that at which the shearing force changes sign.[1] Considering the truss shown in Fig. 131 and taking a cross section mn, Schwedler finds the forces in the three intersected bars by using the three equations of statics. He shows how these forces change with changes in the depth of the truss and defines an optimum (or normal) shape of truss such that its depth is proportional to the bending moment produced by the dead load and the uniformly distributed moving load at every cross section. Such a load will produce no stresses in the diagonals of a normal truss. He shows that the sign of the stress in a diagonal changes with the sign of the shearing force and gives a method of determining the limits of that portion of the truss where two diagonals are required, if these diag-

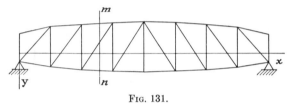

FIG. 131.

onals can work only in tension or only in compression. He also discusses more complicated trusses (*e.g.*, Town trusses) and suggests that, in their analysis, they be divided into simple systems, as shown in Fig. 129.

Ritter simplified the calculation of forces in the bars intersected by a cross section mn (Fig. 131) by using the equations of moments about points of intersection of two of the three intersected bars. In this way we have to solve only one equation with one unknown each time to obtain very simple formulas for the forces in the bars.

Both Schwedler and Ritter used analytical methods. Another simplification was brought about by the introduction of graphic methods into truss analysis. These were proposed by K. Culmann and J. C. Maxwell.

43. K. *Culmann* (1821–1881)

Karl Culmann was born in Bergzabern (Rheinpfalz) where his father was the minister. He was given a very good preparatory education by his father so that, at the tender age of seventeen, he was able to enter the

---

[1] K. Pearson in "History . . . ," vol. 2, p. 613, thinks that Schwedler was not the first to derive Eq. (*a*). But the writer was unable to find it in any earlier publication. Navier, in his book "Résumé des Leçons . . . ," p. 234, 1833, gives an unsatisfactory discussion of the point, for he states that the point where bending moment is a maximum "est toujours situé dans la verticale contenant le centre de gravité des poids dont cette pièce est chargé."

junior class of the Karlsruhe Polytechnicum without spending time in the two preparatory classes. After his graduation, he started his practical work (in 1841) at Hof on the Bavarian railroads. There he had a chance to work on the design and construction of important railway structures. Navier's "Résumé des Leçons . . . " was the accepted book on strength of materials at that time, and Culmann often refers to it in his analysis of structures. Culmann undertook a journey to England and the United States of America in 1849 to better his engineering education. As a result of his travel, he prepared and published an extensive study of English and American bridges.[1] This work greatly influenced the growth of theory of structures and of bridge engineering in Germany. England and America were then ahead of Germany in railway construction, and Culmann found many interesting structures for his study. At the same time, he had had a much better theoretical training than the majority of English and American engineers so that, in studying new structures, he was often able to make some important criticisms.

FIG. 132. K. Culmann.

His writings are of a great historical interest. He begins with a description of American wooden bridges. To his surprise, he finds that they are not merely imitations of European types but that they incorporate many original features. He is especially impressed by the work of S. H. Long, whom he considers to be one of the best American engineers of his day and who has good theoretical knowledge which he uses in selecting the proper dimensions of members of bridge trusses. Long's system of trusses was similar to that of Palladio, but he evidently knew a valid method of calculating stresses in truss members and, in his work,[2] gave very reasonable proportions for all the members of structures of various spans. After describing some of Long's bridges, Culmann remarks that he was unable to find out if Long was still alive and, if so, what he was doing. He remarks: "The American engineers are too practical to be

[1] See *Allgem. Bauztg*, vol. 16, pp. 69–129, 1851; vol. 17, pp. 163–222, 1852. A large number of carefully prepared drawings of American and English bridges is appended to these two articles.

[2] Title of the paper "Improved brace bridge." Specification of a Patent for an Improved Bridge, granted to Lt. Col. S. H. Long in 1839.

interested in thinking about their outstanding men. Each practical engineer considers himself as the highest authority, looks down on the others and pays no attention to them."

In discussing Howe's system, Culmann observes that it introduces nothing new to the theory of structures. Its success is due to some practical improvements in the method of construction used by Long. Regarding Town's bridges, Culmann is very critical and does not recommend them for use in Germany. Finally Culmann discusses the bridges erected by Burr. These he regards as being developed from the European type of arch bridge that was introduced by Gauthey and Wiebeking (page 183). This type (he says) proves very satisfactory in practice but no means of rational analysis has been offered for it since it is difficult to find what proportion of the load is transmitted by the arch and what is carried by the stiffening truss.

Aiming at a more satisfactory means of comparing the various types of wooden bridges, Culmann develops new methods of analyzing these truss systems. He presents a valid discussion of Long's and of Howe's configurations and draws up approximate methods of analysis for such statically indeterminate types as Town's and Burr's bridges. This work and the (previously mentioned) paper by Schwedler (page 189) probably constitute the most complete truss analyses of that time.

Applying his own methods of analysis to the various types of wooden bridges, Culmann is able to demonstrate that the Americans allow much smaller values for the moving loads in their calculations than is required by European specifications and at the same time they use much higher working stresses. This fact prompts Culmann to make some remarks of disapproval regarding the safety of American bridge structures.

The second issue (1852) of Culmann's work treats iron bridges. In England, the author was impressed by Brunel's plate girders (see page 162) and especially by the large tubular bridges, which had just been finished. It seems that Culmann obtained all his preliminary information about these bridges from Fairbairn's book (see page 157), for he is inclined to give most of the credit to that engineer. Now in England, Culmann met the men who continued to work with Stephenson[1] and, under their influence, he evidently changed his mind because, in his writing, he became very critical of Fairbairn's part in the work. He considers that it was a mistake to entrust the preliminary study of the problem to a man who had insufficient theoretical knowledge and who, only after numerous expensive experiments, discovered that iron in compression is weaker than in tension, whereas this could have been learned from existing books.

[1] Fairbairn, as a result of a disagreement with Stephenson, retired from the work on tubular bridges after completion of the first tube of the Conway bridge in 1849.

It seems that Culmann's disparaging remarks are ill considered because little was known at that time about thin-walled structures. In fact, Fairbairn's experiments first drew the attention of engineers to the importance of the stability question in designing compressed iron plates and shells.

At the beginning of his discussion of metal trusses, Culmann remarks that these were originally mere imitations of existing wooden trusses. When in Washington, he visited Rider's shop where the fabrication of parts of Rider's patented bridges was in progress. The author comments that the American inventor seemed to be completely satisfied at getting money from his invention and did not care about further improvements of his construction. The chief defect of Rider's bridges, in Culmann's opinion, is their lack of sufficient rigidity in the upper compressed chord of an open bridge. As a result of this weakness the upper chords in some of those bridges buckled and Culmann describes one such catastrophe, the results of which he was able to examine. According to Culmann, Whipple's truss is much better in this matter of stability, since the upper compressed chord has adequate stiffness in the horizontal plane. This writer is especially disparaging where iron lattice trusses, similar to Town's wooden trusses (Fig. 121), are concerned. He explains that the thin flat compression members of that system cannot withstand the compressive forces and buckle sideways. In this matter of buckling they prove weaker than the wooden members of Town bridges since, where wood is employed, much thicker material is used. He also says that the introduction of vertical stiffeners is bad and shows that such members completely change the conditions under which the lattice system works. Culmann is against the use of lattice trusses in Germany.

At the time of Culmann's visit to America, the majority of American engineers had little confidence in iron trusses, and it appears that fatigue failures were the main cause of their mistrust. They told Culmann that, under the action of repeated impact and vibration, the fibrous iron assumes a crystalline structure and a member might fail suddenly without any warning of the onset of such weakness. Culmann observes (probably correctly) that the cause of the trouble lies in the high stresses which Americans are using. By reducing the working stresses, fatigue failures could be eliminated.

This German engineer was specially impressed by American suspension bridges and remarks that, of all the bridges that he had seen during the course of his travels, the one which impressed him most was the suspension bridge across the Ohio River at Wheeling. At that time, the question of safety of suspension bridges for railway traffic was under discussion by American engineers. Culmann, on the strength of his own analysis, favors the use of such bridges for railroads, provided that a suitable stiffening truss be used to eliminate excessive flexibility. He recommends the use of the same cross-sectional dimensions in the stiffening truss as would be required for a truss bridge the span of which is equal to one-half of the span of the suspension bridge.

In general, Culmann was impressed by American bridge construction and by the courage of American engineers. But, in his opinion, insufficient importance was attributed to preliminary theoretical analysis of structures. He remarks that American engineers conclude that this or that construction is unsatisfactory, not because it is theoretically unsound, but because it fails in service.

After returning home in 1852, Culmann continued his practical work on the Bavarian railways till he was called upon by the newly organized Zürich Polytechnicum to become the professor of the theory of structures in 1855. Culmann liked teaching and threw all his energy into the preparation of the courses in which he stressed the importance of graphic methods in analyzing engineering structures. The construction of the polygon of forces and of the funicular polygon had been known since the time of Varignon[1] and they were utilized by Lamé and Clapeyron in their analysis of arches. Poncelet[2] used them in his theory of retaining walls. But these applications, before Culmann's day, only amounted to a few particular cases where graphic methods were used in the solution of structural problems. Culmann's principal accomplishment lies in his systematic introduction of graphic methods into the analysis of all kinds of structures and in the publication of the first book on graphical statics.[3]

Many original graphical solutions are incorporated in this book. The writer considers projective geometry very important in the development of graphical statics and discusses the projective properties of systems of forces in his introductory chapters.[4] In dealing with applications, Culmann begins with parallel forces in one plane and shows how, by using the funicular polygon, the reactions at the supports of a beam can be deter-

[1] See Varignon, "Nouvelle Mécanique," vol. 1, Paris, 1725.

[2] *Mém. officier génie*, vol. 12, 1835; vol. 13, 1840. Michon, who succeeded Poncelet at the school of Metz, used graphic methods in the theory of arches and in the theory of retaining walls. Culmann refers to Michon's lectures in the preface to the French translation of his book (Paris, 1880).

[3] K. Culmann, "Die graphische Statik," Zürich, 1866. Copies of Culmann's lectures appeared in note form in 1864 and 1865. In 1875 the first volume of the second enlarged edition appeared, but Culmann died before the second volume of that edition was ready. In 1880, a French translation of the second edition was published.

[4] In the preface of his book, Culmann expresses the opinion that projective geometry should be taught as a required course in engineering schools and he assumes that his readers have some knowledge of that science. This idea did not win general approval, since those portions of graphical statics which are of direct practical importance can easily be presented without recourse to the new geometry.

mined. He demonstrates how to construct the bending-moment diagram and how to find that position of a moving load which produces the maximum value of bending moment. Further on, the author gives the graphic method of fixing the position of the centroid of a plane figure and the method of calculating the moment of inertia of a figure with respect to an axis. He shows how the directions of the principal axes of a figure can be found, constructs the ellipse of inertia, and shows how it can be used in finding the core of the cross section of a bar.

In his work on bending of beams, Culmann shows how the stresses at any point A (Fig. 133a) can be analyzed graphically. Considering an infinitesimal element Amn and denoting the stress components acting on the planes through A and perpendicular to the x and y axes by σ_x and τ_{xy}, he demonstrates that the normal and the tangential components of stress acting on any inclined plane mn are given by the coordinates of points on a *circle of stress*. To construct this circle, we have only to take the point

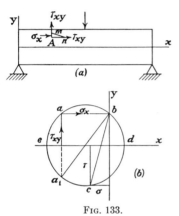

Fig. 133.

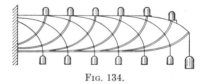

Fig. 134.

a (Fig. 133b) with coordinates τ_{xy}, and σ_x and consider point a_1 symmetrically placed with respect to a about the x axis. Then the line a_1b represents the diameter of the circle of stress, and we can draw this circle. From the equations of equilibrium of the element Amn, Culmann shows that the stress components σ and τ acting on any plane mn are given by the coordinates of the point c, which is obtained by making a_1c parallel to mn. From this construction, he shows that the points d and e at the ends of the horizontal diameter define the magnitudes and the directions of the principal stresses at point A. He also proves that the planes upon which the maximum shear acts bisect the angles between the principal planes and that the maximum shear is given by the magnitude of the radius of the circle of stress. It is easy to see that Culmann's circle of stress is a precursor and particular case of Mohr's circles (page 285) which are so widely used in stress analysis. Knowing the directions of the principal stresses, Culmann discusses stress trajectories and, in illustration of the case of a cantilever, gives the sketch shown in Fig. 134.

Later, Culmann gives an extensive survey of continuous beams, which were widely used in railway bridges at that time. He offers a graphical solution of the problem which was later simplified somewhat by Mohr (page 285). Regarding the inflection points of a continuous beam, he suggests the introduction of intermediate hinges in the structure as shown in Fig. 135a and observes that with such an arrangement bridges of longer spans could be erected. The trajectories of stresses for a continuous beam are given as the curves shown in Fig. 135b, and the author remarks that such forms should be copied in the design of bridges of very large spans. He mentions that a model bridge that was exhibited at the industrial exhibition in London conformed to this notion.

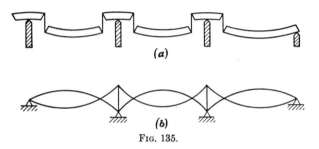

FIG. 135.

Applying graphic methods in analyzing trusses, Culmann uses the *method of sections* and shows that if, in this process, it intersects only three bars each time he can readily obtain the forces in these bars by resolving the known external forces along the directions of the three bars. He also constructed diagrams of internal forces in trusses by considering the conditions of equilibrium of the consecutive joints and drawing the appropriate polygons. Although he was not guided by the reciprocity relations, which were introduced by Maxwell (see page 202), his diagrams coincide with those derived by Maxwell's method almost without exception when he uses the method of joints.

In the last chapters of his book,[1] Culmann develops graphic methods of analyzing arches and retaining walls. The effects of this book were felt in the theory of structures far beyond the confines of the Zürich Polytechnicum. His graphic methods were accepted and widely used by German engineers, and such scientists as E. Winkler and O. Mohr contributed much to the later development of graphical statics. In Italy, this subject was introduced by Cremona, who was a professor of the Polytechnical Institute at Milan. His well-known book[2] on graphical calculus and

---

[1] He planned to revise this part of the book completely in the second edition, but died before that work was finished.

[2] "Le figure reciproche nella statica grafica," Milan, 1872.

reciprocal figures was translated into English by Prof. Thomas H. Beare in 1890.

44. W. J. Macquorn Rankine (1820–1872)

Rankine[1] was born in Edinburgh. His father, after retiring from the army, started to work as an engineer on the construction of railways and was a very well-informed man in the field of engineering. After studying at a Glasgow school, Rankine was privately instructed for some years in Edinburgh. In 1836, he entered Edinburgh University where he attended the natural philosophy course given by Professor Forbes and he won the gold medal for an essay on the "Undulatory Theory of Light." At the same time, he gained some acquaintance with practical engineering, having assisted his father in superintending some work on the Edinburgh and Dalkeith Railway. In 1838, Rankine himself started work on a railway as an engineer under J. B. Macneil and worked for the most part on various surveys and structures on the Dublin and Drogheda railway. He later worked on several schemes promoted by the Caledonian Railway Company.

FIG. 136. W. J. Macquorn Rankine.

Rankine started publishing his scientific work very early. In 1842, his first paper appeared and it was entitled "An Experimental Inquiry into the Advantage of Cylindrical Wheels on Railways." The following year, he presented his paper on fatigue failures of railway axles to the Institute of Civil Engineers (page 163). Rankine mentions in his notes that some of his early publications were based on suggestions made by his father. Beginning in 1848, Rankine became interested in molecular physics and thermodynamics. He published several important papers in physics and was elected a Fellow of the Royal Society in 1853.

In 1855, Rankine was appointed to the chair of engineering at Glasgow University and he worked there to the end of his life (1872). It seems that Glasgow University was the first British university to take up engineering education. Professor John Anderson (1726–1796) (of that uni-

---

[1] A biography of W. J. Macquorn Rankine, written by P. G. Tait, is contained in the volume of "Scientific Papers" of Rankine, London, 1881.

versity) provided for an institution for the instruction of artisans in his will, and the college which bears his name started to offer lectures on natural philosophy and chemistry in 1796. In 1799 George Birkbech (1776–1841) began to give his lectures on mechanics and applied science there, and they led to the foundation of engineering institutes in other cities. In 1840, the first chair of civil engineering in the United Kingdom was established. The Glasgow and West of Scotland Technical College was formed in 1886 as a combination of several institutions and the arts side of Anderson College constituted a part of that combination.

After 1855, most of Rankine's energy was spent on his educational activity and through his lectures, and more especially through his textbooks, he achieved much in the early stages of the growth of engineering science and engineering education at the university level in England. Most of his original work on strength of materials and theory of structures was included in his "Manual of Applied Mechanics," the first edition of which appeared in 1858, and in his "Manual of Civil Engineering," which was published in 1861. Since their appearance, these books have seen many new editions and even now they are not completely superseded. In writing these books, Rankine paid much attention to the portions dealing with the scientific background of the various branches of engineering, and most of his development of these sections was his own original work.

In his presentation of strength of materials in "Applied Mechanics," Rankine begins with the mathematical theory of elasticity. He gives a complete discussion of stress and strain at a point and develops the fundamental equations of equilibrium. Here, perhaps for the first time in English literature, we see such terms as *stress* and *strain* rigorously defined. In his work Rankine prefers to treat each problem first in its most general form and only later does he consider various particular cases which may be of some practical interest. Rankine's adoption of this method of writing makes his books difficult to read, and they demand considerable concentration of the reader.

In discussing problems of theory of structures, Rankine introduces his method of parallel projections in statics[1] and shows some interesting applications of it. He shows "that if a balanced system of forces acting through any system of points be represented by a system of lines, then any parallel projection of that system of lines will represent a balanced system of forces." Applying this principle to the theory of frames, he states: "If a frame whose lines of resistance constitute a given figure, be balanced under a system of external forces represented by a given system of lines, then a frame whose lines of resistance constitute a figure which is a parallel projection of the original figure, will be balanced under a system of forces

[1] "Applied Mechanics," 14th ed., Chap. IV, 1895.

represented by the corresponding parallel projection of the given system of lines; and the lines representing the stresses along the bars of the new frame, will be the corresponding parallel projections of the lines representing the stresses along the bars of the original frame." He gives an interesting application of this theory in the analysis of arches. Taking first a *circular linear arch*[1] (or ring) submitted to a uniformly distributed normal pressure p (see Fig. 137a), he concludes that the thrust T around this ring is the product of the pressure p by the radius r. Making a parallel projection such as is shown in Fig. 137b, we do not change any vertical dimensions, while all the horizontal dimensions are increased in the same ratio (say in the ratio n). In this way an *elliptical linear arch*

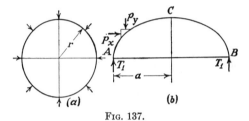

Fig. 137.

is obtained. Since, in this projection, the vertical components of forces do not change in magnitude, while at the same time the distances between them increase in the ratio $n:1$, it can be concluded that the vertical pressure on the arch (Fig. 137b) will be $p_y = p/n$. In a similar way, we find that the horizontal pressure is $p_x = np$. The thrust at the horizontal cross sections A and B of the elliptical arch will be

$$T_1 = p_y \cdot a = \frac{p}{n} rn = pr$$

At the same time the thrust at the vertical cross section C is

$$T_2 = p_x r = npr$$

In the case of a linear arch ACB supporting water pressure (see Fig. 138), at any point M the curve must have a curvature proportional to the depth y, since only under this condition can the resultant of the thrusts acting in the ends of an element of the curve balance the water pressure. To obtain such a curve, we may use a straight elastic wire DCE to the ends of which we attach the bars EF and DG, as shown in the figure by dotted lines. Now by rotating the ends of the wire through 180 deg, we can keep it in its bent condition ACB by applying the horizontal forces H as shown in the figure. The curvature of the elastic line at any point

[1] "Applied Mechanics," 14th ed., Arts. 179–184, 1895.

will be proportional to the bending moment at that point, i.e., proportional to Hy and the curve obtained in this way will satisfy the requirement stated above. Thus a linear arch, having the shape of the curve ACB, will undergo only axial compression without any bending[1] under the action of water pressure. By using parallel projection, Rankine arrives at *geostatic arches* from *hydrostatic arches*. Reducing the horizontal dimensions of the arch ACB in Fig. 138 to some constant ratio without changing the vertical dimensions, he obtains a linear arch for the case where the intensity of the horizontal pressure p_x acting on the arch is only a certain portion of the vertical pressure p_y as it should be for earth pressure.

Treating a plane polygonal frame, Rankine states: "If lines radiating from a point be drawn parallel to the lines of resistance of the bars of a

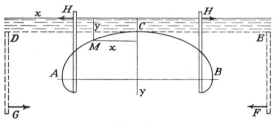

Fig. 138.

polygonal frame, then the sides of any polygon whose angles lie in those radiating lines will represent a system of forces, which, being applied to the joints of the frame, will balance each other."[2] This statement was later[3] extended by Rankine to three-dimensional systems, and it was stated that the polyhedron of forces and the polyhedral frame are reciprocally related in a certain way. This work of Rankine suggested the famous work on reciprocal figures to Maxwell (see page 202).

Regarding suspension bridges, Rankine refers to P. W. Barlow who, "having made some experiments upon models, finds that very light girders, in comparison with what were supposed to be necessary, are sufficient to stiffen a suspension bridge." Assuming a certain shape for the deflection curve of the stiffening truss, Rankine[4] gives an approximate solution of the problem and concludes that "the transverse strength of the stiffening girder should be four twenty-seventh parts of that of a

[1] Such forms of arches were mathematically investigated by Yvon Villarceau, Mémoires sur les voûtes en berceaux cylindriques, *Rev. architect. et travaux publics*, Paris, 1845.
[2] "Applied Mechanics," 14th ed., p. 140, 1895.
[3] *Phil. Mag.*, February, 1864.
[4] "Applied Mechanics," 14th ed., p. 370, 1895.

simple girder of the same span suited to bear a uniform load of the same intensity." It seems that this investigation was the first theoretical analysis of stiffening trusses.

Rankine also made an important addition to the theory of bending of beams by taking into consideration the effect of shearing stresses on the magnitude of deflections.[1] He states that the additional slope of the elastic line produced by shear is

$$\frac{dy_1}{dx} = \frac{VS}{GIb} = \frac{dM}{dx}\frac{S}{GIb}$$

where V is the shearing force, S is the statical moment with respect to the neutral axis of the lower portion of the cross section, and b is the width of the cross section along the neutral axis. For a rectangular, uniformly

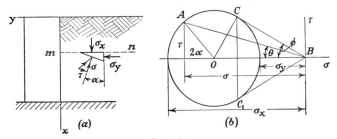

FIG. 139.

loaded, and simply supported beam, Rankine finds that the ratio of the additional deflection to that given by the usual formula is $6Eh^2/5Gl^2$, which he considers to be very small and negligible in practical applications. Naturally, he would have reached a quite different conclusion in considering an I beam having large depth and a thin web.

Perhaps the most important of Rankine's contributions to the theory of structures was his investigation "On the Stability of Loose Earth,"[2] in which he offered a method of finding the proper dimensions of retaining walls. Let us consider the simplest case in which loose earth presses against a vertical wall and is bounded above by a horizontal plane (Fig. 139a). Since the condition of stress at all points of any horizontal plane mn is the same, horizontal planes and the vertical planes parallel to the wall are the principal planes of stress. The corresponding principal stresses may be denoted by σ_x and σ_y (Fig. 139a). The stress components σ and τ for a plane at an angle α to the horizontal can be readily obtained from the Mohr circle shown in Fig. 139b. We draw the radius OA at an

[1] "Applied Mechanics," 14th ed., p. 342, 1895.
[2] *Phil. Trans.*, 1856–1857.

angle 2α to the horizontal and the coordinates of the point A give us the required stress components. The length $\overline{AB}$ represents the total stress acting on the inclined plane and the angle θ is the *angle of obliquity* of that stress. For the equilibrium of loose earth, without any cohesive forces, it is necessary that the obliquity angle should never exceed the angle of friction or the *angle of repose*, which we shall call ϕ. From the Mohr circle, we see that the maximum obliquity is found on the planes corresponding to points C and C_1 and we conclude that the limiting condition of equilibrium of earth is reached when the tangents BC and BC_1 make an angle ϕ with the horizontal. Assuming that we have this limiting condition, the corresponding relation between the principal stresses σ_x and σ_y can be obtained from the Mohr circle. Observing that the radius of the Mohr circle is equal to $(\sigma_x - \sigma_y)/2$ and that the coordinate OB of its center is equal to $(\sigma_x + \sigma_y)/2$, we find from the triangle OBC that

$$\sigma_x - \sigma_y = (\sigma_x + \sigma_y) \sin \phi$$

from which

$$\frac{\sigma_x}{\sigma_y} = \frac{1 + \sin \phi}{1 - \sin \phi} \tag{a}$$

Now considering the retaining wall of Fig. 139a, we have

$$\sigma_x = \gamma x \tag{b}$$

where γ denotes the weight of unit volume of the earth. The minimum value of the horizontal reaction of the wall, consistent with the equilibrium requirement (a), is

$$\sigma_y = \gamma x \frac{1 - \sin \phi}{1 + \sin \phi} \tag{c}$$

Rankine recommends this value of pressure as that to be used when investigating the stability of a retaining wall. The Scottish engineer also arrives at a more general expression for the pressure acting on the wall when the earth is bounded by an inclined plane.

45. J. C. Maxwell's Contributions to the Theory of Structures[1]

We have seen that Culmann used force diagrams in analyzing trusses. But in his diagrams, the same force sometimes appears more than once. The method of drawing diagrams of forces in which the force in each member is represented by a single line was found independently by J. C. Maxwell[2] and by W. P. Taylor.[3] To illustrate Maxwell's method, let us consider a plane triangular frame (Fig. 140a) acting upon which are

---

[1] Maxwell's biography and his work in theory of elasticity are discussed later (see page 268).
[2] *Phil. Mag.*, vol. 27, p. 250, 1864.
[3] See Fleming Jenkin's paper, *Trans. Roy. Soc. Edinburgh*, vol. 25, p. 441, 1869.

three forces P, Q, R in equilibrium. The forces in the bars are obtained by constructing the diagram shown in Fig. 140b. The two figures can be considered as plane projections of two pyramids. Denoting the three sides of the pyramid in Fig. 140a by a, b, c and its base by o, we use the same letters, as shown in Fig. 140b. Thus, for every three lines forming a triangle in Fig. 140a, there is a corresponding point in Fig. 140b, through which pass the three lines parallel to the sides of the triangle. To every apex in Fig. 140a there corresponds a triangle in Fig. 140b, which represents the condition of equilibrium of the forces passing through the apex under consideration. Figures 140a and 140b constitute the simplest example of two reciprocal diagrams. It can easily be seen that, on

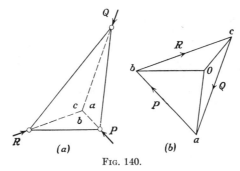

Fig. 140.

the basis of the reciprocity relation described above, and having Fig. 140a, we can construct the reciprocal diagram 140b in a purely geometrical way without consideration of the equilibrium of individual joints.

For the general case, Maxwell formulates his conclusions in the following two sentences: "Two plane figures are reciprocal when they consist of an equal number of lines, so that corresponding lines in two figures are parallel, and corresponding lines which converge to a point in one figure form a closed polygon in the other. If forces represented in magnitude by two lines of a figure be made to act between the extremities of the corresponding lines of the reciprocal figure, then the points of the reciprocal figure will all be in equilibrium under the action of these forces." Such an abstract account of the important property of reciprocal figures is hardly of much help to a practical engineer, and we agree with Professor Jenkin[1] who, after quoting the above two sentences, observes: "Few engineers would, however, suspect that the two paragraphs quoted put at their disposal a remarkably simple and accurate method of calculating the stresses in a framework." After this remark, Jenkin gives several examples of the construction of reciprocal diagrams following rules developed

[1] *Trans. Roy. Soc. Edinburgh*, vol. 25, p. 441, 1869.

by a practical draughtsman, W. P. Taylor, who worked in the office of a contractor. On the Continent the use of reciprocal diagrams became known from Cremona's book, previously mentioned (see page 196), and very often those figures are called Cremona diagrams.

We shall now discuss briefly some general conclusions regarding forces in the members of statically determinate trusses. Maxwell arrived at these conclusions some years later.[1] He states: "In any system of points in equilibrium in a plane under the action of repulsions and attractions, the sum of the products of each attraction multiplied by the distance of the points between which it acts, is equal to the sum of the products of the repulsions multiplied each by the distance of the points between which it acts." To prove this, we note that each point of the system is in equilibrium under the action of a system of repulsions and attractions in one plane. It will remain so if the system is turned through a right angle in the clockwise direction. But if this operation is performed on the systems of forces acting on all the points, then at the extremities of each line joining two points we have two equal forces at right angles to that line and acting in opposite directions. These form a couple whose magnitude is the product of the force between the points and their distance apart, and the direction is clockwise if the force be repulsive and counterclockwise if it be attractive. Now since every point is in equilibrium these two systems of couples are in equilibrium, or the sum of the positive couples is equal to that of the negative couples, which proves the theorem.

In the case of a loaded truss, we have external forces and reactions in addition to forces in the bars. If the rotation of forces described above is performed at all joints, the sum of moments of the rotated external forces and reactions with respect to any point in the plane of the truss must balance the sum of couples formed by the rotated forces in the bars. If, for example, we have a truss with a horizontal lower chord and all external forces and reactions are vertical and are applied to the joints of the lower chord, then the summation of moments of all rotated external forces and reactions vanishes and the sum of the products of the length of all tension members and the tensile forces acting in them must be equal to the sum of the products of length of all compression members and the corresponding compressive forces. This conlcusion can be utilized for checking the results of a truss analysis. It also follows that, if the stresses in all bars are numerically equal, the total volume of all tension members will be equal to the volume of the compression members.

Maxwell did not confine his attention to the analysis of statically determinate trusses, but attacked the problem in a more general form.[2] He

[1] *Trans. Roy. Soc. Edinburgh*, vol. 26, 1870.
[2] *Phil. Mag.*, vol. 27, p. 294, 1864.

shows that if we have a plane framework with n joints, we can write $2n$ equations of equilibrium. Three equations will usually be needed for calculating the reactions at the supports and the remaining $2n - 3$ equations may be used to compute the forces in the bars of the framework if the number of the bars is equal to $2n - 3$. Should the number of the bars be larger than $2n - 3$, the problem is statically indeterminate so that the elastic properties of the bars must be taken into consideration and the deformation of the structure must be investigated for its solution. Regarding the method of doing this, Maxwell says: "I have therefore stated a general method of solving all such questions in the least complicated manner. The method is derived from the principle of Conservation of Energy, and is referred to in Lamé's 'Leçons sur l'Élasticité,' Leçon 7me, as Clapeyron's Theorem; but I have not yet seen any detailed application of it."

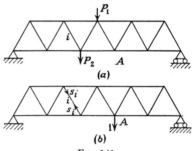

To illustrate Maxwell's method of analysis, let us begin with a calculation of the deflections of a truss such as is shown in Fig. 141a. This truss is statically determinate, and we can readily find the forces in all bars produced by the given loads $P_1, P_2, \ldots$ Let S_i be the force acting in any bar i, let l_i be the length of that bar, and let A_i be its cross-sectional area. Then the elongation of the bar is $\Delta_i = S_i l_i / E A_i$. We now have the geometrical problem of finding the deflection of any joint, say A, knowing the elongations of all bars of the truss. To solve this, Maxwell considers an auxiliary problem, shown in Fig. 141b, in which he again takes the same truss, but instead of applying the actual loads $P_1, P_2, \ldots$, he applies a unit load at joint A, the deflection of which we have to calculate. This problem is again statically determinate, and we can find the force s_i which the unit load induces in any bar i. Let us now calculate the deflection δ_a' of the hinge A resulting from the elongation Δ_i' of one bar i. For this purpose we assume that all bars of the truss except the bar i are absolutely rigid. Then the work done by the unit load in moving through the deflection δ_a' will be transformed into strain energy of the bar i and, assuming proportionality between the force and the deflection, we obtain the equation

$$\tfrac{1}{2} \cdot 1 \cdot \delta_a' = \tfrac{1}{2} s_i \Delta_i'$$

From this it follows that

$$\delta_a' = \Delta_i' \frac{s_i}{1}$$

This relation holds for any small value of Δ_i'. If instead of Δ_i' we write $S_i l_i / E A_i$, we shall find the deflection in the actual case (Fig. 141a) due to elongation of the one bar i. To get the total deflection δ_a in the actual case, we have only to sum up the deflections brought about by the deformations of all bars. This gives

$$\delta_a = \sum_{i=1}^{i=m} \frac{S_i s_i l_i}{E A_i} \qquad (a)$$

By using Eq. (a), some statically indeterminate problems can be solved. Consider, for example, the truss in Fig. 142a with one redundant support.

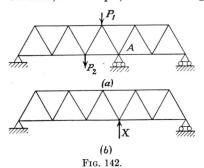

(a)

(b)

Fig. 142.

To obtain the reaction X at the point A, we assume first that this support is removed and calculate the consequent deflection of the joint A by using Eq. (a). Now, separately as in Fig. 142b, we calculate the deflection at A brought about by the reaction X acting alone. Using the results obtained for the case depicted in Fig. 141b, we see that the force in any member i of the truss produced by this reaction X is equal to $-s_i X$, so that the deflection δ_1 at A produced by the reaction X is, from Eq. (a),

$$\delta_1 = -\sum_{i=1}^{i=m} \frac{X s_i^2 l_i}{E A_i}$$

Since in the actual case of Fig. 142a, the deflection at A vanishes, the equation for determining X becomes

$$\delta + \delta_1 = \sum_{i=1}^{i=m} \frac{S_i s_i l_i}{E A_i} - \sum_{i=1}^{i=m} \frac{X s_i^2 l_i}{E A_i} = 0 \qquad (b)$$

It is seen that the solution of the statically indeterminate problem of Fig. 142a is reduced to the two statically determinate problems of calculating forces S_i and s_i. In a similar manner, trusses with redundant bars can be treated.

In his work on deflections of trusses, Maxwell discovered the existence of a very important relation between the deflections of a truss produced by two different kinds of loading. Let us consider the two cases of load-

Strength of Materials in the Evolution of Railway Engineering

ing shown in Figs. 143a and 143b. In the first one, the load P is applied at the hinge B, and it is required to find the deflection δ_a at A. In the second, the load P acts at A, and it is required to find the deflection δ_b. Following the method illustrated by Fig. 141, we consider the two auxiliary cases shown in Figs. 143c and 143d. Let us use the notation S_i and s_i for the forces in the ith bars of Figs. 143a and 143c, respectively, and S_i' and s_i' for the forces in Figs. 143b and 143d. Then, from Eq. (a), we know that

$$\delta_a = \sum_{i=1}^{i=m} \frac{S_i s_i l_i}{A_i E} \qquad \delta_b = \sum_{i=1}^{i=m} \frac{S_i' s_i' l_i}{A_i E} \qquad (c)$$

But comparing Figs. 143c and 143b, we see that $S_i' = s_i P$ and, from Figs.

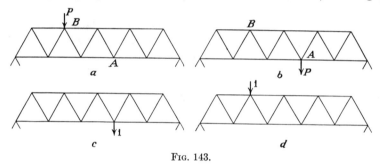

Fig. 143.

143d and 143a, we find $S_i = s_i' P$. Substituting into Eqs. (c), we arrive at

$$\delta_a = \delta_b = P \sum_{i=1}^{i=m} \frac{s_i s_i' l_i}{A_i E} \qquad (d)$$

Thus we see from the two cases of loading, shown in Figs. 143a and 143b, that when the load is moved from joint B to joint A, the deflection of A moves to B. This is the reciprocity theorem as it was obtained by Maxwell in its simplest form.

Later on, as we shall see, the theorem was generalized and became very important in the analysis of statically indeterminate structures.

Maxwell presented his method of indeterminate frameworks in an abstract form without any illustrating figures and, as far as we can judge from the contemporary literature, his work remained completely unnoticed by engineers.[1] Ten years later, O. Mohr rediscovered[2] Maxwell's

---

[1] The discussion of Maxwell's work and some corrections to Maxwell's example are given in the recent paper by A. S. Niles, *Engineering*, Sept. 1, 1950.
[2] *Z. Architek. u. Ing. Ver. Hannover*, 1874, S.223, 509; 1875, S.17.

equation (*a*) and showed various applications of it in structural analysis. Since Mohr arrived at his results without knowledge of Maxwell's work and by using a quite different process of derivation, and since the method found practical application only after Mohr's publication, it is usually called the *Maxwell-Mohr method.*

46. Problems of Elastic Stability. Column Formulas

With the introduction of iron structures, questions of the elastic stability of slender struts and thin webs became of great practical importance. At the beginning of the nineteenth century, Duleau showed, by his compression tests of iron bars (see page 82), that Euler's formula gives satisfactory values for the buckling load if the end conditions that are assumed in the theory are fulfilled. Duleau used comparatively thin bars whereas the struts used in trusses were not so slender. In the latter case, Euler's formula predicts exaggerated values for the buckling load, and toward finding a more reliable formula, recourse had to be had to experimentation.

In the middle of the nineteenth century, the formulas that were derived on the basis of Hodgkinson's tests (page 128) were widely used. But in most of Hodgkinson's experiments, the bars had flat ends and the end conditions were somewhat indefinite. Again sometimes rounded ends were used, but the surfaces of contact were not spherical so that any small buckling introduced some eccentricity. As a result, Hodgkinson's formulas left something to be desired. Furthermore, it was by no means easy to apply them in any practical problem so that some simplified formulas were introduced later by Love (see page 123) and by Lewis Gordon.[1]

A very important theoretical investigation of lateral buckling was made at that time by E. Lamarle.[2] He was the first to introduce a definite limit up to which Euler's formula should be used and beyond which experimental data should be relied upon. Having the formula

$$\sigma_{cr} = \pi^2 E \frac{i^2}{l^2} \qquad (a)$$

for a bar with hinged ends in which $l:i$ is the slenderness ratio, he correctly concludes that it can give satisfactory results only so long as the value of σ_{cr} does not exceed the elastic limit of the material. For iron, he evidently assumed the existence of an elastic limit equivalent to the yield-point stress and obtained the value

$$\frac{l^2}{i^2} = \frac{\pi^2 E}{\sigma_{yp}} \qquad (b)$$

---

[1] See Rankine's "A Manual of Civil Engineering," 20th ed., p. 236, 1898.

[2] E. Lamarle, Mémoire sur la flexion du bois, *Ann. travaux publics Belgique*, vol. 3, pp. 1–64, 1845; vol. 4, pp. 1–36, 1846. The buckling problem is discussed in the second part of the memoir.

for the limiting slenderness ratio. For slenderness ratios larger than this, Lamarle recommends Euler's formula, and for smaller values he proposes that the constant value σ_{yp} be taken for σ_{cr}.

Apparently this reasonable proposition remained unnoticed by engineers, and throughout Rankine's books we find Gordon's formula. Rankine gives an interesting history of this formula. It seems that, before Hodgkinson did his experiments, English engineers used a formula due to Tredgold which was derived along the following lines. Assume that a strut of a rectangular cross section (of sides $b \times h$) and with hinged ends is deflected by an amount δ under the action of an axial compressive force P. Then the maximum compressive stress is

$$\sigma_{\max} = \frac{P}{bh} + \frac{6P\delta}{bh^2} \qquad (c)$$

Now for a given material, the deflection is proportional to $M_{\max}l^2/I$ and the maximum bending stress is proportional to $M_{\max}h/I$. Hence for a given bending stress, δ is proportional to l^2/h. Using this value of δ in Eq. (c), Tredgold concluded that the compressive stress at which the strut fails can be calculated from the following equation

$$\sigma_{\max} = \frac{P}{bh}\left(1 + a\frac{l^2}{h^2}\right) \qquad (d)$$

where a is a certain constant. Rankine states that formulas based on such a consideration "having fallen for a time into disuse, were revised by Mr. Lewis Gordon, who determined the values of the constants contained in them by a comparison of them with Mr. Hodgkinson's experiments." In this manner Gordon obtained

$$\sigma_{ult} = \frac{36{,}000}{1 + \frac{1}{12{,}000}\frac{l^2}{h^2}}. \qquad (e)$$

for wrought iron. In the case of a strut with built in ends the factor $\frac{1}{3{,}000}$, instead of $\frac{1}{12{,}000}$, was proposed. Recommending a formula analogous to (e) for nonrectangular struts, Rankine uses the slenderness ratio l/i instead of l/h and accordingly changes the magnitude of the numerical factor in the denominator.

Rankine also takes up several other buckling problems in his manuals. Considering the compression of a built-up square iron tube and assuming that the thickness of the wall is not smaller than $\frac{1}{30}$ of the side of the cell, he recommends the use of formula (e) with a stress 27,000 lb per in.² instead of 36,000 lb per in.² in calculating σ_{ult}.

Regarding the bending of built-up I beams, Rankine discusses the possibility of lateral buckling of the compressed flange of the beam. Denoting the width of the flange by b, he calculates the ultimate stress for the flange with the formula

$$\sigma_{ult} = \frac{36{,}000}{1 + (1/5{,}000)(l^2/b^2)} \qquad (f)$$

In determining the correct thickness of the web of an I beam, Rankine considers the possibility of buckling of that web due to a compressive stress acting at the neutral axis at an angle of 45 deg to the horizontal. To calculate the ultimate value of that stress, he offers the formula

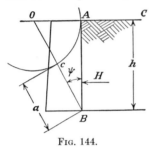

FIG. 144.

$$\sigma_{ult} = \frac{36{,}000}{1 + (1/3{,}000)(s^2/t^2)} \qquad (g)$$

where t denotes the thickness of the web and s is the distance measured along a line, inclined at 45 deg to the horizontal, between two adjacent vertical stiffeners.

A more rational method of analyzing short struts was proposed by H. Scheffler.[1] He assumes that, due to unavoidable inaccuracies in applying the compressive forces, we always have some eccentricity of loading at the ends. He derives a formula for the maximum stress in terms of the eccentricity so that, for any assumed value of the latter, the value of the dangerous load can be calculated. Scheffler selects eccentricities, for struts of various materials, of such magnitudes that his values of the ultimate load agree satisfactorily with Hodgkinson's experimental results. Apparently Scheffler was the first to use an assumed eccentricity in calculating the dangerous compressive load. The value of this method was not appreciated at that time, and engineers continued to use Rankine's formula up to the end of the nineteenth century.

47. Theory of Retaining Walls and Arches between 1833 and 1867

During this period, engineers continued to use Coulomb's theory (page 61) in investigating the stability of retaining walls. Progress in this field mainly consisted in the development of graphic methods for determining the magnitude of the earth pressure. Poncelet[2] was the first to show that the earth pressure on the retaining wall (as calculated by Coulomb) can readily be obtained by a graphic method. In the simplest case of a vertical wall and of earth having a horizontal surface AC (Fig. 144) he finds

[1] Hermann Scheffler, "Theorie der Festigkeit gegen das Zerknicken . . . ," Brunswick, 1858.
[2] *Mém. officier génie*, vol. XIII, pp. 261–270, 1840.

Strength of Materials in the Evolution of Railway Engineering   211

the following formula for the horizontal earth pressure per unit length of the wall by neglecting friction between the wall and the earth:

$$H = \frac{\gamma}{2} a^2 \quad (a)$$

where γ is the weight per unit volume of earth and a is the length obtained from Fig. 144 as follows:

$$a = \overline{Bc} = \overline{OB} - \overline{OA} = \frac{h}{\cos \psi} - h \tan \psi$$

ψ being the angle of natural slope. Then, from (a), he obtains

$$H = \frac{\gamma h^2}{2} \left(\frac{1}{\cos \psi} - \tan \psi \right)^2$$

This result coincides with Coulomb's formula (a) if we substitute expressions (b) and (c) (page 61) into it.

Poncelet also developed a simple graphical solution for the more general case in which the wall is inclined, the earth is bounded by a polygonal surface, and in which friction between the wall and the earth is taken into consideration. In this work, Poncelet derives first (on the basis of Coulomb's theory) an analytical expression for the earth pressure and then shows how this expression can be evaluated by a graphical construction.

While engineers continued to use Coulomb's theory in their analysis of retaining-wall stability, some attempts were made to determine the true distribution of stresses in pulverized materials supported by a retaining wall. These were made, as we have seen (see page 20), by Rankine and also by Scheffler;[1] but the work of these two men remained unnoticed by engineers for a long time and it had little influence on the design of retaining walls in the period under consideration.[2]

In the field of arch design, engineers continued to consider stone arches as systems of absolutely rigid stone blocks although, as we have seen (page 149), Bresse gave a complete solution for an elastic arch with built-in ends. The notions of a *line of pressure* and a *line of resistance* were introduced into arch analysis about 1830. F. J. Gerstner[3] must apparently be credited with the first investigation of lines of pressure. He starts with an analysis of suspension bridges, treats the catenary, and

[1] Scheffler, *Crelle's J. Baukunst*, 1851.
[2] For the history of the theory of retaining walls, see G. C. Mehrtens, "Vorlesungen über Ingenieur-Wissenschaften," vol. 3, part 1, 1912.
[3] "Handbuch der Mechanik," edited by Anton Gerstner, vol. 1, p. 405, Prague, 1831.

gives tables for the construction of that curve. Then he indicates that the same curve, inverted, is the proper one for an arch of constant cross section. Such an arch will work in compression only under the action of its own weight. Since circular and elliptic arches were in common use, he investigates the question of how the intensity of load should be distributed such that circles or ellipses are pressure lines. He shows that, in practice, the load distribution differs from that obtained theoretically for the ideal case. This means that in point of fact, we not only have compression but also bending present. He also shows that the problem is statically indeterminate and that it is possible to construct an infinite number of pressure lines which satisfy the conditions of equilibrium and pass through various points of the keystone and of the abutments. To each of these curves, there corresponds a value of the horizontal thrust H. To make the problem definite, Gerstner uses finally some arbitrary assumptions regarding the position of the true pressure curve.

Some general idea regarding the position of the true pressure line in an arch was given by the English scientist Henry Moseley (1802–1872). Moseley was born in the vicinity of Newcastle. He studied at the city grammar school and later at Abbeville. Finally, he went up to St. John's College, Cambridge, and graduated in 1826. In 1836, Moseley became professor of natural philosophy and astronomy at King's College, London. There he gave the courses of mechanics, with its application to engineering, and published two books: "A Treatise on Mechanics Applied to the Arts" (1839) and "The Mechanical Principles of Engineering and Architecture" (1843). From the preface to the second of these, we learn that Moseley was greatly influenced by the work of French scientists and he often refers to the publications of Navier, Dupin, and especially to the book "Mécanique Industrielle" by Poncelet. The chief merit of Moseley's work is that it introduced the methods of French engineers into England, so that his writings represented considerable progress in English engineering literature of that time. Moseley became interested in statically indeterminate problems and for their solution introduced *the principle of least resistance*.[1] Moseley develops a method of applying this principle to the analysis of arches in his paper "On the Theory of the Arch," written in 1839 and published in the first volume of "The Theory, Practice, and Architecture of Bridges" edited by John Weale in 1843. It is shown for the first time that the pressure line and the resistance line are two different curves. To illustrate this, a sketch (shown in Fig. 145) is used. The arrow A represents the force acting on the cross section 1-2 at point a. Adding the weight of the stone 1234 to this force, the force B acting on the cross section 3-4 at point b is obtained. Continuing in this

[1] *Phil. Mag.*, October, 1833.

way, the forces $C, D, E, \ldots$ applied at points $c, d, e, \ldots$ are obtained. The polygon formed by the intersections of forces $A, B, C, D, E, \ldots$ gives us, in the limit, the line of pressure and the polygon $a, b, c, d, e, \ldots$ similarly gives the line of resistance.

In the construction of the pressure line for a symmetrical arch (Fig. 146), Moseley begins with an arbitrarily chosen point C and applies a thrust H of such magnitude that the pressure line becomes tangent to the intrados of the arch at a certain point B. By changing the position of the starting point C, an infinite number of pressure lines can be obtained. To select the true pressure curve from all those possible, Moseley used his principle of least resistance and states that the required one is that

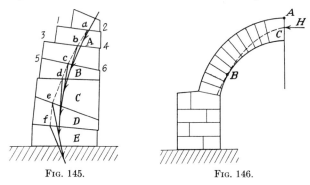

FIG. 145.   FIG. 146.

curve corresponding to the minimum value of H. It can be seen that by moving the point C upward, we shall diminish H, so that finally Moseley finds that the true pressure line passes through the points A and B. This conclusion is in agreement with the basic assumption of Coulomb's theory which, as Moseley states, "was unknown to me until my researches had long been completed, and which, after an oblivion of more than sixty years among the pages of the *Mémoires des Savants Étrangers*, undisturbed by the many and fierce discussions to which the question had been subjected since that time in this country and elsewhere, has recently been exhumed and made the subject of valuable researches by M. M. Navier, Lamé and Clapeyron and Garidel."[1]

Moseley's work attracted the interest of German engineers and his book "The Mechanical Principles of Engineering and Architecture" was translated into German by Scheffler (1844) who later tried to improve upon Moseley's theory by introducing a consideration of the strength of the material of the arch.[2] He notes that if the pressure line passes through

[1] See *Mém. officier génie*, vol. 12, p. 7, 1835.
[2] Scheffler, "Theorie der Gewölbe, Futtermauern und eisernen Brücken," Brunswick, 1857.

points A and B (Fig. 146), the stress must become infinitely large at those points. To eliminate the difficulty, Scheffler proposes that the pressure line be moved into the arch and considers that line for which the maximum stress at A and B reaches the values of the ultimate strength of the material as the true pressure line.

Further progress in arch analysis was made by the introduction of the graphic method of investigation. Poncelet was responsible for this.[1] He takes Coulomb's theory as the basis of his analysis, and shows how the position of the cross section of rupture can be determined graphically. He proves (in his investigation) the theorem that the point on the intrados of an arch, corresponding to the cross section of rupture, is that at which the tangent to the intrados passes through the point of intersection of the horizontal thrust, acting at the keystone, and the gravity force of the portion of the arch between the extreme arch stone and the cross section of rupture.[2]

Another improvement in the graphical analysis of arches was due to K. Culmann.[3] Assuming that the material of an arch cannot withstand tensile stresses, he deduces that the extreme position of the pressure line is that for which it passes through the upper third point at the keystone and the lower third point at the cross section of rupture. Using these two points, he constructs the funicular polygon for the gravity forces of the stone blocks and of the external load and determines the force acting on (and the corresponding stresses in) each cross section of the arch in this way. With the work of Culmann, that period in the development of the theory of arches during which no regard was given to elastic deformation was brought to a conclusion. Further progress in this field, as we shall see, was made by treating the arch as an elastic curved bar and by applying the theory developed by Navier (page 77) and by Bresse (page 147).

In conclusion, it would be well to discuss briefly the well-known work of Yvon Villarceau[4] (1813–1883) on arch bridges. The son of a merchant of Vendôme,[5] he did not show much interest in study during his youth and spent much of his time playing with various mechanical devices and acquired great skill as a mechanic. He then became interested in music, participated in the activity of the local orchestra, and in 1830 moved to Paris where he studied at the Conservatoire de Musique. At that time,

[1] See *Mém. officier génie*, vol. 12, pp. 151–213, 1835.
[2] This theorem had been proved before, as we have seen, by Lamé and Clapeyron (see page 84).
[3] See "Grafische Statik," pp. 435–469, 1866.
[4] See his memoir "Sur l'établissement des arches de pont," presented to the French Academy of Sciences in 1845 and published in *Mém. présentés par divers savants*, vol. 12, pp. 503–822, 1854.
[5] See Joseph Bertrand, "Éloges Académiques," Paris, 1890. See also "Histoire de l'École Centrale des Arts et Manufactures," p. 333, by Francis Pothier, Paris, 1887.

he became interested in the socialistic theories of Saint-Simon. He joined the mission of Enfantin, spent several years in Egypt, and not until 1837 did he return to Paris where he decided to study engineering at the École Central des Arts et Manufactures. While studying in this school, he showed his outstanding ability and did some original work in geometry. He became especially interested in arch analysis, and after his graduation (1840) he published several articles on this subject in the *Revue de l'architecture et des travaux publics*. Finally (in 1845), he presented his famous memoir on arches to the Academy. Later, Villarceau became interested in astronomy and started to work in the Paris Observatory under Arago. There he wrote his famous works in astronomy. He was elected to membership of the Academy of Sciences in 1867.

In his work on arches, Villarceau is mainly interested in the selection of the most favorable shape of the center line of an arch. He clearly sees that the problem is statically indeterminate and that its complete solution requires consideration of the elastic deformations. But he concludes that there is insufficient knowledge of the elastic properties of arch materials and therefore decides to treat the problem by considering the arch stones as being absolutely rigid. He believes that the forms found in this way would be sufficiently good when the elastic deformations are taken into consideration. This method of attack leads him to consider cases in which the pressure curve coincides with the center line of the arch. We have already mentioned the simplest instances of this kind before (see page 199) in which the thickness of the arch is neglected. Villarceau goes further and considers not only external pressure, which he assumes acting normally to the extrados (or external surface), but also the weight of the arch. Under these assumptions, the differential equation of the pressure line becomes involved and its integration requires the use of elliptic functions. To render the solution useful in practice, Villarceau did not hesitate to calculate numerous tables which enabled practical engineers to obtain the required center line and the necessary thickness of the arch in each particular case. With these tables, Villarceau applied his theory to many existing bridges and showed that, often, the thickness of the arch could be reduced without increasing stresses by using his shape of the center line.

In conclusion, he discusses the effect of the elasticity of materials (which was neglected in his theory) and considers the increase which must be given to the height of the arch in order to get the correct shape of the center line after deflection. Villarceau's idea of taking the funicular curve constructed for the assumed external loads for the center line is widely used. By this means, the preliminary shape of an arch and its thickness are usually established and the selected dimensions are later checked by means of the elastic theory of arches.

CHAPTER VIII

The Mathematical Theory of Elasticity between 1833 and 1867

48. The Physical Elasticity and "The Elastic Constant Controversy"[1]

Navier assumed that an ideal elastic body consists of molecules between which forces appear on deformation in his derivation of the fundamental equations of the theory of elasticity (see page 105). These forces were proportional to changes in the distances between the molecules and acted in the directions of lines joining them. In this way, Navier was able to establish relations between the deformations and the elastic forces for isotropic bodies by using only one elastic constant. Cauchy used initially two constants (see page 110) to write the relations between stress and strain in the case of isotropy. In the most general case of nonisotropic bodies, Poisson and Cauchy both assumed that each of the 6 stress components can be represented as a homogeneous linear function of 6 strain components (the generalized Hooke's law). For this, 36 elastic constants were introduced. Under the assumptions of the above-mentioned molecular theory they were able to reduce this number to 15 in the general case. They showed also that isotropy makes further reduction possible and that finally only the single constant used by Navier is required in expressing the relations between stress and strain components.

Poisson showed that in the simple tension of an isotropic bar the ratio of lateral contraction to axial elongation should be equal to $\frac{1}{4}$ and that, if E denotes the modulus in simple tension or compression, the modulus in shear will be

$$G = \frac{E}{2(1 + \frac{1}{4})} = 0.4E \qquad (a)$$

Again, if a body is subjected to a uniform pressure p the unit volume change e will be given by the formula

$$e = \frac{3p(1 - 2 \cdot \frac{1}{4})}{E} = \frac{3}{2}\frac{p}{E} \qquad (b)$$

[1] The history of this subject is discussed in detail in Appendixes III and V of Navier's "Résumé des Leçons . . . " 3d ed., 1864. See also "Note Finale du No. 16" of Saint-Venant's translation of the book by Clebsch, "Théorie de l'élasticité des corps solides," Paris, 1883.

The Mathematical Theory of Elasticity between 1833 and 1867   217

The idea that the elastic properties of an isotropic body can be completely defined by one constant (say the modulus in tension E) was generally accepted in the early stages of the development of the theory of elasticity. Navier, Cauchy, Poisson, Lamé, and Clapeyron all agreed with this.

Great changes were brought about in this subject by the work of George Green, who offered a derivation of the equations of elasticity without using any hypothesis regarding the behavior of the molecular structure of elastic bodies.

George Green (1793–1841) was the son of a miller in Nottingham and he was given no chance to obtain an education in his youth. He acquired his great mathematical knowledge in later life from studying books. In 1828, his first paper "An Essay on the Application of Mathematical Analysis to the Theories of Electricity and Magnetism" was published.[1] From the preface of the paper, we see that Green had drawn heavily from the French mathematical literature. He mentions the works of Laplace, Poisson, Cauchy, and Fourier, and remarks: "It must certainly be regarded as a pleasing prospect to analysts, that at a time when astronomy, from the state of perfection to which it has attained, leaves little room for further applications of their art, the rest of the physical sciences should show themselves daily more and more willing to submit to it; . . ." and further on: "It is hoped the difficulty of the subject will incline mathematicians to read this work with indulgence, more particularly when they are informed that it was written by a young man, who has been obliged to obtain the little knowledge he possesses, at such intervals and by such means, as other indispensable avocations which offer but few opportunities of mental improvement, afforded." Green's first paper is considered to be his most important contribution to science. In it, he uses a mathematical function, or "potential," that had already been used by Laplace and it has proved to have universal utility in all branches of physics. Green's publications attracted the interest of mathematicians, and in 1833 Green, who was forty years old, had the opportunity to enter Gonville and Caius College of Cambridge University. In January, 1837, he took his degree of bachelor of arts as the fourth wrangler. He was elected to a fellowship of his college in 1839, but did not long enjoy this position, as he died in 1841.

Green took up problems of elasticity in his work "On the Laws of the Reflexion and Refraction of Light at the Common Surface of Two Noncrystallized Media."[2] He did not wish to make any assumptions regarding the ultimate constitution of the luminiferous ether or on the mutual action of molecules but assumed only that the properties of the ether

---

[1] See "Mathematical Papers of the late George Green," edited by N. M. Ferrers. London, 1871.

[2] "Mathematical Papers," p. 245.

must comply with the principle of conservation of energy. He states: "If . . . we are so perfectly ignorant of the mode of action of the elements of the luminiferous ether on each other . . . it would seem a safer method to take some general physical principle as the basis of our reasoning, rather than assume certain modes of action, which, after all, may be widely different from the mechanism employed by nature; more especially if this principle includes in itself, as a particular case, those before used by M. Cauchy and others, and also leads to a much more simple process of calculation. The principle selected as the basis of the reasoning contained in the following paper is this: In whatever way the elements of any material system may act upon each other, if all the internal forces exerted be multiplied by the elements of their respective directions, the total sum for any assigned portion of the mass will always be the exact differential of some function." Now, if we denote that function by ϕ and combine D'Alembert's principle with that of virtual displacements, the equations of motion in the absence of external forces could be obtained from the equation

$$\iiint \rho \, dx \, dy \, dz \left(\frac{\partial^2 u}{\partial t^2} \delta u + \frac{\partial^2 v}{\partial t^2} \delta v + \frac{\partial^2 w}{\partial t^2} \delta w \right) = \iiint \delta \phi \, dx \, dy \, dz \qquad (a)$$

Assuming that displacements u, v, w are small, Green concludes that the function ϕ must be a homogeneous function of second degree of six strain components

$$\epsilon_x = \frac{\partial u}{\partial x} \qquad \epsilon_y = \frac{\partial v}{\partial y} \qquad \epsilon_z = \frac{\partial w}{\partial z}$$
$$\gamma_{xy} = \frac{\partial u}{\partial y} + \frac{\partial v}{\partial x} \qquad \gamma_{xz} = \frac{\partial u}{\partial z} + \frac{\partial w}{\partial x} \qquad \gamma_{yz} = \frac{\partial v}{\partial z} + \frac{\partial w}{\partial y} \qquad (b)$$

In the most general case such a function contains 21 coefficients, and these define the elastic properties of a material. Using this function we can obtain from Eq. (a), the three equations of motion which must prevail at each point of an elastic medium. In an isotropic body, the function ϕ can be as Green shows greatly simplified. It then contains only two coefficients and the corresponding equations of motion contain only two elastic constants, as Cauchy assumed in his initial equations (see page 110).

Green's paper started a controversy and gave rise to two schools in the development of the theory of elasticity. The scientists who followed Navier and Cauchy, and accepted their ideas regarding the molecular structure of elastic bodies, used 15 constants for defining the elastic properties of the general material and 1 constant for isotropy, while the followers of Green used 21 constants and 2 constants for the same purposes. Naturally, many attempts were made to settle this argument by direct tests, and several physicists became interested in the experimental determination of elastic constants.

W. Wertheim[1] (1815–1861) was especially interested in these determinations and his results are still very often quoted in physics textbooks. At first, he accepted the uni-constant hypothesis and his first paper gives the tensile modulus for various materials.[2] For this he not only uses static tensile tests but also experiments with longitudinal and lateral vibrations. He finds that (1) for the same metal, every process (say hammering, rolling) which increases the density also increases the modulus; (2) the modulus, as obtained from vibration experiments, is higher than that obtained from static tests, and he notes that the difference can be used for finding the ratio of the specific heat at constant stress to the specific heat at constant volume. He also studies the effect of temperature on the modulus and finds that generally the tensile modulus decreases continuously as the temperature rises from -15 to $200°C$. However, steel is an exception to this last rule, for its modulus increases with temperature from -15 to $100°C$ and then begins to decrease so that at $200°C$ it is smaller than at $100°C$ and sometimes smaller than at room temperature. The experiments show that there is no real elastic limit and Wertheim takes this limit as the stress at which the permanent set in tension reaches the value 0.00005 per unit length.

Later on Wertheim, in collaboration with E. Chevandier, made extensive investigations of glass[3] and various kinds of woods.[4] In 1848 he presented the "Mémoire sur L'équilibre des corps solides homogènes"[5] to the Academy in which he adopted a suggestion by Regnault and used glass and metallic cylindrical tubes for his tests and measured changes of the internal volumes of tubes produced by axial extension. In this way the lateral contraction of material could be calculated. The experimental results did not agree with Poisson's theoretical ratio of $\frac{1}{4}$, and the lateral contraction only could be explained by using two elastic constants for the (isotropic) bodies. But Wertheim continued to accept the uni-constant theory and suggested the adoption of the value $\frac{1}{3}$ for Poisson's ratio, although that value boasted neither a theoretical basis nor the ability to give satisfactory agreement between the theory and his experiments.

Wertheim was interested in the electroelastic and magnetoelastic properties of materials and made an extensive study of the effect of an electric current on the tensile modulus of a conducting wire. He also investigated the influence on the longitudinal deformations of an iron bar of an electric current in a surrounding solenoid.

[1] W. Wertheim was born in Vienna in 1815. There, he obtained his doctor's degree in medicine in 1839. In 1840 he moved to Paris where he obtained the degree of doctor of science in 1848. After 1855 he was one of the examiners of those seeking admission to the École Polytechnique. He committed suicide in 1861.
[2] See *Ann. chim.*, vol. 12, pp. 385–454, Paris, 1844.
[3] *Ann. chim.*, vol. 19, pp. 129–138, Paris, 1847.
[4] "Mémoire sur les propriétés mécaniques du bois," Paris, 1848.
[5] *Ann. chim.*, vol. 23, pp. 52–95, Paris, 1848.

In studying the optical properties of elastic bodies,[1] Wertheim drew up a very complete table of colors by the use of which he could measure stresses in a stretched transparent bar. He found that, if a material does not follow Hooke's law exactly, then the double refraction is proportional to strain, but not to stress.

Lastly, mention should be made of a very extensive memoir by Wertheim on torsion.[2] Experiments were conducted with prisms of circular, elliptical, and rectangular cross sections and, in some cases, tubular specimens were used. The materials were steel, iron, glass, and wood. From his tests, Wertheim again concluded that the ratio of lateral contraction is different from $\frac{1}{4}$ and is closer to $\frac{1}{3}$. By measuring the inner volume of the tubes during torsion, Wertheim finds that it diminishes with increase of the angle of twist (as it should if the helical shape assumed by longitudinal fibers is considered). In discussing torsion of bars of elliptical and rectangular cross sections, this writer does not use Saint-Venant's theory although the experimental results are in good agreement with it. Instead, he applies Cauchy's unsatisfactory formula (see page 110) with some correction factor. Regarding torsional vibrations, Wertheim noticed that the frequency of vibration is higher for smaller amplitudes, indicating that for very small stresses the magnitude of the modulus must be higher than it is for higher stresses.

Although Wertheim's work contributed much to physical elasticity, the fundamental question concerning the necessary number of elastic constants remained unsolved. Whenever experiments showed that Poisson's ratio differed from the value $\frac{1}{4}$, it was possible to show that the material used in the specimens was not perfectly isotropic.

Another physicist of this period who did much work in physics of elasticity was A. T. Kupffer (1799–1865). The Russian government founded the Central Laboratory of Weights and Measures in 1849 and Kupffer was installed as the director of that institution. He was interested in the physical properties of metals in so far as they can affect standards of measurement, and the results of his work were published in *Comptes rendus annuel* of the Observatoire Physique Central for the years 1850–1861. Regarding this work, Todhunter and Pearson[3] observe: "Probably no more careful and exhaustive experiments than those of Kupffer have ever been made on the vibrational constants of elasticity and the temperature-effect."

Kupffer begins with torsion tests from which he determines the shear modulus G. Multiplying G by $\frac{5}{2}$ he should, in accordance with the uni-constant hypothesis, obtain the modulus in tension E. But he finds that

[1] *Ann. chim. et phys.*, vol. 40, pp. 156–221, Paris, 1854.
[2] *Ann. chim. et phys.*, vol. 50, pp. 195–321 and 385–431, Paris, 1857.
[3] See "History . . . ," vol. 1, p. 750.

the values obtained in this way differ considerably from those obtained by tensile or bending tests. Thus his results do not uphold the uniconstant hypothesis. In studying torsional vibration, he investigates damping and shows that only a part of it can be attributed to air resistance and that the remainder is due to viscosity of the material. He takes up the problem of elastic afterstrain and shows that in steel bars deflections do not vanish immediately upon removal of loads and that a continuous decrease of the deflection takes place over a period of several days after unloading. The writer points out that this elastic after effect gives rise to damping of vibrations also. It is not proportional to deformation so that vibrations are not truly isochronous.

Kupffer examined the question of influence of temperature on the modulus of elasticity with great care and presented a paper on this subject to the Russian Academy of Sciences in 1852.[1] He shows that for small changes in temperature (from $t = 13$ to $t = 25°R$) the tensile modulus can be represented by the formula

$$E_{t_1} = E_t[1 - \beta(t_1 - t)]$$

where β is a constant depending on the material. He also shows that, with increase of temperature, the damping due to the afterstrain effect increases. Kupffer won a prize offered by the Royal Society of Göttingen (in 1855) for this work.

In 1860 Kupffer published a book[2] in which his numerous experimental investigations on the flexure and transverse vibrations of bars were collected. In its preface, Kupffer stresses the importance of having a national institution for studying the elastic properties and strength of structural materials. He states that, by publishing information concerning the properties of metals that are produced by various companies, very useful data can be supplied to the designing engineer by such an organization. This practice can have a beneficial influence toward improving the quality of materials, since the companies will try to improve their products in order to expand their markets.

In Germany, Franz Neumann and his pupils were also interested in physical elasticity and they also determined the elastic constants from experiments. From the correspondence[3] between Neumann and Kupffer, we learn that the former assumed that the ratio of the longitudinal extension to the lateral contraction does not remain constant, but depends upon

[1] See *Mém. acad. . . . St. Petersburg, sci. math. et phys.*, vol. 6, pp. 397–494, 1857.

[2] "Recherches expérimentales sur l'élasticité des métaux faites à l'observatoire physique central de Russie," vol. 1 (all published), pp. 1–32 and 1–430, St. Petersburg, 1860.

[3] "History . . . ," vol. II, p. 507.

the nature of the material. Neumann was the first to attach small mirrors to the sides of a rectangular bar under flexure and to show by means of them that a cross section becomes trapezoidal during bending. From the angle of relative rotation made by the two sides of the bar, Poisson's ratio can be calculated.

Kirchhoff (Neumann's pupil) used circular steel cantilevers in his experiments.[1] To the free ends of these, he applied a transverse load with a certain eccentricity so that bending and torsion were produced simultaneously. The angle of torsion and the angle which the tangent at the end of the cantilever made with the horizontal were measured optically by using a mirror attached at the end of the cantilever. From these very carefully performed tests, Kirchhoff found that Poisson's ratio for steel is 0.294, while for brass he gave the value 0.387. But he made the reservation that the bars of brass which he used could not be considered as isotropic.

All these experimental results were in conflict with the uni-constant hypothesis for isotropic bodies. Again, this hypothesis became increasingly incompatible with the current views regarding the constitution of matter. Thus, in the subsequent development of the theory of elasticity, the method proposed by Green whereby the stress-strain relations are derived from a consideration of strain energy prevailed and it is now commonly used.

49. Early Work in Elasticity at Cambridge University

The brilliant work of French mathematicians at the École Polytechnique greatly affected the development of mathematical instruction in other countries and was, without doubt, an important factor in the revival of scientific activity at Cambridge University which took place in the first quarter of the nineteenth century.[2] In 1813, a group of Cambridge students under the leadership of Charles Babbage (1792–1871), George Peacock (1791–1858), and John Frederick Herschel (1792–1871) formed The Analytical Society. They held meetings, read papers, and conducted the famous struggle for the introduction of the Continental notation of the infinitesimal calculus into Cambridge.[3] To facilitate the introduction of this notation, Babbage (with the collaboration of Herschel and Peacock) prepared a translation of the French book on differential calculus by S. F. Lacroix (1816). Later on, the "Collection of Examples of the Application of the Differential and Integral calculus" was published (1820) by Peacock.

[1] *Pogg. Ann. Physik u. Chem.*, vol. 108, 1859. See also *Ges. Abhandl.*, p. 316.

[2] See "History of the Study of Mathematics at Cambridge" by W. W. Rouse Ball, Cambridge, 1889.

[3] Interesting information about Cambridge of that time can be found in Babbage's book, "Passages from the Life of a Philosopher," 1864.

While traveling on the Continent, Babbage established contacts with many scientists. In the autumn of 1828, he attended the annual congress of German naturalists and wrote an account of it to Prof. D. Brewster. Later Brewster, in collaboration with John Robison and William V. Harcourt, undertook the organization of a similar society called the British Association for Advancement of Science. At the second meeting of this association, Babbage strongly advocated that "attention should be paid to the object of bringing theoretical science in contact with the practical knowledge on which the wealth of the nations depends." He was himself interested in practical applications and participated in discussions of various technical questions connected with the work on English railways. He organized the early experimental work on the Great Western Railway and constructed a special experimental car in which instruments were installed for recording the locomotive's tractive force and the train's velocity. He was interested in the construction of large tubular bridges and corresponded with Fairbairn on the experiments with model tubes.

William Whewell (1794–1866) and George Biddell Airy (1801–1892) were mainly responsible for the early work in applied mathematics at Cambridge. Whewell graduated in 1816, was elected a Fellow of Trinity College in 1817, and started teaching mechanics. In 1819, his book "An Elementary Treatise on Mechanics" was published. This book promoted the introduction of Continental mathematics to Cambridge since Whewell freely used the differential and integral calculus in it. In time, Whewell published several books of a more advanced character. To his elementary statics in the above-mentioned book, he added "Analytical Statics" (1833), which contains a chapter on the equilibrium of an elastic body. An elementary theory of bending of beams is given in this chapter and references are made to the works of Hodgkinson and Barlow. To enable students to get some knowledge of Newton's work, Whewell published (in 1832) his book, "On the Free Motion of Points, and on Universal Gravitation, Including the Principal Propositions of Books I and III of the Principia." Two years later his book "On the Motion of Points Constrained and Resisted, and on the Motion of a Rigid Body" appeared. The best known of Whewell's books are his "History of the Inductive Sciences" (1837) and "The Philosophy of the Inductive Sciences" (1840). The first of these was reedited in the United States (1859) and was also translated into German and French. Whewell was also interested in higher education. He published a book "On the Principles of English University Education" (1837) and the views expressed in it can be seen to have guided mathematical instruction at Cambridge University.[1]

[1] Much interesting information about Whewell and about Cambridge University in Whewell's time can be found in I. Todhunter's book "William Whewell, Master of Trinity College, Cambridge," 1876.

George Biddell Airy entered Trinity College, Cambridge, in 1819. As a student, his career was brilliant and he graduated as senior wrangler in 1823.[1] In 1826, he was appointed Lucasian professor of mathematics and published the book "Mathematical Tracts on the Lunar and Planetary Theories, The Figure of the Earth, Precession and Nutation and the Calculus of Variations." In the second edition (1828) a chapter on the undulatory theory of optics was added. This book became widely used at Cambridge as an introduction to the application of mathematics in solving problems of astronomy and theoretical physics.

Regarding this period of his activity at Cambridge, Airy makes the following remark in his autobiography (p. 73): "There had been no lectures on Experimental Philosophy (Mechanics, Hydrostatics, Optics) for many years. The university in general, I believe, looked with great satisfaction to my vigorous beginning: still there was considerable difficulty about it. There was no understood term for the Lectures: no understood hour of the day: no understood lecture room." He was always ready to use mathematics in various branches of theoretical physics and opposed those who would study pure mathematics exclusively. In a letter to Stokes (1868) regarding the Smith prizes, he states: "I regard the university as the noblest school of the mathematical education in the world. And, in spite of the injury that is done by a too mechanical use of the analysis, it is the most exact. There is no accuracy of teaching like that of Cambridge: and in the applied sciences that are really taken up there, of which I would specially mention Astronomy, the education is complete and excellent. But with this, there is an excessively strong tendency to those mathematical subjects which are pursued in the closet, without the effort of looking into the scientific world to see what is wanted there."

In 1828, Airy was elected Plumian professor of astronomy, and after that time most of his energy was absorbed in the founding of the Cambridge observatory and in astronomical work. He was appointed Astronomer Royal in 1835 and moved to Greenwich. But he continued to keep in contact with Whewell and Peacock and his views regarding mathematical instruction must have been regarded with respect at Cambridge.

Airy was always interested in the application of mathematics to the solution of engineering problems. He concerned himself in the construction of the large tubular bridges and showed Fairbairn how the necessary cross-sectional dimensions of those bridges could be determined from the results obtained from model tests (page 159). He took up the theory of bending of beams and, in 1862, presented a paper on this subject to the Royal Society.[2] Considering the rectangular beam as a two-dimensional

[1] "The Autobiography of Sir George Biddle Airy" was edited in 1896 by Wilfrid Airy.
[2] *Phil. Trans.*, vol. 153, 1863.

problem, Airy obtains the differential equations of equilibrium

$$\frac{\partial \sigma_x}{\partial x} + \frac{\partial \tau_{xy}}{\partial y} = 0 \qquad \frac{\partial \sigma_y}{\partial y} + \frac{\partial \tau_{xy}}{\partial x} = 0 \qquad (a)$$

and shows that both of these equations will be satisfied if the expressions for the stress components are derived from a function ϕ in the following manner:

$$\sigma_x = \frac{\partial^2 \phi}{\partial y^2} \qquad \tau_{xy} = -\frac{\partial^2 \phi}{\partial x\, \partial y} \qquad \sigma_y = \frac{\partial^2 \phi}{\partial x^2} \qquad (b)$$

Airy takes the function ϕ in the form of a polynomial and selects the coefficients of that polynomial in such a way as to satisfy the conditions at the boundary. He does not consider the fact that ϕ must also satisfy the compatibility equation; thus his investigation is not complete. But this was the first time that a *stress function* was used and it can be considered as the birth of a very useful method of solving elasticity problems.

50. Stokes

The golden age of theoretical physics at Cambridge began with the scientific work of George Gabriel Stokes (1819–1903).[1] Stokes was a member of the large family of the Rector of Skreen, a village south of Sligo in Ireland. Stokes received his elementary training in arithmetic

FIG. 147. G. G. Stokes.

from the parish clerk and in 1832, at thirteen years of age, he was sent to the Rev. R. H. Wall's school in Dublin. There, his solutions of geometrical problems attracted the attention of the mathematics master. In 1835, he went to Bristol College, where he studied mathematics under Francis Newman and, at his final examination (1837), he won a prize "for eminent proficiency in mathematics."

In the same year, Stokes entered Pembroke College, Cambridge. Regarding the work at Cambridge, Stokes later writes:[2] "In those days boys coming to the University had not in general read so far in math-

---

[1] The "Memoirs and Scientific Correspondence" of Stokes was edited by Joseph Larmor, Cambridge, 1907. A short biography of Stokes, written by Lord Rayleigh, is to be found in the fifth volume of Stokes's "Mathematical and Physical Papers."

[2] See Glazebrook's article in "Good Words," May, 1901.

ematics as is the custom at present; and I had not begun the Differential Calculus when I entered College, and had only recently read Analytical Sections. In my second year I began to read with a private tutor, Mr. Hopkins, who was celebrated for the very large number of his pupils who obtained high places in the University examinations for Mathematical Honours. In 1841, I obtained the first place among the men of my year as Senior Wrangler and Smith's Prizeman, and was rewarded by immediate election to a Fellowship in my College. After taking my degree I continued to reside in College and took private pupils. I thought I would try my hand at original research; and, following a suggestion made to me by Mr. Hopkins while reading for my degree, I took up the subject of Hydrodynamics, then at rather a low ebb in the general reading of the place, not withstanding that George Green, who had done such admirable work in this and other departments, was resident in the University till he died . . . . In 1849, when thirty years of age, I was elected to the Lucasian Professorship of Mathematics, and ceased to take pupils. The direction of the Observatory was at that time attached to the Plumian Professorship; and the holder of that office, Professor Challis, used to give lectures in Hydrodynamics, and Optics, as had been done by his predecessor Airy. On my election Challis, who was desirous of being relieved of his lectures on Hydrostatics and Optics, so as to be free to take up the subject of Astronomy, which was more germane to his office at the Observatory, agreed with me that I should undertake the course of lectures which he himself had previously given."

The first papers of Stokes deal with hydrodynamics. But in his paper "On the Theories of the Internal Friction of Fluids in Motion, and of the Equilibrium and Motion of Elastic Solids,"[1] presented to the Cambridge Philosophical Society in 1845, he devotes considerable attention to the derivation of the fundamental differential equations of elasticity for isotropic bodies. Stokes considers that the theory of elasticity must be based on the results of physical experiments and not upon theoretical speculation regarding the molecular structure of solids. He remarks: "The capability which solids possess of being put into a state of isochronous vibration shews that the pressures called into action by small displacements depend on homogeneous functions of those displacements of one dimension.[2] I shall suppose moreover, according to the general principle of the superposition of small quantities, that the pressures due to different displacements are superimposed, and consequently that the pressures are linear functions of the displacements." Following these assumptions, he finally establishes equations of equilibrium containing

[1] See Stokes, "Mathematical and Physical Papers," vol. 1, p. 75.
[2] We have seen (see page 20) that Hooke had already mentioned isochronous vibration as a physical fact proving the law of proportionality of stress with strain.

two elastic constants. He mentions india rubber[1] and jelly as materials for which the ratio of lateral contraction to axial elongation differs greatly from the prediction of uni-constant hypothesis. He then describes experiments (involving tension, torsion, and uniform compression) for finding the ratio of the two elastic constants (E and G) for any particular material. He states: "There are two distinct kinds of elasticity; one, that by which a body which is uniformly compressed tends to regain its original volume, the other, that by which a body, which is constrained in a manner independent of compression tends to assume its original form." Considering the second kind of elasticity, Stokes gives an interesting comparison of solids with viscous fluids. He remarks: "Many highly elastic substances, as iron, copper, etc. are yet to a very sensible degree plastic. The plasticity of lead is greater than that of iron or copper, and, as appears from experiment, its elasticity less. On the whole it is probable that the greater the plasticity of a substance the less its elasticity. When the plasticity of a substance is still further increased, and its elasticity diminished, it passes into a viscous fluid. There seems no line of demarcation between a solid and a viscous fluid."

From his early work in hydrodynamics, Stokes turned to optics. Considering the luminiferous ether as a homogeneous uncrystallized elastic medium (such as an elastic solid), he again had to work with the equations of theory of elasticity. In his paper "On the Dynamical Theory of Diffraction"[2] he states: "Equations may be obtained by supposing the medium to consist of ultimate molecules, but they by no means require the adoption of such a hypothesis, for the same equations are arrived at by regarding the medium as continuous."

In this work, he establishes two theorems which have proved very important in the theory of vibration of elastic bodies. Let us examine these theorems in the simplest case of vibration of a system with one degree of freedom. The displacement x from the equilibrium position is then given by the expression

$$x = x_0 \cos pt + \frac{\dot{x}_0}{p} \sin pt \qquad (a)$$

and we see that the part of this expression which is due to the initial displacement x_0, is obtained from the part due to the initial velocity by differentiation with respect to t and replacing $\dot{x}_0$ by x_0. This holds in the most general case and Stokes concludes that, in an elastic medium, "the part of the disturbance which is due to the initial displacements may be obtained from the part which is due to the initial velocities by differentiat-

[1] Here he refers to the memoir of Lamé and Clapeyron (see page 115).
[2] "Mathematical and Physical Papers," vol. 2, p. 243.

ing with respect to t, and replacing the arbitrary functions which represent the initial velocities by those which represent the initial displacements." In this way, the problem of finding the displacements is limited to that of finding the displacements due to initial velocities only.

The second theorem deals with the disturbance due to a given variable force acting in a given direction at a given point of the medium. Again considering the system with a single degree of freedom and denoting the disturbing force per unit mass of the system by $f(t)$, we find that, due to an impulse $f(t)\,dt$ communicated to the system at the instant t, an increment $d\dot{x}$ in the velocity equal to $f(t)\,dt$ will be produced. Then, from the form of the second term in expression (a), we conclude that the displacement at an instant t_1, due to the increment of the velocity $f(t)\,dt$ communicated at the instant t, will be

$$\frac{f(t)\,dt}{p} \sin p\,(t_1 - t)$$

Considering now the action of the disturbing force in the interval from $t = 0$ to $t = t_1$, we obtain the expression[1]

$$x = \frac{1}{p}\int_0^{t_1} f(t)\,\sin p\,(t_1 - t)\,dt \qquad (b)$$

for the displacement of the system at the instant t_1. Similar reasoning can be applied to the general case and the displacements produced by a disturbing force can be calculated if the free vibrations of the system are known.

In the same year (1849) that Stokes presented his work on diffraction, he also made the investigation on dynamical deflections of bridges which has been discussed before (page 176).

In 1854, Stokes became secretary of the Royal Society. The work in this new position absorbed much of his energy and Lord Rayleigh, in Stokes's obituary notice, notes: "The reader of the Collected Papers can hardly fail to notice a marked falling off in the speed of production after this time. The reflection suggests itself that scientific men should be kept to scientific work, and should not be tempted to assume heavy administrative duties, at any rate until such time as they have delivered their more important messages to the world." Regarding Stokes's exper-

[1] It seems that the idea of subdividing the action of a continuous disturbing force into infinitesimal intervals and of obtaining the forced motion by summing up the motions produced during those intervals was first used by M. Duhamel in "Mémoire sur les vibrations d'un système quelconque de points matériels," *J. école polytech.* (*Paris*), cahier 25, pp. 1–36, 1834. It was applied by Saint-Venant in his study of the forced vibrations of bars. See Saint-Venant's translation of the book by Clebsch, p. 538.

imental work and his lectures Lord Rayleigh states: "His experimental work was executed with the most modest appliances.[1] Many of his discoveries were made in a narrow passage behind the pantry of his house, into the window of which he had a shutter fixed with a slit in it and a bracket on which to place crystals and prisms. It was much the same in lectures. For many years he gave an annual course on Physical Optics, which was pretty generally attended by candidates for mathematical honours. To some of these, at any rate, it was a delight to be taught by a master of his subject, who was able to introduce into his lectures matter fresh from the anvil."

Stokes was always interested in the instruction of theoretical subjects at Cambridge and often acted as an examiner for the mathematical tripos. After he became the Lucasian professor of mathematics, it was his duty to set one of the papers in the examination for the Smith's prizes. These examination papers, together with his mathematical tripos problems, are included in the fifth volume of the "Mathematical Papers" in the form of an appendix.

From 1885 to 1890, Stokes was the president of the Royal Society. The most remarkable testimony of the estimation in which Stokes was held by his scientific contemporaries was the celebration of the jubilee of his professorship. Many outstanding scientists from all parts of the world came to Cambridge to do him homage at that time.

50a. Barré de Saint-Venant

Saint-Venant was born in 1797 in the castle de Fortoiseau (Seine-et-Marne).[2] His prowess in mathematics was noticed very early and he was given careful coaching by his father who was a well-known expert in rural economy. Later he studied at the lycée of Bruges and in 1813, when sixteen years of age, he entered the École Polytechnique after taking the competitive examinations. Here he showed his outstanding ability and became the first in his class.

The political events of 1814 had a great effect on Saint-Venant's

FIG. 148. Barré de Saint-Venant.

---

[1] There was no physics laboratory at Cambridge. The organization of the famous Cavendish Laboratory was effected by Maxwell in 1872.

[2] Saint-Venant's biography was written by M. J. Boussinesq and M. Flamant and published in *Ann. ponts et chaussées*, 6th series, vol. XII, p. 557, 1886.

career. In March of that year, the armies of the allies were approaching Paris and the students of the École Polytechnique were mobilized. On March 30, 1814, they were moving their guns to the Paris fortifications when Saint-Venant, who was the first sergeant of the detachment, stepped out from the ranks with the exclamation:[1] "My conscience forbids me to fight for an usurper. . . . " His schoolmates resented that action very much and Saint-Venant was proclaimed a deserter and never again allowed to resume his study at the École Polytechnique. The famous mathematician Chasles, who was one of Saint-Venant's schoolmates, later judged him indulgently with the words: "If St. Venant, as he said, believed that his conscience was concerned . . . you can cut him to pieces without getting the smallest satisfaction. The accusation of cowardice was absurd. On the contrary, he was very courageous and showed, on the 30th of March, a hundred times more resoluteness and vigour in provoking the indignation of his commanders and the contempt of his comrades, than in exposing himself to the bayonets of the cossacks."[2] During the eight years following this incident, Saint-Venant worked as an assistant in the powder industry. Then, in 1823 the government permitted him to enter the École des Ponts et Chaussées without examination. Here, for two years Saint-Venant bore the protests of the other students, who neither talked to him nor sat on the same bench with him. He disregarded these unpleasant actions, attended all the lectures,[3] and graduated from the school as the first of his class. We do not know how much all this influenced Saint-Venant's subsequent career, but it can be seen that his progress in the outside world as an engineer was not as rapid as we should expect, knowing his outstanding capability, his energy, and his great love of hard work. Whereas nothing is said about Saint-Venant's outstanding accomplishments in the Encyclopaedia Britannica and very little in the French Grande Encyclopédie, we read, on the title page of Todhunter and Pearson's "History of Elasticity," "To the memory of M. Barré de Saint-Venant the foremost of modern elasticians the editor dedicates his labour on the present volume."

After graduating from the École des Ponts et Chaussées, Saint-Venant worked for some time on the channel of Nivernais (1825–1830) and later, on the channel of Ardennes. In his spare time he did theoretical work and, in 1834, he presented two papers to the Academy of Sciences, one dealing with some theorems of theoretical mechanics and the other with fluid dynamics. These papers made him known to French scientists and, in 1837–1838, during the illness of Professor Coriolis, he was asked to give

---

[1] See J. Bertrand, *Éloges Académiques*, New Series, p. 42, Paris, 1902.

[2] See Bertrand's eulogy.

[3] At this time Saint-Venant attended Navier's lectures and became a great admirer of that outstanding engineer and scientist.

lectures on strength of materials at the École des Ponts et Chaussées. These lectures were lithographed[1] and are of great historical interest, since some problems are mentioned which were later to become the objects of the author's scientific researches. At that time, the most advanced book on strength of materials was Navier's "Résumé des Leçons. . . ." Although Navier himself established the fundamental equations of the theory of elasticity, he did not present them in his course and developed the theory of tension, compression, bending, and torsion of prismatical bars, assuming that cross sections of bars remain plane during deformation. Saint-Venant was the first to try to bring the new developments in the theory of elasticity to the attention of his pupils. In the introduction to his course, he discusses the hypotheses regarding the molecular structure of solids and the forces acting between the molecules. Using this hypothesis, he defines the notion of stress. He speaks of shearing stress and shearing strain and shows that tension in one direction combined with compression of the same magnitude in the perpendicular direction is equivalent to pure shear. In treating the bending of beams, Saint-Venant attends to shearing stresses but does not yet know how they are distributed over cross sections. He assumes a uniform distribution. Combining these shearing stresses with the tensile and compressive stresses in longitudinal fibers of a beam, he calculates the principal stresses. In discussing the safe dimensions of a beam, he assumes that the maximum strain must be considered as the basis for selecting permissible stresses.

At that time, the theory of elasticity had no rigorous solutions for problems of practical importance and there were engineers who did not expect much from such theoretical investigations and who preferred to select the safe dimensions of structures by using empirical formulas.[2] Saint-Venant does not believe that any progress in engineering can be made in this way and expresses the opinion that our knowledge can be improved only by combining experimental work with theoretical study.

While lecturing at the École des Ponts et Chaussées Saint-Venant also undertook practical work for the Paris municipal authorities. But some of his plans were disapproved, and in protest he abandoned the work for the city. Very early Saint-Venant became interested in hydraulics and its applications in agriculture. He published several papers in that subject for which he was awarded the gold medal of the French Agricultural Society. For two years (1850–1852) he delivered lectures on mechanics at the agronomical Institute in Versailles. These occupations did not distract Saint-Venant from his favorite work and he continued his

[1] A copy of these lectures can be seen in the library of the École des Ponts et Chaussées.
[2] See Vicat's criticism of the theory of bending (page 83).

research in the field of theory of elasticity. In 1843 he presented a memoir on bending of curved bars to the Academy, while his first memoir on torsion appeared in 1847. However, his final ideas on the treatment of torsional and bending problems were formulated later in the two famous memoirs which were published in 1855 and 1856. These will be discussed in the next articles.

Saint-Venant was not only interested in the analysis of stresses produced by statically applied forces, but he also studied the dynamical action of loads moving along a beam and the impact action of a load falling down onto a bar and producing lateral or longitudinal vibration therein. Several important papers dealing with these problems will be discussed later.

Saint-Venant was also interested in the derivation of the fundamental equations of the theory of elasticity and actively participated in the controversy regarding the required number of elastic constants. He always took the side of those scientists who favored the smaller number of constants on the basis of the assumed molecular structure of solids. His ideas on that subject were discussed in many papers and were finally set down very completely in the fifth Appendix (pp. 645–762) to his edition of Navier's "Résumé des leçons. . . . "

The history of the equations of elasticity is very completely presented by Saint-Venant in Moigno's book "Leçons de Mécanique Analytique, Statique," 1868, the last two chapters of which (pp. 616–723) were written by him. Moigno makes some very interesting remarks, in the preface of the book, regarding these chapters. He wanted the portion on the statics of elastic bodies to be written by an expert in the theory of elasticity, but every time he asked for the collaboration of an English or a German scientist, he was given the same answer: "You have there, close to you, the authority *par excellence*, M. de Saint-Venant, consult him, listen to him, follow him." One of them, M. Ettingshausen, added: "Your Academy of Sciences makes a mistake, a great mistake when it does not open its doors to a mathematician who is so highly placed in the opinion of the most competent judges." In conclusion Moigno observes: "Fatally belittled in France of which he is the purest mathematical glory, M. de Saint-Venant enjoys a reputation in foreign countries which we dare to call grandiose."

In 1868, Saint-Venant was elected a member of the Academy of Sciences and was the authority in mechanics at that institution to the end of his life. He continued to work actively in the field of mechanics of solids and was especially interested in vibration problems and in problems of plastic deformation. His attention was drawn to these latter problems by Tresca's experimental investigations of plastic flow of metals under great pressure.[1] This was a completely new field of research at that time and Saint-Venant was the first to set up the fundamental equations of plasticity and to use them in (several) practical problems.

Saint-Venant never presented his numerous investigations in the field of theory of elasticity in book form, but he edited Navier's "Résumé des leçons . . . " (1864) and he translated and edited "Théorie de l'élasticité des corps solides" by Clebsch (1883). In the first of these two books, Saint-Venant's additional notes are so numerous that the inital material of Navier constitutes only one-tenth of the volume! The book by Clebsch was increased threefold in volume by editorial notes. These two publications, without any doubt, are the most important books for all who are interested in the history of the development of theory of elasticity and of strength of materials.

Saint-Venant continued his work up to the last days of his life. On Jan. 2, 1886, his last article appeared in *Comptes rendus*. On Jan. 6, 1886, this great scientist died. The president of the French Academy, in announcing Saint-Venant's death, used the following words: "Old age was kind to our great colleague. He died, advanced in years, without infirmities, occupied up to the last hour with problems which were dear to him and supported in the great passage by the hopes which had supported Pascal and Newton."

51. *The Semi-inverse Method*

In 1853, Saint-Venant presented his epoch-making memoir on torsion to the French Academy. The committee, composed of Cauchy, Poncelet, Piobert, and Lamé, were very impressed by this work and recommended its publication.[2] The memoir contains not only the author's theory of torsion but also gives an account of all that was known at that time in the theory of elasticity together with many important additions developed by its author. In the introduction Saint-Venant states that the stresses at any point of an elastic body can be readily calculated if the functions representing the components u, v, w of the displacements are known. Substituting the known functions $u(x,y,z)$, $v(x,y,z)$, and $w(x,y,z)$ into the known expressions (see page 107) for the strain components, viz.,

$$\epsilon_x = \frac{\partial u}{\partial x} \qquad \epsilon_y = \frac{\partial v}{\partial y} \qquad \epsilon_z = \frac{\partial w}{\partial z}$$
$$\gamma_{xy} = \frac{\partial u}{\partial y} + \frac{\partial v}{\partial x} \qquad \gamma_{xz} = \frac{\partial u}{\partial z} + \frac{\partial w}{\partial x} \qquad \gamma_{yz} = \frac{\partial w}{\partial y} + \frac{\partial v}{\partial z} \qquad (a)$$

---

[1] See Tresca, Mémoires sur l'écoulement des Corps Solides, *Mém. presentés par divers savants*, vol. 20, 1869.

[2] *Mém. acad. sci. savants étrangers*, vol. XIV, pp. 233–560, 1855.

he obtains these strain components by differentiation and the components of stress can then be calculated by using Hooke's law. Having expressions for the stress components he obtains, from the differential equations of equilibrium [Eq. (b), page 109] and from the boundary conditions [Eq. (a), page 109], the forces which must be applied to the body to produce the assumed displacements. Proceeding in this way and assuming expressions for u, v, and w, there is little chance, says Saint-Venant, of arriving at solutions of practical interest. In the inverse case, where the forces are known, we have to integrate the differential equations of equilibrium and methods of integration have not been found to enable us to solve this problem in its general form.

Saint-Venant then proposes the *semi-inverse method* by which he assumes only some features of the displacements and of the forces and determines the remaining features of those quantities so as to satisfy all the equations of elasticity. He remarks that an engineer, guided by the approximate solutions of elementary strength of materials, can obtain rigorous solutions of practical importance in this way. He then gives solutions for the torsion and bending of prismatical bars of various shapes of cross section.

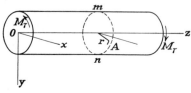

Fig. 149.

To explain the method more clearly let us consider the case of simple torsion of a bar by couples M_t applied at the ends (Fig. 149). Placing the origin of coordinates at the centroid of the left-hand end of the bar and taking the positive directions of the coordinate axes as shown in the figure, let us consider the components of the displacement of a point A in any cross section mn taken at a distance z from the left end of the bar. We assume that that end is prevented from rotating during torsion and denote the angle of twist per unit length of the shaft by θ. The cross section mn will rotate through the angle θz and any point A with coordinates x and y will be displaced, owing to this rotation, by the amounts

$$u = -\theta z y \qquad v = \theta z x \qquad (b)$$

In a circular shaft, the cross sections remain plane during torsion and displacements w, in the axial direction, vanish. Using expressions (b), the torsional stress in circular shafts can be readily calculated. Attempts had been made to use the hypothesis of plane cross sections for noncircular shafts also, but results obtained in this way did not agree with experiments.[1] For this reason, Saint-Venant assumes that warping of the cross

---

[1] See the description of Duleau's experiments, page 82.

sections will occur and that it will be the same for all the cross section so that w is independent of z. Thus he puts

$$w = \theta \cdot \phi(x,y) \qquad (c)$$

where ϕ is some function of x and y to be determined later.

Equations (b) and (c) represent the assumptions that Saint-Venant makes concerning the displacements brought about by torsion. Regarding the forces, he assumes that there are no body forces and that no forces are applied to the curved surface of the shaft. Concerning the forces applied at the ends, he assumes that they are statically equivalent to the given torque M_t. Substituting expressions (b) and (c) into Eqs. (a), Saint-Venant finds that only two strain components, γ_{xz} and γ_{yz}, are different from zero and that their expressions are

$$\gamma_{xz} = \theta\left(\frac{\partial \phi}{\partial x} - y\right) \qquad \gamma_{yz} = \theta\left(\frac{\partial \phi}{\partial y} + x\right)$$

The corresponding stress components are

$$\sigma_x = \sigma_y = \sigma_z = \tau_{xy} = 0$$
$$\tau_{xz} = G\theta\left(\frac{\partial \phi}{\partial x} - y\right) \qquad \tau_{yz} = G\theta\left(\frac{\partial \phi}{\partial y} + x\right) \qquad (d)$$

Substituting these into the equations of equilibrium (page 109), he finds that they will be satisfied if the function ϕ satisfies the equation

$$\frac{\partial^2 \phi}{\partial x^2} + \frac{\partial^2 \phi}{\partial y^2} = 0 \qquad (e)$$

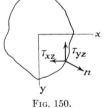

Fig. 150.

At the boundary of the cross section (Fig. 150), the component of shearing stress in the direction of the normal n to the boundary must vanish, since otherwise, according to the theorem of Cauchy (page 109), shearing forces must act on the curved surface of the shaft. That would contradict the initial assumption that the lateral surface is free from traction. This consideration leads to the boundary condition

$$\left(\frac{\partial \phi}{\partial x} - y\right)\cos(nx) + \left(\frac{\partial \phi}{\partial y} + x\right)\cos(ny) = 0 \qquad (f)$$

In this manner, the problem of torsion of a prismatical bar having any shape of cross section is reduced to that of finding a solution of Eq. (e) in each particular case such that it satisfies the boundary condition (f).

Saint-Venant solves the problem for a variety of shapes. The case for an elliptic shaft is specially simple. In this case, Eqs. (e) and (f) are satisfied by taking

$$\phi = \frac{b^2 - a^2}{a^2 + b^2} xy \tag{g}$$

where a and b are the lengths of the semiaxes of the ellipse (Fig. 151). The stress components (d) are then

$$\tau_{xz} = -G\theta \frac{2a^2 y}{a^2 + b^2} \qquad \tau_{yz} = G\theta \frac{2b^2 x}{a^2 + b^2} \tag{h}$$

The corresponding torque is

$$M_t = \iint (\tau_{yz} x - \tau_{xz} y) \, dx \, dy = \frac{\pi G \theta a^3 b^3}{a^2 + b^2} \tag{i}$$

and

$$\theta = \frac{M_t (a^2 + b^2)}{\pi G a^3 b^3} \tag{j}$$

From expressions (h) we see that for any line Oc lying in the plane of the cross section and passing through its centroid (Fig. 151), the ratio $\tau_{xz}:\tau_{yz}$

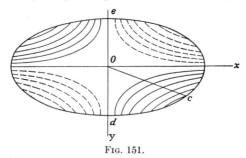

Fig. 151.

remains constant for all points along it. This indicates that the resultant shearing stresses at these points are parallel and must be parallel to the tangent at point c of the boundary. We also conclude that the maximum stress occurs at the boundary at the points d and e, the ends of the shorter axis of the ellipse. Equations (c) and (g) define the form of the warped cross sections and we see that in an ellipse the contour lines of the warped surface are hyperbolas, as shown in Fig. 151, in which full lines indicate points above the xy plane (w is negative for these points). Equation (j) can be written in the form

$$\theta = \frac{M_t}{C} \tag{k}$$

where
$$C = \frac{\pi G a^3 b^3}{a^2 + b^2} = \frac{GA^4}{4\pi^2 I_p} \qquad (l)$$
$$A = \pi ab \qquad I_p = \frac{\pi a^3 b}{4} + \frac{\pi a b^3}{4}$$

C is the *torsional rigidity* of the bar.

From his calculations of the torsional rigidity of solid bars of various cross section, Saint-Venant concludes that formula (l) gives the value of C with good accuracy in all cases.[1] Thus, the torsional rigidities of solid bars can be taken as being equal to the torsional rigidity of a bar having an elliptical cross section with the same area A and the same polar moment of inertia I_p. The torsional rigidity evidently varies inversely as the polar moment of inertia and not directly as in the old theory.

The solutions obtained by Saint-Venant give the same stress distribution in all cross sections and therefore require that the external forces applied at the ends be distributed in the same way. Only under this condition are Saint-Venant's results rigorous solutions of the problems. But, by applying his principle (see page 139) concerning the stresses produced by statically equivalent systems of forces, Saint-Venant concludes that his solutions give the stresses with sufficient accuracy when the distribution of the forces at the ends is different from that assumed in his theory, provided that the points under consideration are at a sufficient distance from the ends. The stresses near the ends naturally depend upon the distribution of the external forces; but this distribution is usually unknown, so that any further investigation of the stresses near the ends has little practical value.

Saint-Venant also applied the semi-inverse method to the bending of a cantilever[2] by a force applied at the end. Assuming that normal stresses on any cross section are given correctly by the elementary beam theory, he shows that it is possible to find a distribution of shearing stresses such that all the equations of elasticity are satisfied. In this way, he finds rigorous solutions for the bending of prismatical bars of various cross sections. He does not limit himself to a general solution of the problem but (in order to facilitate the application of his results) prepares tables from which the maximum shearing stress in rectangular beams can be readily calculated. These calculations show that, when a beam of rectangular cross section is bent in its plane of maximum rigidity, the maximum

[1] See *Compt. rend.*, vol. 88, pp. 142–147, 1879.

[2] The bending problem is briefly treated in the memoir on torsion. It is discussed in more detail in the memoir on bending published in the *J. math. Liouville*, 2d series, vol. 1, pp. 89–189, 1856. At the beginning of this memoir, a very interesting review of old theories of bending is given.

shearing stress approximates very closely that given by Jourawski's elementary theory (see page 142).

Having solutions for torsion and bending of prismatical bars, Saint-Venant goes on to consider combined bending and torsion.[1] He not only calculates the stresses distributed over a cross section, but also finds the principal stresses and computes the maximum strain. He recommends that, in the designing of beams, the dimensions should be selected in such a way as to keep the maximum strain within limits established for each structural material by direct test.

The impact of Saint-Venant's memoirs, giving rigorous solutions in practical cases of torsion and bending, upon strength of materials may be clearly seen. From the time of their appearance, there was a tendency to introduce the fundamental equations of the theory of elasticity into engineering books on strength of materials. Saint-Venant himself worked in that direction in the numerous notes in his edition of Navier's book. Rankine devotes considerable space in his manual of applied mechanics to the theory of elasticity. Grashof and Winkler both tried to develop formulas of strength of materials without using the hypothesis of plane cross sections and basing their derivations on the equations of exact theory. This method of presenting strength of materials has since fallen into disuse[2] and it is the custom now to teach the subject from a more elementary standpoint. More refined work, based upon the theory of elasticity, is generally reserved for those engineers who are especially interested in stress analysis, nowadays.

52. The Later Work of Saint-Venant

After finishing the famous papers on torsion and bending, Saint-Venant continued his researches in theory of elasticity and published many articles, principally in the *Comptes rendus* of the French Academy. The most important results of this work were presented later in the form of notes to his edition of Navier's "Résumé des Leçons . . . " and in his translation of the book by Clebsch. His appendixes to the first of these two works contain a complete theory of elasticity for aeolotropic bodies and a very good account of the question regarding the necessary number of elastic constants.[3] Up to the end of his life, Saint-Venant retained the opinion that one can proceed legitimately only by considering a molecular structure of solids and by assuming the existence of molecular forces. He favored the reduction of the number of elastic constants from 21 to 15 in

---

[1] See Chap. XII of the memoir on torsion.

[2] The courses of strength of materials based on the theory of elasticity, such as were given by Grashof at Karlsruhe and by some professors in Russian engineering schools, proved too difficult for the average student and were gradually discontinued.

[3] See the fifth appendix.

the general case, and from 2 to 1 in isotropic systems. His investigations have not only been of great value to elasticians, but have also proved an asset to molecular physicists.

Saint-Venant's notes to Clebsch's book are also of great value, especially his notes dealing with vibrations of bars and with the theory of impact. In discussing the lateral impact of beams, we have already mentioned Saint-Venant's important contribution to this subject (page 179). Assuming that the body striking a simply supported beam remains in contact with the beam, he discusses the problem of impact as a problem of the vibration of a beam with a mass attached to it. He investigates the first seven modes of the system, calculates the corresponding frequencies of vibration, and finds the shapes of the corresponding curves for various values of the ratio of the weight of the beam to the weight of the striking body. Assuming that the beam is initially at rest while the attached mass has a certain velocity, Saint-Venant calculates the amplitude of each mode of vibration. Summing up the deflections corresponding to these component vibrations, he is able to obtain the deflection curve of the beam at various values of the time t, and to find the maximum deflection and the maximum curvature.[1] From this analysis, the translator concludes that the elementary theory of lateral impact, as developed first by Cox (page 178), gives satisfactory values for the maximum deflections, but is not accurate enough for calculating the maximum stress.[2]

Saint-Venant was also interested in the longitudinal impact of bars. Considering a horizontal bar which is fixed at one end and subjected to an axial blow at the other, he again discussed the impact problem as that of the vibration of a bar with a mass attached at its end. He assumes that initially $(t = 0)$ the bar is at rest and that the attached mass has a given velocity in the axial direction. He takes a solution in the form of a trigonometric series and by summing the first few terms reaches a satisfactory conclusion regarding the motion of the end of the bar. But in calculating stresses, Saint-Venant finds that his series does not converge rapidly enough to allow the computation of an accurate answer. He tried later to use some closed expression for displacements instead of infinite series. The problem was finally solved, at about the same time, by Boussinesq and by two officers of artillery, Sébert and Hugoniot. They gave the solution in terms of discontinuous functions.

[1] The numerous curves which Saint-Venant used in this analysis were published after his death by his pupil M. Flamant, *J. école polytech.* (Paris), Cahier 59, pp. 97–128, 1889.

[2] It seems that Saint-Venant was unaware of Cox's elementary solution, and gives his own derivation together with a very complete discussion of its consequences. See Saint-Venant's translation of Clebsch's book, p. 576.

It was characteristic of Saint-Venant's work that he was never satisfied by giving only a general solution. He always wanted (by calculating tables and preparing diagrams) to present his results in such a form that engineers could use them without difficulty in practical applications. Using Boussinesq's solution Saint-Venant, in collaboration with Flamant, prepared diagrams which illustrate various phases of longitudinal impact for various values of the ratio r of the mass of the struck bar to that of the striking mass.[1] For instance, Fig. 152 shows a diagram from which the compressive stress at the end of the bar in contact with the striking body can be calculated. In this diagram, l denotes the length of the bar, α is the velocity of sound along the bar, and V is the velocity of

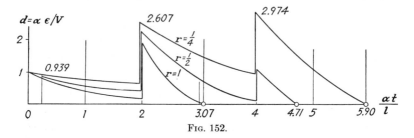

FIG. 152.

the striking body. For abscissas he takes the ratio $\alpha t/l$ and for ordinates he uses the quantity $d = \alpha\epsilon/V$, which is proportional to the unit compression ϵ at the end of the bar. The three curves in the diagram correspond to three different values of the ratio r of the weight of the bar to that of the striking body. It is seen that the compressive strain, which is produced at the instant of impact, at the plane of contact of the bar with the striking body gradually diminishes up to the time $t = 2l/\alpha$. At that time the compression wave, reflected from the fixed end of the bar, reaches the plane of contact and the compressive strain suddenly jumps to a higher value. The gradual decrease in compressive strain then begins afresh. When $r = 1$, as we see, the compressive strain vanishes at $t = 3.07l/\alpha$ so that contact between the striking body and the bar is broken. For small values of r the contact continues beyond this time, and at $t = 4l/\alpha$, we have the second abrupt change in compressive strain. After this, the strain decreases and the contact ends at $t = 4.71l/\alpha$ for $r = \frac{1}{2}$, and at $t = 5.90l/\alpha$ for $r = \frac{1}{4}$. The diagram shows that the duration of impact increases with decrease of the ratio r, and the abrupt changes in maximum strain explain why the solution did not give satis-

[1] This work was published in *Compt. rend.*, vol. 97, pp. 127, 214, 281, and 353, 1883. It also appeared, in the form of an appendix, in the French translation of the book by Clebsch.

factory results when in the form of a series. Diagrams similar to that of Fig. 152 are constructed by Saint-Venant for several cross sections of bar and, in this way, he shows that the maximum stress during impact occurs at the fixed end.

This theory of impact assumes that contact takes place at the same instant over the whole surface of the end of the bar. This condition cannot be realized in practice and experiments made by W. Voigt[1] (see page 345) did not agree with the theory.

Saint-Venant was not only interested in impact but also in the forced vibration of bars and in the "note du no. 61" (about 150 pages long) of his translation of the book by Clebsch he gives a very complete discussion of the vibration produced in a bar by a force which varies with time. He also considers the forced vibrations which are produced by a specified movement of one of the points of the bar and discusses in great detail the case when the middle point of a simply supported prismatical bar performs a given harmonic movement. In the same note, he devotes a considerable amount of space to the problem of bending of a bar under a moving load. He refers to the work of Willis and Stokes (page 174), in which the mass of the beam was neglected. Then he derives the general equations of the problem and discusses the work of Phillips[2] who tried to give an approximate solution of those equations. He shows that Phillips's solution is not complete, and he finally derives an approximate solution by using Willis's mode of reasoning but adding the inertia forces corresponding to the uniformly distributed weight Q of the beam to the inertia force of the moving load W. In this way, he finds that the maximum bending moment produced by both the load W and the load Q taken together is[3]

$$M_{\max} = \frac{Wl}{4}\left(1 + \frac{1}{\beta}\right) + \frac{Ql}{8}\left(1 + \frac{5}{4}\frac{1}{\beta}\right)$$

where, as before (page 176),

$$\frac{1}{\beta} = \frac{16\delta_s v^2}{gl^2}$$

The difference from Phillip's solution lies in the second term of the expression for $M_{\max}$ in which Phillips gives $\frac{3}{4}1/\beta$ instead of $\frac{5}{4}1/\beta$.

[1] See W. Voigt, *Ann. Physik*, vol. 19, p. 44, 1883; vol. 46, p. 657, 1915. A review of the literature on impact is given in the article by T. Pöschl, "Handbuch der Physik," vol. 6, p. 525, Berlin, 1928.

[2] *Ann. mines*, vol. 7, 1855.

[3] A similar result was obtained by Bresse in his "Cours de mécanique appliquée," 1st ed., 1859. Bresse also gave the solution for the case in which a uniformly distributed load moves along a beam.

In 1868, Tresca presented two notes dealing with the flow of metals under great pressures to the French Academy.[1] Saint-Venant, who had to prepare a report on that work, became interested in the plastic deformation of ductile materials. He later published several papers in which he developed the fundamental equations of plasticity on the assumptions that (1) the volume of material does not change during plastic deformation, (2) the directions of principal strains coincide with the directions of the principal stresses, and (3) the maximum shearing stress at each point is equal to a specific constant. Using these hypotheses, Saint-Venant solves several simple problems such as that of torsion of circular shafts, pure bending of rectangular prismatic bars,[2] and the plastic deformation of hollow circular cylinders under the action of internal pressure.[3] In this way, Saint-Venant initiated the study of a completely new field of mechanics of materials. He called the new subject *plasticodynamics;* recently it has been the object of considerable research.

53. Duhamel and Phillips

J. M. C. Duhamel (1797–1872) was born in Saint-Malo, entered the École Polytechnique in 1814, and graduated in 1816. After this schooling, he studied law for some time at Rennes before returning to Paris where he taught mathematics in several schools. In 1830, he succeeded Coriolis as the teacher of calculus at the École Polytechnique and he was connected with that school to the end of his teaching career (1869). His position was an influential one and many of his plans were adopted by the famous school. His textbooks on calculus[4] and on theoretical mechanics[5] were widely used in French schools. In his original work, Duhamel owed much to his former teacher Fourier and to Poisson. His was the time of rapid growth in the application of mathematical analysis to physics and his memoirs followed that trend.

After publishing several important memoirs on heat flow in solid bodies, he presented his "Mémoir sur le calcul des actions moléculaires développées par les changements du température dans les corps solides"[6] to the Academy of Sciences. This paper constitutes Duhamel's chief contribution to the theory of elasticity. In the introduction, he states that Fourier discussed various distributions of temperatures in solid bodies in his famous book "Théorie analytique de la Chaleur," but that

---

[1] "*Mém. presentés par divers savants,*" vol. 20, 1859.
[2] *J. mathématiques,* vol. 16, pp. 373–382, 1871.
[3] *Compt. rend.,* vol. 74, pp. 1009–1015, 1872.
[4] "Cours d'analyse de l'École Polytechnique," 2 vols., 2d ed., 1847.
[5] "Cours de mécanique," 2 vols., 2d ed., 1853.
[6] See *Mém. acad. sci. savants étrangers,* vol. 5, pp. 440–498, 1838.

he did not consider the deformations produced by temperature changes. The elements, into which a solid body may be divided, cannot expand freely under the effects of the temperature changes and stresses will ensue. In studying these stresses, Duhamel follows the method introduced by Navier (see page 105) and develops differential equations of equilibrium, similar to Eqs. (g) (page 106). However, in addition to the components X, Y, Z of body forces, terms of the form $-K(\partial T/\partial x)$, $-K(\partial T/\partial y)$, $-K(\partial T/\partial z)$ appear. These are proportional to the rate of temperature change in the x, y, z directions. He also finds the conditions at the boundary of the body and shows that the thermal stresses can be determined in the same manner as the stresses produced by body forces and by forces applied at the surfaces. Duhamel then shows that the stresses produced by forces and stresses due to temperature changes can be calculated separately and that the total stresses can be obtained by superposition. It seems that here, for the first time, we find the use of superposition in stress analysis. The writer indicates that a uniform distribution of temperature produces no stresses in a body.

Duhamel proceeds to apply his fundamental equations to several specific cases and obtains solutions of practical interest. He begins with a hollow sphere the temperature of which is a given function of the distance from the center. He shows that the change in length of the outer and inner radii depends only upon the average value of the temperatures of the wall of the shell. He extends this to a shell that is made of two concentric layers of different materials. A cylindrical tube, having its temperature a given function of the radial distance, is also discussed in the paper. Finally Duhamel investigates motions of a spherical shell produced by variation of temperature. In all this work, the author assumes that the elastic constant is independent of temperature. He discusses the changes in temperature produced by deformations and the difference in specific heat at constant volume and at constant pressure in his second memoir,[1] which is of primary importance in the theory of heat.

Duhamel worked in the theory of vibration of elastic bodies. The free vibration of strings and of bars of uniform cross section had already received plenty of attention. Duhamel considered more complicated cases. He took up, for example, the vibration of a string to which concentrated masses are attached and not only gave a complete solution of the problem but also conducted numerous experiments which gave results in satisfactory agreement with the theory.[2] He gave a general method for analyzing the forced vibration of elastic bodies.[3] Using the principle of superposition, he showed that the displacements produced by a variable

[1] *J. école polytech.* (*Paris*), cahier 25, vol. 15, 1837.
[2] *J. école polytech.* (*Paris*), cahier 29, 1843.
[3] *J. école polytech.* (*Paris*), cahier 25, pp. 1–36, 1843.

force can be obtained in form of an integral (see page 228). This method was used later by Saint-Venant in studying lateral forced vibration of bars (see page 241).

Phillips (1821–1889) was born in Paris. In 1840 he entered the École Polytechnique and afterwards studied at the École des Mines from which he graduated in 1846. After several years of work for the government, he entered the service of the French railways. His first scientific investigations were connected with his work as a railway engineer. He was put in charge of the rolling stock of the company and became interested in the design of railroad springs. Little was known in that field and Phillips developed the complete theory of leaf springs; he designed the springs according to his theory and was later given the chance to make tests to prove that his theory was accurate enough for practical use. The work was presented to the Academy of Sciences and the committee recommended its publication in the *Mémoires des savants étrangers*.[1] The paper was very valuable to designers of springs and it clearly showed that an engineer having a thorough training in mathematics in addition to practical experience can be of great service in new fields of engineering. The analysis is based upon the elementary theory of bending of straight beams. Assuming that during bending contact is maintained between the laminas

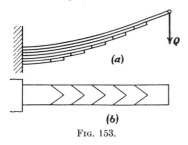

FIG. 153.

(Fig. 153a), Phillips derives formulas for the change of curvature of the spring in each section and shows that this change can be made continuous along the length of the spring by tapering the ends of the laminas as shown in Fig. 153b. By integration he obtains formulas for the deflection of the springs. Differentiation gives him expressions for the pressure between the laminas. Having this pressure, friction forces can be calculated and upon these depend the damping characteristics of the spring during vibration. Phillips applies all these results to the design of actual railway springs and gives numerical examples of such a design. Considerable space is given to the testing of the mechanical properties of the material of the laminas and from the results of these tests he makes recommendations regarding allowable stresses.

A very complete review of this work can be found in Pearson's " History . . . " (vol. 2, part I, p. 330). At the end of this review, Pearson remarks: "The memoir is a striking example of how a very simple elastic

[1] In its complete form, the paper was published in *Ann. mines*, 5th series, vol. 1, 1852.

theory—sufficiently accurate for the range of facts to which it is applied—can be made to yield most valuable results. Phillips' theory of springs such as are employed in the ordinary rolling stock of railways is one of those excellent bits of work which can only be produced by the practical man with a strong theoretical grasp."

In connection with his work for the railway, Phillips became interested in the problem of the dynamical action of a moving load on a bridge.[1] He gave an approximate solution of the problem in which not only the moving mass but also the mass of the bridge was considered. Later on, this solution was simplified by Saint-Venant, who made some corrections to it (see page 241).

Phillips also took up the forced longitudinal and lateral vibration of bars and gave solutions[2] to such problems as that of the longitudinal vibration of a bar, one end of which is subjected to the action of a periodic force.[3] In treating lateral vibrations Phillips examined the stresses in a side rod, all points on the axis of which describe circles of the same radius. He also considered the vibrations of a string having one end fixed and the other attached to a tuning fork which performs harmonic vibration. The methods developed by Phillips in treating lateral vibration of bars were later applied by Saint-Venant in discussing particular cases of lateral vibration in the "note du no. 61" of his translation of the book by Clebsch (see page 241).

Phillips retired from practical work in 1864 and started teaching mechanics, first at the École Central (1864–1875) and later also at the École Polytechnique (1866–1879). In 1868, he was elected to membership of the French Academy of Sciences.

In the latter part of his life, Phillips became interested in the theory of deformation of spiral springs such as are used in watches. He published[4] several memoirs on the subject in which he discussed the problem in great detail. He showed that one of the conditions for satisfactory operation is that the center of gravity of the spiral spring of a watch must be retained on the axis of the balance staff. He also investigated the influence of temperature and of friction on the oscillation of the balance and gave a number of experimental results which were of great value in the watchmaking industry an example of the successful application of theoretical analysis to the solution of an important practical problem.

[1] See *Ann. mines*, vol. 7, pp. 467–506, 1855.
[2] *J. mathématiques*, vol. 9, pp. 25–83, 1864.
[3] This type of problem has recently become of great practical importance in oil fields, where rods of great length are used.
[4] *J. mathématiques*, 2d series, vol. 5, pp. 313–366, 1860; *Ann. mines*, vol. 20, pp. 1–107, 1861.

54. Franz Neumann

Franz Neumann[1] (1798–1895) was born near Joachimsthal, in the province of Brandenburg. His father, Ernst Neumann, was a superintendent of an estate. The Napoleonic Wars presented a very difficult time to Germany, and Neumann's childhood was one of great misery. He received his first instruction in the school at Joachimstahl until he reached ten years of age when he was placed in Werder's *Gymnasium* in Berlin. He lived with the family of a carpenter, again with very limited means. When Prussia was freed from the French occupation in 1813, Neumann volunteered for service in the German Army, but he was too young and did not succeed with this plan until 1815. He joined Blücher's army and took part in the battle of Ligny (June 16, 1815), which preceded the battle of Waterloo. He was severely wounded and was left for dead on the battlefield until the next day when he was found to be alive and was taken to an ambulance. It took several months for him to recover sufficiently to rejoin his regiment. In the fall he participated in the siege of Givet and remained there up to the end of the campaign.

In February, 1816, the volunteers were returned to Berlin and only then was Neumann able to continue his interrupted study at the high school. He graduated in 1817 and in the same year entered Berlin University. Life continued to be difficult for him, since his father was unable to support him and he had to make do with what little money he could earn by giving lessons. At the beginning of his university career Neumann attended lectures in theology and laws, but very soon he changed to the natural sciences and became very interested in mineralogy. Although keen on mathematics, he was unable to get much instruction in that subject since little was offered. He gleaned his mathematical knowledge later from books and the work of Fourier played a large part in his study.

Fig. 154. Franz Neumann.

[1] The biography of Franz Neumann, written by his daughter Luise Neumann, was published in 1904. See also the book by A. Wangerin, "Franz Neumann und sein Wirken als Forscher und Lehrer," 1907, and the article by W. Voigt, Zur Erinnerung an F. E. Neumann, *Nachr. Ges. Wiss. Göttingen, Math.-physik. Klasse*, 1895.

In the fall of 1820, Neumann made a trip to Silesia in order to collect minerals and fossils for Berlin's museum of natural sciences. This excursion brought him into close contact with the professor of mineralogy at Berlin, E. C. Weiss, who arranged an assistantship for Neumann at the mineralogical institute. Neumann's work in mineralogy resulted in the publication of a book called "Krystallonomy," in which a new projective method was described for analyzing crystal structures. The book was a great success, and Neumann was asked to give a series of lectures to a group of men who were specially interested in mineralogy. These lectures presented him with an opportunity to develop his new methods. He also started working for a doctor's degree, and this was awarded to him in the spring of 1826. The same year, he accepted an offer to go to the University of Königsberg to teach mineralogy in the capacity of lecturer. In Königsberg, Neumann met the famous astronomer Bessel and some younger scientists, the physicist Dove and the mathematician Jacoby.

The German policy of academic freedom allowed Neumann to expand the scope of his activity and he started to teach in various branches of theoretical physics such as geophysics, theory of heat, theory of sound, optics, and electricity. Thus, in the first three years of his teaching, he covered all the subjects of theoretical physics. He was rapidly promoted to associate professorship (1828) and to professorship (1829). In 1834, Neumann, in collaboration with Jacoby, organized a seminar in theoretical physics and mathematics. This form of teaching had only existed before in the school of humanitarian sciences and its application to physics was an experiment which can be interpreted as the first attempt to bring some degree of organization into the postgraduate study of the theoretical sciences. This new way of training graduate students rapidly found favor in Germany and proved very successful. It did much toward the progress made in the physical sciences in Germany during the second half of the nineteenth century.

Each of the students who attended the seminar had to prepare a paper for each meeting in which some advanced problem in physics or the scientific work done by some of the students under the direction of the professor was discussed. Later, Neumann divided the students into two groups according to their level of knowledge. For those who had had less training, the seminar was usually run in conjunction with the course of lectures given in the preceding semester. The papers presented in such meetings often consisted of more detailed discussions of topics which had only been briefly touched upon in the lectures. Sometimes, these more elementary discussions contained descriptions of experiments done by students for the verification of theories. Neumann paid great attention to this kind of work since, in it, students acquired experience in handling scientific instruments, in the technique of measurement, and in the presentation of their results in mathematical form. Such were the seminars dealing with problems of theoretical mechanics, problems of capillarity, of heat conduction, acoustics, optics, and electricity. In those for the more advanced students, the students' original work was usually discussed.

The seminar showed itself to be an excellent training ground. Neumann's students often became very good teachers in other universities and, in turn, organized their own seminars. In this way, Neumann's influence on physics in Germany was very important, and, looking over the development of the theory of elasticity during the second half of the nineteenth century, we can say that the principal contributions to that science in Germany were made by his pupils. Borchardt, Clebsch, Kirchhoff, Saalschütz, and Voigt, all of them spoke in the physics seminar at Königsberg and their work in elasticity was initiated by Neumann.[1]

Neumann's original work in the theory of elasticity was started while Navier, Cauchy, and Poisson were still active and when the principal application of this theory was to optics. In his paper on double refraction,[2] Neumann considers an elastic solid the structure of which has three perpendicular planes of symmetry and, following Navier's method (page 105), he develops equations of equilibrium containing six elastic constants and investigates wave propagation in such a medium. Later on he became immediately interested in the elastic properties of crystals with three perpendicular planes of symmetry[3] and showed what kind of experiments must be made to obtain the values of the six constants by direct test. He derived (for the first time) a formula for calculating the modulus in tension for the material of a prism cut out from a crystal with any arbitrary orientation. In these early publications Neumann takes as the basis of his investigations the theory of the molecular structure of elastic bodies and so uses the reduced number of elastic constants as did Poisson, and later Saint-Venant.

However, his constant search for agreement between theory and experiments soon caused Neumann to reject Navier's and Poisson's assumptions. He finally established the necessary number of constants for various kinds of crystal without taking recourse to the molecular theory. He proposed various methods of testing prisms cut from crystals, such that the necessary elastic constants could be computed from measurements. The experiments were made by Neumann's pupils. The work

[1] More information about Neumann's seminar and his pupils can be found in Wangerin's book.
[2] *Pogg. Ann. Physik. u. Chem.*, vol. 25, pp. 418–454, 1832.
[3] *Pogg. Ann. Physik. u. Chem.*, vol. 31, pp. 177–192, 1834.

The Mathematical Theory of Elasticity between 1833 and 1867

of this kind by W. Voigt was especially important[1] since it definitely showed that the reduction in the number of elastic constants, brought about by the assumption of central elastic forces acting between the molecules, was incompatible with test results and that in the most general case 21 elastic constants and not 15, as Poisson's theory indicated, are needed. For isotropic bodies, we need 2 constants, and not 1, as was assumed by Navier, Poisson, and Saint-Venant. So long as the adherents of the multiconstant theory brought forward such examples as cork, rubber, and jelly, which definitely show values different from $\frac{1}{4}$ for Poisson's ratio, it was always possible to argue that those materials were not isotropic. But Voigt's experiments definitely showed that the rariconstant theory gives unsatisfactory results when applied to prismatical bars cut out from perfect crystals.

The most important contributions made by Neumann to the theory of elasticity are included in his great memoir dealing with double refraction.[2] Brewster[3] and Seebeck[4] noticed that a nonuniformly heated plate of glass becomes double refractive. Brewster discovered that glass also acquires a similar property under the action of stresses[5] and suggested the exploitation of this phenomenon with the useful object of investigating stresses in structures by the use of glass models.

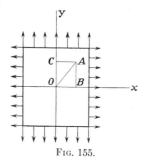

Fig. 155.

Brewster also experimented with jelly and showed that if this substance is dilated and allowed to dry and harden in that state it retains the double refractive property when the dilating force is removed. The double-refractive property of glass under compressive stresses was also studied by Fresnel.[6] Experimenting with triangular prisms, he showed that under axial compression they assume the same doubly-refracting properties as crystals have.

[1] *Ann. Physik. u. Chem.*, vol. 31, p. 701, 1887; vol. 34, p. 981, 1888; vol. 35, p. 642, 1888; vol. 38, p. 573, 1889.
[2] *Abhandl. preuss. Akad. Wiss., Math.-Naturer. Klasse*, 2d part, pp. 1–254, 1843.
[3] *Phil. Trans.*, 1814, p. 436; 1815, p. 1.
[4] Seebeck's work is mentioned in the letter by Maxwell to William Thomson. See "Origins of Clerk Maxwell's Electric Ideas," edited by Sir Joseph Larmor, p. 31, Cambridge, 1937.
[5] *Phil. Trans.*, 1815, p. 60; 1816, p. 156.
[6] *Ann. chim. et phys.*, vol. 20, p. 376, 1822. See also "Oeuvres d'Augustin Fresnel," vol. 1, p. 713, 1866. Regarding Fresnel's work, Maxwell makes the following statement: "D. Brewster discovered that mechanical stress induces temporarily in transparent solids directional properties with respect to polarized light, and Fresnel has identified these properties with the double refraction of crystal." (See the letter of Maxwell mentioned in footnote, 4, above.)

In his memoir Neumann develops the theory of double refraction in stressed transparent bodies. In the simplest case of a homogeneously stressed plate (Fig. 155), this theory states that, if a beam of polarized light perpendicular to the plate passes through a point O and OA represents the amplitude of shear vibration of the aether, this vibration will be resolved into two components OB and OC parallel to the x and y axes. These components will be propagated through the plate with two different velocities. The difference between these velocities $(v_x - v_y)$ is proportional to the difference between the two principal strains, $\epsilon_x - \epsilon_y$, that is, it is proportional to the maximum shearing strain γ_{xy}. By using an analyzer, we can make the two component rays interfere, and by observing the colored picture of the plate on a screen, we can find the relation between the color of the image and the magnitude of shearing strain γ_{xy} for a given plate. Having this information, we can analyze the distribution of shearing strain in a nonhomogeneously stressed plate by again examining the colored pattern on a screen. If, instead of white light, a beam of monochromatic polarized light is used, we shall obtain an image of the plate on the screen with a system of dark fringes. The conclusion regarding the distribution of shearing strains can be reached from the configuration of this. This technique is now widely used by engineers in *photoelastic stress analysis*.

Neumann draws up a theory for the general case of a three-dimensional stress distribution. He shows how, from simple tests, the optical constants can be obtained. By using these, the colored pattern can be predicted for a given stress distribution and a given material. He applied this theory to the particular cases of a circular shaft in torsion and that of stresses radially symmetrical in a sphere.

Next Neumann applies his theory to the study of the colored patterns which Brewster observed in the nonuniformly heated glass plates and points out that the double refractive property of such a plate is due to stresses produced by the nonuniform temperature distribution. To investigate these stresses, Neumann develops equations of equilibrium, similar to those obtained by Duhamel (page 242) and containing terms to allow for thermal expansion. Applying these equations to the case of a sphere with a temperature distribution depending only on distance from the center, Neumann computes the thermal stresses and also makes photoelastic tests which show that the colored fringes that appear when polarized light is passed through the sphere agree satisfactorily with his theory.

Neumann also discusses thermal stresses in a plate having a nonuniform temperature distribution but with the temperature constant through the thickness at any point. He derives the necessary equations and applies them to a circular plate and a circular ring. He considers the circular ring as having a very small thickness in the radial direction and discusses bending of such a ring if its temperature is a function of the distance s measured along the axis of the ring only. The case of two plates of differing materials cemented together, as in Breguet's bimetallic thermometer, is taken up by Neumann, who also attacks the problem of bending of such plates when the temperature is uniformly distributed.

In the last chapter of his great memoir, Neumann takes up the problem of the residual stresses which are left in a body after removal of external forces when plastic deformation takes place during loading. His theory is based on the assumption that the directions of the principal plastic strains coincide with those of the principal elastic deformations and that their magnitudes are linear functions of the principal elastic strain components. Neumann applies his theory in his investigation of the residual stresses set up in a rapidly cooled glass sphere. It seems that Neumann was the first to study residual stresses.

Some of Neumann's original research in elasticity is included in his course of lectures.[1] Neumann usually discussed the problems in which he was especially interested in his class and some of these talks provided subjects for the scientific work of his pupils. Others were never published in the form of memoirs and appeared in print for the first time only when the lectures were published; *i.e.*, perhaps more than thirty years after the work was done and when the results no longer held any novelty.

Let us look briefly at the contents of Neumann's book. We note that, in the first five chapters, he develops the fundamental equations of elasticity for isotropic bodies by introducing the notion of stress and strain components and establishes the relations between them using two elastic constants. His notation for stress components was later accepted by many authors; for example, A. E. H. Love adopted it. In the following three chapters, the derivation of the fundamental equations is given under the hypothesis of the molecular structure of solids. The works of Navier and of Poisson are discussed. The derivation of the equations for non-uniform distribution of temperature is given and a discussion of the theorem on the uniqueness of solutions of the elasticity equations appears. The following portion of the book deals with the applications of the equations to particular problems. The chapter dealing with the deformation of prisms cut out from crystals is especially interesting, as it represents Neumann's original work and served as the basis for the experimental work of several of his pupils who were interested in determining the elastic constants of crystals of various materials.

[1] F. Neumann, "Vorlesungen über die Theorie der Elasticität der festen Körper und des Lichtäthers," edited by Dr. O. E. Meyer, Leipzig, 1885.

Next we come to a portion of the book which deals with propagation of waves in an elastic medium. However, this application of elasticity to optics belongs properly to the history of the undulatory theory of light. The last chapter covers the vibrations of strings, membranes and prismatical bars. In treating longitudinal vibrations of a circular shaft, Neumann sets up the necessary equations in a more complete form than has appeared before by considering not only longitudinal but also radial displacements of particles. He gives an approximate method[1] of solving these equations and applies the results to the longitudinal impact of cylindrical bars. In this respect he was the first to point out that, in applying the principle of conservation of energy in studying longitudinal impact, the vibrations in the bars should be considered. This problem was taken up later by Saint-Venant, as we have seen (page 239).

55. G. R. Kirchhoff

Gustave Robert Kirchhoff (1824–1887)[2] was the son of a lawyer and was born in Königsberg. After leaving high school in 1842, he entered Königsberg University and, during the period 1843–1845, he attended Neumann's lectures and his seminar in theoretical physics. Neumann noticed Kirchhoff's outstanding ability and, in his report to the secretary of education, he recommended his pupil as a very promising young scientist. He also allowed Kirchhoff to publish several papers in 1845–1847 which had been started in the seminar under his direction. In 1848, Kirchhoff obtained his doctor's degree and started teaching at Berlin University. He did not stay long at Berlin, for, in 1850, he was invited by the University of Breslau to join the teaching staff in the capacity of an associate professor of physics. Here he met the famous chemist Bunsen (1811–1899) with whom he later worked in close cooperation for many years. In 1854, Bunsen moved to the University of Heidelberg, and when, in 1855, the chair of physics there fell vacant, he succeeded in drawing Kirchhoff to Heidelberg. In 1858, Helmholtz joined them. It

Fig. 156. G. R. Kirchhoff.

[1] A complete solution of problems of that kind was offered later by L. Pochhammer, *J. Math.* (Crelle), vol. 81, p. 324, 1876.

[2] A brief biography of Kirchhoff can be found in L. Boltzmann's book "Populäre Schriften," Leipzig, 1905.

The Mathematical Theory of Elasticity between 1833 and 1867   253

was thus that a great scientific era was started at Heidelberg University. The lectures of the three outstanding professors attracted students from other German universities and from foreign countries. Kirchhoff and Bunsen worked together on spectrum analysis, and in 1859 the former published his famous papers in this field. He was not only a very good lecturer and an expert in theoretical physics, but he was also an experimenter of no mean ability so that his students received a very thorough laboratory training. In 1868, Kirchhoff injured his leg in an accident and this greatly affected his health. He could no longer work so hard in his laboratory and was forced to restrict himself to theoretical work. He moved to the University of Berlin in 1875 and there he occupied the chair of theoretical physics and was freed from supervising students' laboratory work. In 1876, his famous book on mechanics appeared as the first volume[1] of his lectures on theoretical physics. Kirchhoff published[2] the volume of his collected papers in 1882. His health continued to deteriorate, and as a result he had to discontinue his lectures in 1884. He died at the age of sixty-three (in 1887).

Being a pupil of F. Neumann, Kirchhoff soon became interested in theory of elasticity. In 1850, he published his important paper on the theory of plates[3] in which we find the first satisfactory theory of bending of plates. At the beginning of the paper, Kirchhoff gives a brief history of the problem. He mentions the first attempts of Sophie Germain to obtain the differential equation for bending of plates and also Lagrange's correction of her mistake. He does not mention Navier's work of deriving the equation for plates, using some hypothesis regarding molecular forces (page 121). He discusses the work of Poisson and shows that that writer's three boundary conditions (page 113) cannot be met simultaneously in general and that the problem of a vibrating circular plate had been correctly solved by the French elastician only because the symmetrical modes of vibration which he discussed met with one of the three boundary conditions automatically.

Kirchhoff based his theory of plates on two assumptions which now are generally accepted. The hypotheses were (1) that each line which is initially perpendicular to the middle plane of the plate remains straight during bending and normal to the middle surface of the deflected plate and (2) that elements of the middle plane of the plate do not undergo stretching during small deflections of plates under lateral load. These assumptions approximate closely the hypothesis regarding plane cross sections used in the elementary theory of bending of plates nowadays. Using his two conditions, Kirchhoff established the correct expression for

[1] "Vorlesungen über mathematische Physik, Mechanik," 1st ed., Leipzig, 1876.
[2] "Gesammelte Abhandlungen," Leipzig, 1882.
[3] *J. Math. (Crelle)*, vol. 40, 1850.

the potential energy V of a bent plate, viz.,

$$V = \frac{1}{2} D \iint \left[\left(\frac{\partial^2 w}{\partial x^2}\right)^2 + \left(\frac{\partial^2 w}{\partial y^2}\right)^2 + 2\mu \frac{\partial^2 w}{\partial x^2} \frac{\partial^2 w}{\partial y^2} \right.$$
$$\left. + 2(1 - \mu) \left(\frac{\partial^2 w}{\partial x \, \partial y}\right)^2 \right] dx \, dy \quad (a)$$

In this expression $D = Eh^3/12(1 - \mu^2)$ is the flexural rigidity of the plate and w is the deflection of its mid-surface. To obtain the differential equation of bending, Kirchhoff now uses the principle of virtual work, which states that for any virtual displacements the work done by the load q distributed over the plate must be equal to the increment in the potential energy of the plate. This gives

$$\iint q \, \delta w \, dx \, dy = \delta V \quad (b)$$

Substituting expression (a) for V and performing the indicated variation, Kirchhoff derives the well-known equation for bending of plates:

$$D \left(\frac{\partial^4 w}{\partial x^4} + 2 \frac{\partial^4 w}{\partial x^2 \, dy^2} + \frac{\partial^4 w}{\partial y^4}\right) = q \quad (c)$$

Further, he shows that there are only two boundary conditions and not three, as was supposed by Poisson.

Kirchhoff applies his equations to the theory of vibration of a circular plate with a free edge. He investigated not only the symmetrical modes (the nodal lines of which are concentric circles) but also the modes for which the nodal lines are diameters of the circle and for which Poisson's edge conditions cannot be applied. Having arrived at his general solution, he undertakes much numerical work and gives a table of frequencies calculated for various modes of vibration. He uses these numerical results to analyze the experimental results on plate vibration obtained by Chladni (page 119) and by Strehlke. From these results, he wanted to find the correct value of Poisson's ratio μ. However, since the frequencies are not influenced much by the magnitude of μ, these experiments proved unsuitable for an accurate determination. Later, as mentioned before (page 222), Kirchhoff carried out his own experiments for the same purpose.

In his lectures,[1] Kirchhoff subsequently extended his theory of plates to cover the case where the deflections are not very small. The advent of this theory of plates was a very great step forward in the theory of elasticity, and it has become especially important lately owing to its wide application in the design of various kinds of thin-walled structures.

Another important contribution made by Kirchhoff to the theory of

[1] See his "Mechanik," 2d ed., p. 450, 1877.

The Mathematical Theory of Elasticity between 1833 and 1867 255

elasticity was his theory of deformation of thin bars.[1] He derived the general equations of equilibrium for the three-dimensional deflection curve of a bar, when deflections are not small. He then demonstrated that, when forces are applied only to the ends of the bar, these equations are identical with the equations of motion of a rigid body about a fixed point. In this way the solutions, which were already known in the dynamics of a rigid body, could be applied directly to the deformation of thin bars. This is known as *Kirchhoff's dynamical analogy*. As a simple example of this analogy, let us compare the lateral buckling of a compressed bar AB(Fig. 157a) with the oscillation of a mathematical pendulum (Fig. 157b). In both cases we have the same differential equation and we can state the following relation between the two problems: if a point M moves along the curve AB with uniform velocity such that it passes from A to B in a time equal to the half period of the pendulum, and if M begins to move away from A at the instant when the pendulum is in its extreme position and the tangent to the curve at A makes an angle equal to that defining the extreme position of the pendulum with the vertical, then at any intermediate position of M the direction of the tangent to the curve coincides with that of the pendulum. A similar statement can be made relating the deflection of a thin wire into a helix and the regular precession of a gyroscope.

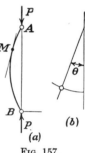

Fig. 157.

Kirchhoff's theory aroused much discussion, and this cleared up many difficulties, led to simplified derivations, and at the same time confirmed Kirchhoff's results. In more recent times, this theory has been used in the analysis of elastic stability problems such as that of the buckling of a uniformly compressed circular ring and the lateral buckling under uniform bending of a curved bar of narrow rectangular cross section.

In conclusion, mention should be made of Kirchhoff's paper on the vibration of bars of variable cross section.[2] The general equation for lateral vibration of such bars was already known, and Kirchhoff shows that, in certain cases, it can be integrated exactly. In particular, he considers bars having the form of a thin wedge and of a very sharp cone and calculates the frequency of the fundamental mode of vibration for both.

56. A. Clebsch

A. Clebsch (1833–1872) was born in Königsberg. At an early age he entered Königsberg University where he received his training in theoret-

---

[1] *Borchardt's J.*, vol. 56, 1858. See also "Gesammelte Abhandlungen," p. 285.
[2] Kirchhoff, "Gesammelte Abhandlungen," pp. 339–351.

ical physics from Neumann. Under Neumann's direction, he wrote his doctoral thesis, which dealt with the motion of an ellipsoid in an incompressible fluid. After obtaining this degree (1854), Clebsch remained at Königsberg as a lecturer. He later worked at the University of Berlin in the same capacity. Very soon, his scientific work began to attract attention and in 1858, at the age of twenty-five, he was given a professorship at the Karlsruhe Polytechnicum, an important engineering school, where he was put in charge of theoretical mechanics. While there, he wrote his famous book "Theorie der Elasticität fester Körper" (1862) and it represents his principal contribution to the theory of elasticity. As time went on, Clebsch's scientific interest swung into the direction of pure mathematics.[1] He stopped teaching at the engineering school, and in 1863 he became a professor of pure mathematics at the University of Giessen. In 1868, he was elected to the chair at Göttingen. Here he showed his outstanding talent both as a scientist and as a teacher. With his friend Carl Neumann, the son of Franz Neumann, he founded a new mathematical journal, the *Mathematische Annalen* (1868), and it became the leading periodical in mathematics. His lectures attracted many students, for he knew how to foster their interest in mathematics and how to inspire them to creative work in that field. In 1872, he was elected Rector of Göttingen University but did not live long in this capacity, for in the same year he died suddenly from diphtheria at the age of thirty-nine.

In 1861, Clebsch, then twenty-eight, started to write his book on elasticity. At that time only one book existed on the subject, that by Lamé. The French author was interested in the application of the theory of elasticity to theoretical physics, in such fields as acoustics and optics. Clebsch, who was teaching mechanics to engineers, wanted to write a book which would embrace the engineering application of the subject. His introductory treatment of stress and strain is considerably shorter than that of Lamé. He assumed from the beginning that structural materials are isotropic and he derived all the necessary equations on that assumption. He completely eliminates all problems of optics and instead incorporates Saint-Venant's theories of torsion and bending and Kirchhoff's analysis of deformations of thin bars and thin plates. But as a book for engineers Clebsch's work never enjoyed much popularity, since the writer's scientific interests were always too remote to produce simplicity. He pays more attention to devising mathematical methods for solving problems than to explaining the physical and technical significance of the results which he obtains. Again his originality leads him to discuss

[1] Clebsch's work in that field is discussed by his pupil F. Klein. See his book "Vorlesungen über die Entwicklung der Mathematik im 19 Jahrhundert." See also Clebsch's biography in *Math. Ann.*, vol. 6, pp. 197–202; vol. 7, pp. 1–55.

questions in which he is especially interested at the moment, and as a result of this the content of the book is not well balanced. The book is of great interest as a collection of the original works of Clebsch, but it is not a book suitable for studying the theory of elasticity, especially for those who have no special interest in the mathematical treatment of problems. Saint-Venant, in his French translation of Clebsch's book, added many important notes, and this annotated edition is still one of the most complete books in the theory of elasticity and its engineering applications.

The introduction of Clebsch's book contains a condensed presentation of the fundamental equations of elasticity, and as early as page 50 the author begins with applications and discusses the stresses and deformations of a hollow sphere under the action of a uniformly distributed external or internal pressure. There is nothing new in this problem, but Clebsch uses it as an introduction to the theory of radial vibrations of a sphere and gives an original study of the roots of the frequency equation and a mathematical proof that all the roots are real and positive. He also uses this case in proving that the state of equilibrium of elastic bodies is completely determined if the acting forces are given and the body is fixed in such a manner that it cannot move as a rigid body.

In the next portion, Clebsch discusses Saint-Venant's problem. He omits Saint-Venant's physical considerations, which were used in applications of the semi-inverse method and puts the problem into a purely mathematical form which can be stated as follows: Find the forces which should be applied at the ends of a prismatical bar if there are no body forces, if there are no forces acting on the lateral surface of the bar, and if on the lateral surfaces of longitudinal fibers only shearing stresses in the axial direction are acting. In this way, Clebsch can treat the problems of axial tension, of torsion, and of bending jointly. The presentation becomes more involved than that of Saint-Venant, while the physical aspect of the problem is lost in this method of approach and the treatment becomes too abstract to promote the interest of an engineer. Clebsch is not interested in the numerous applications which Saint-Venant discussed in considering beams having various shapes of cross section. He takes as examples a solid elliptical bar and a hollow bar the cross section of which is formed by two confocal ellipses. The problems are of little practical interest, but Clebsch uses them to introduce a novelty into the mathematical treatment by applying conjugate functions in discussing Saint-Venant's problem for the first time.

The next chapter deals with two-dimensional problems and it is, perhaps, the most valuable part of Clebsch's work in the theory of elasticity. Two-dimensional problems have since received great attention, and the results obtained have found important practical applications. The writer's approach to this problem is again purely mathematical. Considering a cylindrical bar, he assumes that the forces and stress components which vanish in Saint-Venant's problem are now different from zero while the stresses and forces which Saint-Venant's problem retains now vanish. In such a system no stresses act on cross sections perpendicular to the axis of the bar and any portion of the bar between two adjacent cross sections will represent a plate stressed by forces distributed over its cylindrical bounding surface and parallel to its faces. This is *plane stress*. Clebsch investigates the conditions which the distribution of forces at the boundary must satisfy in plane stress and gives general expressions for the displacements. Afterwards he applies these general expressions to a circular plate for which the radial displacements at the boundary are specified. As his next example, Clebsch discusses the case of a very thin plate and suggests that, instead of the previously considered stresses, the average values of those stresses through the thickness be used. He shows

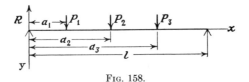

Fig. 158.

that then the problem can be solved for any force distribution along the boundary and gives the general solution for a circular plate. As a particular case of the general form of plane stress, Clebsch treats the bending of a plate by couples distributed along the boundary and solves the problem for a circular disk, the deflections of which are prescribed along the edge.

In the next portion of his book, we find problems concerning the deformation of thin bars and thin plates. He gives Kirchhoff's theory of thin bars (page 255) in somewhat modified form. He also extends the theory of bending of plates considerably by writing equations for the case where deflections are not small.[1] In conclusion, he applies the theory of small deflections to the bending of a circular plate having its edge clamped and being loaded at any point by a force perpendicular to its surface.

The last part of the book contains an elementary discussion of problems in strength of materials. In finding the deflection of a beam loaded by several lateral concentrated forces (Fig. 158), he shows that the determination of the constants of integration can be simplified if the integrals of the differential equations of equilibrium for the consecutive portions of the beam

[1] See the footnote on p. 264 of his book.

$$EI\frac{d^2y}{dx^2} = -Rx$$

$$EI\frac{d^2y}{dx^2} = -Rx + P_1(x - a_1)$$

$$EI\frac{d^2y}{dx^2} = -Rx + P_1(x - a_1) + P_2(x - a_2)$$

. . . .
. . . .

are put in the following form:

$$EI\frac{dy}{dx} = -\frac{Rx^2}{2} + C$$

$$EI\frac{dy}{dx} = -\frac{Rx^2}{2} + \frac{P_1(x-a_1)^2}{2} + C_1$$

$$EI\frac{dy}{dx} = -\frac{Rx^2}{2} + \frac{P_1(x-a_1)^2}{2} + \frac{P_2(x-a_2)^2}{2} + C_2$$

. . . .
. . . .

$$EIy = -\frac{Rx^3}{6} + Cx + D$$

$$EIy = -\frac{Rx^3}{6} + \frac{P_1(x-a_1)^3}{6} + C_1x + D_1$$

$$EIy = -\frac{Rx^3}{6} + \frac{P_1(x-a_1)^3}{6} + \frac{P_2(x-a_2)^3}{6} + C_2x + D_2$$

. . . .
. . . .

From the condition that, at the points of application of the loads, that is, for $x = a_1, x = a_2, x = a_3, \ldots$, the two adjacent portions of the deflection curve must have a common tangent and a common deflection, it follows that $C = C_1 = C_2 \ldots$ and $D = D_1 = D_2 \ldots$ This means that, independently of the number of forces, we have only to calculate two constants of integration in each case. Clebsch extends his method to problems where, in addition to concentrated forces, a distributed load also acts. Further, he uses it in analyzing uniformly loaded continuous beams. For these he gives very simple formulas for the reactions at the supports where the spans are equal.

In the last article of the book Clebsch gives a method of analyzing trusses. Here, for the first time, this problem is discussed in its general form. He shows that if instead of the forces in the bars we select the displacements of hinges as our unknown quantities, we shall always be confronted with as many linear equations as there are unknowns and so the problem can be solved. Clebsch demonstrates this with two simple problems: (1) the case of a single hinge attached to the foundation by several bars (Fig. 159a) and (2) the case of a beam reinforced by a system of three bars, as shown in Fig. 159b.

Clebsch's contribution to the theory of elasticity is mainly of a mathematical character. He, like Cauchy, was principally a pure mathematician and he applied his mathematics to developing new methods of treating problems. His book, especially the French translation with Saint-Venant's notes, occupies an important place in the history of our science.

Some original papers of Clebsch, dealing with optics, are also of interest to elasticians. This particularly applies to his investigation of the vibratory motion of an elastic sphere by which displacements vanish at the surface.[1] He uses spherical functions in his solution of the problem and adds considerably to the theory of those functions. In the particular case where the acceleration vanishes, we obtain the solution of the statical problem which was discussed later by Lord Kelvin (see page 265).

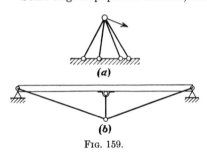

FIG. 159.

57. Lord Kelvin (1824–1907)

William Thomson, Lord Kelvin, was born in Belfast in 1824 of Scottish descent.[2] His father, James Thomson, was professor of mathematics at the Royal Belfast Academical Institution. In 1832, the family moved to Glasgow, where James Thomson occupied the chair of mathematics at the University of Glasgow. In 1834, at the age of ten years, William Thomson matriculated in the University of Glasgow, where he studied classic languages, mathematics, and natural philosophy. The lectures on natural philosophy that were given by Prof. J. P. Nichol attracted young Thomson's interest to mathematics and it was Nichol who, in 1840, drew his attention to the famous book, "Théorie analytique de la Chaleur," by Fourier. The contents of that book greatly influenced Thomson's early scientific work. He later wrote[3] "The origin of my devotion to these problems is that after I had attended in 1839 Nichol's Senior Natural Philosophy Class, I had become filled with the utmost admiration

---

[1] See *J. reine u. angew. Math.*, vol. 61, pp. 195–262, 1863.

[2] An extensive biography of Lord Kelvin is given in the book in two volumes, "The Life of William Thomson, Baron Kelvin of Largs," by Silvanus P. Thompson, London, 1910.

[3] See p. 14 of the above-mentioned book by S. P. Thompson.

for the splendour and poetry of Fourier. . . . I asked Nichol if he thought I could read Fourier. He replied 'perhaps.' He thought the book a work of most transcendent merit. So on 1st of May, 1840, the very day, when the prizes were given, I took Fourier out of the University library; and in a fortnight I had mastered it—gone right through it."

To help his childrens' study of foreign languages, Thomson's father took his family to Paris in the summer of 1839, and in the summer of 1840 they traveled in Germany. William was, at that time, absorbed in mathematics and was not interested in the German language. Reminiscently he says: "Going that summer to Germany with my father and brothers and sisters, I took Fourier with me. My father took us to Germany, and insisted that all work should be left behind, so that the whole of our time should be given to learning German. We went to Frankfort where my father took a house for two months. . . . I used in Frankfort to go down to the cellar surreptitiously every day to read a bit of Fourier. When my father discovered it he was not very severe upon me."[1]

Fig. 160. Lord Kelvin.

In April, 1841, William Thomson left the University of Glasgow and entered St. Peter's College, Cambridge. Here, he continued to be interested in Fourier's work, and in November, 1841, his first scientific paper, dealing with Fourier's series, appeared in the *Cambridge Mathematical Journal*. Two more papers of a more advanced character appeared in the same journal during the course of the following year. In mathematical aptitude and knowledge William was far ahead of other students of his class and it was expected that he would become senior wrangler for the year 1845. But to his and his father's great disappointment he won only the place of the second wrangler in the final examination. His biographer states: "That devotion to scientific activity of the highest order . . . acted as a hindrance rather than a help to University honours. . . . To secure a leading place in the Senate-house (final examination) the candidate must have at his fingertips the solutions of known problems, polishing up his methods of writing out the solutions, so as to be able to get through an immense quantity of

[1] S. P. Thompson's biography of Lord Kelvin, p. 17.

bookwork in a short time. Drilled for months by his coach or his tutor in this species of gymnastic, as if he were being trained for a race, the candidate indeed acquired facility in dealing with particular classes of problems in the particular mode then in vogue. But such a training was not adapted either to cultivate originality or to advance mathematical science. It set up a false ideal of attainment, and not infrequently caused the coveted position of Senior Wrangler to be awarded, not to the best mathematician of his year, but to the one who had been best groomed for the race."[1]

After graduating from Cambridge, Thomson decided to continue his study and for that purpose he went to Paris. France, at that time, offered the best facilities for studying mathematics and its applications to the various branches of physics. He there met Liouville, Sturm, and Cauchy, the leading mathematicians of the day, and succeeded in drawing their interest to Green's work (see page 217), a copy of which he brought from Cambridge.

He also met several French physicists. In order to acquire a better experimental technique, he entered the physics laboratory of the Collège de France where he worked with Professor Regnault, helping him in his famous experimental researches upon the laws of heat. During this period, Thomson became acquainted with the famous paper by Clapeyron, "Mémoires sur la puissance motrice du feu," in which the writer explains Carnot's cycle. In his early scientific work, Thomson owed much to the writing of Carnot, Fourier, and Green.

After four months of study in Paris, he returned to Cambridge in May, 1846, and, in the fall of the same year, obtained a fellowship at St. Peter's College. He became a lecturer in mathematics in the college and began to coach pupils in mathematics. He also undertook the management of the *Cambridge and Dublin Mathematical Journal*. Several of his papers dealing with the application of mathematics to theoretical physics were published in this periodical.

In the fall of 1846, at the age of twenty two, William Thomson was elected professor of natural philosophy at Glasgow University, where he taught this subject for fifty-three years. His biographer says: "For the purposes of systematic teaching he became a bad expositor, profound as was his grasp, and accurate as was his phraseology. He could not lecture on the most ordinary and elementary parts of physics, to the most ordinary of university students, without his thoughts traveling ever and anon to the recondite border-regions of science, where only a few could follow him. These original digressions, says a former student of his class, containing the priceless jewels of his discourse, were simply flung

[1] S. P. Thompson's biography of Lord Kelvin, p. 96.

away on all except the abler and wiser scholars, who listened with rapt attention to the flashing torrent, the impetuous cataract of his genius. His imagination was vivid: in his intense enthusiasm he seemed to be driven, rather than to drive himself. The man was lost in his subject, becoming as truly inspired as is the artist in the act of creation."[1]

From the very beginning of his teaching, Thomson recognized the importance to students of experimental work in physics. He was not content merely to give experimental demonstrations during his lectures, but organized a laboratory in which he himself and his students could investigate the properties of matter. This was the first laboratory of its kind for the physical sciences in Britain. The most important contribution made to physics by Thomson during the first years of his work at Glasgow were in the field of thermodynamics, but he also obtained a considerable amount of experimental data in strength of materials and in the theory of elasticity.[2] These results were later used in the preparation of articles which appeared in the ninth edition of the Encyclopaedia Britannica and they became widely read and greatly valued.[3]

Discussing (in these articles) the fundamental assumptions on which the theory of elasticity is based, Thomson explains that the properties of actual materials sometimes deviate noticeably from those considered. He observes that structural materials are not perfectly elastic and, in investigating this imperfection, he introduces the notion of internal friction, which he studies by examining the damped vibrations of elastic systems. From his experiments, he concludes that this friction is not proportional to velocity, as in fluids. Regarding moduli of elasticity, the writer strongly criticizes the rariconstant theory (see page 216), which was then in vogue among many French scientists. In his argument against this theory he cites the example of cork, jelly, and india rubber. Thomson employs Thomas Young's definition of the modulus in simple tension and talks about the *weight-modulus* and *length of modulus* (see page 92). He shows that the length of modulus has a simple physical interpretation—the velocity of transmission of longitudinal vibrations along a bar is equal to the velocity acquired by a body in falling from a height equal to half the length of modulus.

Experiments with elastic bodies led Thomson into the borderland between elasticity and thermodynamics. He studied the temperature changes produced in elastic bodies which are subjected to strain[4] and

[1] See Sylvanus P. Thompson's book, p. 444.
[2] See the papers A Mathematical Theory of Elasticity, *Trans. Roy. Soc. (London),* (A), 1856. On the Elasticity and Viscosity of Metals, *Proc. Roy. Soc. (London),* (A), 1865.
[3] These articles were published later in a book, "Elasticity and Heat," 1880.
[4] See "Elasticity and Heat," p. 19.

showed that the magnitude of the modulus depends upon the manner in which stress is set up in a specimen. Considering the results of a tensile test, let OA (Fig. 161) represent the diagram for the sudden stretching of a specimen within the elastic limit. The diagram for slow application of the tensile force usually has a smaller slope, as shown by the straight line OB. In the first instance, there is no heat exchange between the specimen and its surroundings, and we have an *adiabatic extension*. In the second, we assume that the deformation proceeds so slowly that, due to heat exchange, the temperature of the specimen remains practically constant so that we have an *isothermal extension*. From the diagram we conclude that Young's modulus for sudden stretching is larger than it is for slow extension. The difference, as far as steel is concerned, is very small (about $\frac{1}{3}$ of 1 per cent) and in most practical applications it can be disregarded. A specimen that is subjected to a sudden extension usually becomes cooler than its surroundings and, due to the equalization of the temperature, some increment of extension will ensue, as represented by the distance AB in Fig. 161. If now the tensile force is suddenly removed, the specimen contracts and its state is represented in Fig. 161 by point C. Due to shortening, the temperature of the specimen rises so that a return to the initial state (shown in Fig. 161 by point O) only takes place after cooling to the temperature of the surroundings. The area $OABC$ then represents the mechanical work lost during one cycle.

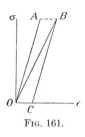

Fig. 161.

There are some peculiar materials, as, for example, india rubber under a certain temperature, which show a heating effect when pulled and a cooling effect on removal of the extending force. In such specimens (explains Thomson), the material must contract when heated, and become elongated when cooled, since otherwise the reasoning illustrated by Fig. 161 leads to the absurd conclusion that a possibility exists of gaining mechanical energy during the course of each cycle. In his experiments, Thomson used an india-rubber band stretched vertically by a weight suspended from its lower end, and he showed that the weight rose when a red-hot body was brought near the band and fell when it was removed.

Later Thomson conducted a general investigation of the thermal changes which occur during deformation of an elastic body. By considering the energy of the body[1] he gives the first logical proof of the existence of a strain-energy function depending only upon the strain measured with respect to some standard state and not upon the manner in which the strain is reached. This work constitutes one of the most important of William Thomson's additions to the theory of elasticity.

[1] See "Mathematical and Physical Papers," vol. 1, pp. 291–313.

The Mathematical Theory of Elasticity between 1833 and 1867 265

In 1860, P. G. Tait was elected to the chair of natural philosophy in Edinburgh and very soon the two men, Thomson and Tait, were working together. Both felt a want of books which could be recommended to students interested in the various branches of theoretical physics. The two scientists decided that they would themselves write the necessary books and thus, in 1861, they started work on their "Treatise on Natural Philosophy." The writing was very much delayed by Thomson's participation in important work on the Atlantic telegraph, which required plenty of theoretical and experimental investigation. It was not until 1867 that the first volume of the famous treatise was published. This volume deals with the mechanics of rigid, elastic, and fluid bodies and contains much original work in the theory of elasticity due to Thomson.

Earlier in his career, Thomson had given[1] a solution of the equations of elasticity for an infinite homogeneous isotropic elastic solid. He now uses that solution to solve such important problems as that of a force applied uniformly to a spherical portion of an infinite homogeneous solid and that of a force applied to an infinitely small part of that body. He also gives a complete solution for a spherical shell and a solid sphere subjected either to given surface stresses or to given surface shifts.[2]

The author uses this solution in his investigation of the rigidity of the earth. At that time the question was raised: "Does the earth retain its figure with practically perfect rigidity, or does it yield sensibly to the deforming tendency of the moon's and sun's attractions on its upper strata and interior mass? It must yield to some extent, as no substance is infinitely rigid: but whether these solid tides are sufficient to be discoverable by any kind of observation, direct or indirect, has not yet been ascertained." Assuming perfect elasticity, Thomson estimates the effect of the elastic strain of a solid earth of uniform density upon the superficial water tides and finds that if the earth were as rigid as steel its elastic yielding would reduce the height of the tide to about $\frac{2}{3}$ of its value as calculated from a theory in which the earth is supposed to be absolutely rigid. In the second edition of the treatise, an additional discussion of the subject was added by G. H. Darwin who arrived at the following conclusion: "On the whole we may fairly conclude that, whilst there is some evidence of a tidal yielding of the earth's mass, that yielding is certainly small, and that the effective rigidity is at least as great as that of steel."[3]

[1] "Mathematical and Physical Papers," vol. 1, p. 97.
[2] The same problems had already been solved by Lamé, who used another method (see page 117).
[3] "Treatise on Natural Philosophy," part 2, p. 460.

In his treatment of the theory of thin bars, Thomson gives a very complete discussion of Kirchhoff's dynamical analogy (see page 255) and uses it in the calculation of deflections of helical springs. In developing the theory of bending of thin plates, he explains, in simple terms, why Kirchhoff's elementary theory is accurate enough only if deflections are small in comparison with the thickness of the plate. A very instructive consideration of boundary conditions is given. Kirchhoff had already shown that there must be only two boundary conditions at an edge and not three as demanded by Poisson. Thomson gives a physical interpretation of this reduction and, by utilizing Saint-Venant's principle (see page 139) regarding the elastic equivalence of statically equivalent systems of load, he shows that a distribution of twisting couples along an edge of a plate may be replaced by a statically equivalent distribution of shearing forces. Thus we need only consider the two conditions concerning shearing forces and bending moments at an edge.[1] In the particular case of twisting couples uniformly distributed along the edges of a rectangular plate and producing a uniform anticlastic curvature, replacement of the twisting couples by shearing forces gives rise to the interesting system of a rectangular plate subjected to normal forces P, P at the ends of one diagonal and normal forces $-P$, $-P$ at the ends of the other. Such a system of forces can easily be applied to a plate and the deflections produced in this way may be used for the experimental determination of the flexural rigidity of the plate.

In treating Saint-Venant's torsional problem, Thomson uses the method of conjugate functions that had been introduced by Clebsch, and applies it in calculating the stresses and the angle of twist in a bar with cross section in the form of an annular sector.

W. Thomson's discussion of the theory of elasticity in the "Treatise on Natural Philosophy" constitutes the first systematic presentation of that subject in the English language. It exerted great influence on future workers in the theory of elasticity in England.

In 1866, telegraphic connection was established between England and America. The success of this important enterprise was due to a great extent to Thomson's scientific advice and to his active and energetic cooperation. The honor of knighthood was conferred upon him in appreciation of this work and of his high position in the scientific world. He was recognized not only as the greatest physicist in England but also as an inventor and research engineer. From this time on, a considerable

---

[1] See "Treatise on Natural Philosophy," pp. 724–729. Thomson gives there a mathematical proof justifying the replacement of one system of forces by another one statically equivalent to it. This is the first instance of a proof of Saint-Venant's principle. Later, the same subject was discussed by Maurice Levy, *J. mathématiques*, vol. 3, pp. 219–306, 1877.

portion of his time was spent in the design of instruments and in technical consultations.

The broad outline of the "Treatise on Natural Philosophy," which was to include all branches of theoretical physics, was never completely fulfilled. The first volume appeared in a second edition, Part I in 1879 and Part II in 1883. However, Kelvin continued to be interested in the properties of matter. Under the influence of Helmholtz's paper on vortex motion,[1] he developed the theory of vortex rings. He showed that in "a perfect medium vortex-rings are stable, and in many respects they possess the qualities essential to the properties of material atoms, permanence, elasticity, and the power to act on one another through the intervening medium."

Kelvin's ideas regarding the structure of matter were further crystallized in the famous "Baltimore lectures" which were delivered before a distinguished audience of physicists and mathematicians in 1884. A considerable portion of those lectures is devoted to the theory of elasticity, to the propagation of light in isotropic and aeolotropic elastic solids, and to propagation in an elastic medium in which an infinite number of gyrostatic molecules are embedded. He does not accept the electromagnetic theory of light and tries to improve the "elastic" theory. He often repeated: "I never satisfy myself until I can make a mechanical model of a thing." He invented several kinds of molecular structures in order to explain optical phenomena. The notes on the Baltimore lectures were prepared and reproduced by papyrograph in 1884, but Kelvin retained an interest in the subject so that, after many revisions and additions, the lectures were finally published in printed form twenty years later.

The book "Baltimore Lectures" together with the relevant chapters of the "Treatise on Natural Philosophy" was very widely read by elasticians, and we shall see that, during the latter part of the nineteenth century and the beginning of the twentieth century, several outstanding English scientists became interested in this subject as a consequence.

There were several attempts made to bring Lord Kelvin to Cambridge to occupy the Cavendish professorship, but he preferred to remain in Glasgow to the end of his teaching activities in 1899. At that time the celebration of the jubilee of Lord Kelvin was organized.[2] Many scientists from all parts of the world came to that celebration to present their congratulations to the great man. In 1899, Lord Kelvin retired from his professorship, but he applied to the Senatus Academicus to be appointed a research student so that he would have the right to pursue

[1] *J. Math. (Crelle)*, 1858; English translation, *Phil. Mag.*, 1867.
[2] At that time the book "Lord Kelvin" was published, in which Lord Kelvin's biography and the description of the jubilee celebration are given.

investigations in the Natural Philosophy Laboratory. On Dec. 17, 1907, Lord Kelvin died.

58. James Clerk Maxwell (1831–1879)[1]

J. C. Maxwell was born in Edinburgh, but the greater part of his childhood was passed in the country house owned by his parents. In 1841, the boy was taken to Edinburgh and sent to school at the Edinburgh Academy, where he studied until 1847, spending his summer vacations in the country.

FIG. 162. James Clerk Maxwell.

James's interest in schoolwork was slow to mature, but when classes in geometry were started in 1844, his mathematical aptitude was soon noticed. He devoted much of his energy to studying mathematics and, in 1845, he received the medal for mathematics as a result of his prowess in the examination room.

Maxwell's father frequently visited the Academy and, on these occasions, he took his son to the meetings of the Edinburgh Society of Arts and the Royal Society. At one of these meetings, the forms of Etruscan urns were discussed and the problem of how to draw a perfect oval arose.

The boy became interested in the problem and evolved an ingenious solution of it. He also devised a simple mechanical means for producing oval curves by wrapping a thread around pins. The work was presented to the Royal Society by Professor Forbes in 1846 and it was published in the *Proceedings of the Royal Society of Edinburgh*.[2] In the spring of 1847, James was taken to the laboratory of Nicol, the inventor of the polarizing prism, and, from that time, he became very keen on experimenting with polarized light. To produce the polarization, he used the reflection from a glass mirror. Later on, he was presented with prisms by Nicol and with the gift he constructed an apparatus for photoelastic stress analysis.

In the fall of 1847, Maxwell entered the University of Edinburgh, where he could work without any pressure and follow his inclination in selecting subjects for study. He attended the lectures on natural philosophy

---

[1] For a complete biography of J. C. Maxwell, see the book "The Life of James Clerk Maxwell" by L. Campbell and W. Garnett, London, 1882. See also the preface to "The Scientific Papers," edited by W. D. Niven, Cambridge University Press, 1890.

[2] *Proc. Roy. Soc. Edinburgh*, vol. 2, pp. 89–93.

The Mathematical Theory of Elasticity between 1833 and 1867   269

given by Professor Forbes, continued his experiments with polarized light, and spent a considerable amount of time in studying books on mechanics and physics. In a letter of March, 1850, he wrote to a friend: "I have been reading Young's *Lectures*, Willis's *Principles of Mechanism*, Moseley's *Engineering and Mechanics*, Dixon on *Heat*, and Moigno's *Répertoire d'Optique* . . . I have a notion for the torsion of wires and rods, not to be made till the vacation; of the experiments on the action of compression on glass, jelly, etc., numerically done up; on the relations of optical and mechanical constants, their desirableness, etc., and suspension-bridges, and catenaries, and elastic curves." We see that, by that time, he had already become interested in theory of elasticity.

In 1850 Maxwell's paper "On the Equilibrium of Elastic Solids" was read to the Royal Society of Edinburgh. This paper begins with a criticism of the rariconstant theory and reference is made to a paper by Stokes.[1] The writer derives the equations of equilibrium of isotropic bodies having two elastic constants. He then uses his equations in discussing several particular problems. Most of these had already been solved previously by other authors, but no one had paid such attention to experimental verification of the theoretical results. He begins with the case "of a hollow cylinder, of which the outer surface is fixed, while the inner surface is made to turn through a small angle $\delta\theta$, by a couple whose moment is M." Using the equations of equilibrium in polar coordinates, he easily shows that shearing stresses are produced and that their magnitude is inversely proportional to the square of the distance from the axis of the cylinder.

In order to verify this result, he applies the photoelastic method and uses, for that purpose, a jelly of isinglass poured between two concentric cylindrical forms while hot. When cold, it formed a convenient solid for experiments. He says: "If the solid be viewed by polarized light, transmitted parallel to the axis, the difference of retardation of the oppositely polarized rays at any point in the solid will be inversely proportional to the square of the distance from the axis of the cylinder. . . . The general appearance is therefore a system of coloured rings arranged oppositely to the rings in uni-axial crystals, the tints ascending in the scale as they approach the center, and the distance between the rings decreasing towards the centre. The whole system is crossed by two dark bands inclined 45° to the plane of primitive polarization." Maxwell observed this colored picture and copied it in water colors.[2] His biographers say: "Maxwell had a great faculty for designing and would frequently amuse

---

[1] *Trans. Cambridge Phil. Soc.*, vol. 8, part 3; paper read, April, 1845.

[2] The pictures were attached to his paper. They are reproduced in Maxwell's biography by Campbell and Garnet, but they are omitted from the "Scientific Papers."

himself by making curious patterns, for wool-work. His designs are remarkable for the harmony of the coloring."

As a second problem, Maxwell considers the torsion of circular bars and uses his analysis in an experimental determination of the modulus of shear. As his third and fourth examples, the author takes up Lamé's problems of the stresses in hollow cylinders and hollow spheres induced by uniform pressure. Maxwell uses his solutions to examine some experimental results dealing with a determination of the compressibility of fluids. He remarks: "Some who reject the mathematical theories, as unsatisfactory, have conjectured that if the sides of the vessel be sufficiently thin, the pressure on both sides being equal, the compressibility of the vessel will not affect the result. The following calculations show that the apparent compressibility of the liquid depends on the compressibility of the vessel, and is independent of the thickness when the pressures are equal."

The fifth problem is that of pure bending of rectangular beams and the writer makes an interesting addition to the elementary theory by considering pressures between the longitudinal fibers which ensue as a result of the curvature of the beam. Maxwell discusses (as his sixth case) the bending of uniformly loaded circular plates. This investigation was made to examine the possibility of making a concave silvered glass mirror by bending. Maxwell calculates the radius of curvature at the center of the plate and observes that a telescope made according to this principle would serve as an aneroid barometer, the focal distance varying inversely as the pressure of the atmosphere.

Case 7 concerns the torque which is due to the stretching of longitudinal fibers of a twisted circular cylinder, and it is shown that the formula for torque contains a term proportional to the cube of the angle of twist, as had been indicated before by Thomas Young.

In case 8, Maxwell considers the important problem of stresses produced by centrifugal force in a rotating thin circular disk, and gives the correct formula for tangential stresses. This is the first satisfactory treatment of this problem.[1]

Cases 9 to 11 concern thermal stresses in hollow cylinders and hollow spheres. Duhamel's solutions (see page 242) of these problems appear to be unknown to Maxwell.

In case 12, Maxwell calculates the deflection of a simply supported beam and gives a formula in which the effect of shearing forces on the magnitude of the deflection is taken into account. This is done under the assumption that these stresses are uniformly distributed over cross sections of the beam.

[1] There is a misprint in Eq. (59), p. 61, of the "Scientific Papers."

The Mathematical Theory of Elasticity between 1833 and 1867   271

The stresses which are produced when two cylindrical surfaces, whose axes are perpendicular to the plane of an infinite elastic plate, be equally twisted in the same direction are treated in case 13. The resultant stress at any point is then obtained by using superposition and adding the stresses produced by each of the two rotating cylinders together. Reference is had to case 1 for this. The isochromatic curves that he obtained by computation are shown in Fig. 163, and Maxwell verified the theoretical results experimentally. He says: "I have produced these curves in the jelly of isinglass described in Case 1. They are best seen

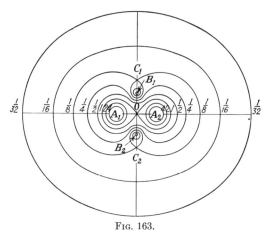

Fig. 163.

by using circular polarized light, as the curves are then seen without interruption, and their resemblance to the calculated curves is more apparent."

In case 14, Maxwell deals with stresses in a triangular plate of unannealed glass. The problem is more complicated, and we have no theoretical solution upon which to base calculations. Maxwell determines the stresses experimentally by applying the photoelastic method and he obtains the isochromatic curves by using circularly polarized light. Then by transmitting plane-polarized light, he determines the directions of principal stresses and draws the system of isoclinic lines shown in Fig. 164. "From these curves others may be formed which shall indicate, by their own direction, the direction of the principal axes at any point. These curves of direction of compression and dilatation are represented in Fig. 165; the curves whose direction corresponds to that of compression, are concave toward the centre of the triangle, and intersect at right angles the curves of dilatation." Having this system of *trajectories* of stresses, Maxwell now shows how the principal stresses can be calculated. Considering an elemental curvilinear rectangle bounded by two adjacent lines of compression and two adjacent lines of tension, he uses the equation of equilibrium

$$q - p = r\frac{dp}{ds} \qquad (a)$$

where q is the principal compressive stress, p is the principal tensile stress, r is the radius of curvature of the trajectory of compression, and ds is the distance between the two consecutive trajectories of compression. Thus Eq. (a) is the same as that which was used by him in discussing Lamé's

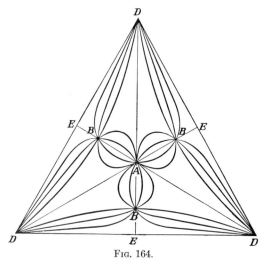

Fig. 164.

problem, where r denoted the radial distance of a point and ds was equal to dr. Since $q - p$ is known from the photoelastic test and the quantities r and ds can be taken from the drawing of the trajectories, he is able to find the values of p using step-by-step integration of Eq. (a) and starting from the boundary where the values of p are known.

We see that Maxwell completely developed the technique of photoelastic stress analysis which is at present widely used in studying two-dimensional problems. He also noticed a property exhibited by some transparent materials, which now is used in three-dimensional photoelasticity. For, in his experiments with torsion (case 1), he noticed: "By continuing the force of torsion while the jelly is allowed to dry, a hard plate of isinglass is obtained, which still acts in the same way on polarized light, even when the force of torsion is removed. . . . Two other uncrystallized substances have the power of retaining the polarizing structure developed by compression. The first is a mixture of wax and resin pressed into a thin plate. . . . The other substance, which possesses similar properties, is gutta percha. This substance in its ordinary state, when cold, is not transparent even in thin films; but if a thin film be drawn out gradually, it may be extended to more than double its length. It then possesses a powerful double refraction, which it retains so strongly that it has been used for polarizing light."

Maxwell did all this important work in theory of elasticity including that in the photoelastic method of stress analysis before he was nineteen

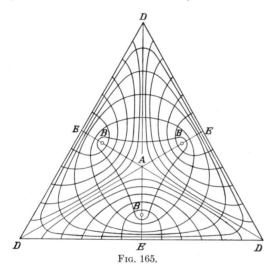

Fig. 165.

years of age. In the autumn of 1850, he departed for Cambridge, where he continued his study.

Concerning Maxwell's life in Cambridge, we have the following statement of Prof. P. G. Tait, who was with him in the same class at Edinburgh University and who left that university to enter Cambridge in 1848: "He [Maxwell] brought to Cambridge, in the autumn of 1850, a mass of knowledge which was really immense for so young a man, but in a state of disorder appalling to his methodical private tutor. Though that tutor was William Hopkins, the pupil to a great extent took his own way, and it may safely be said that no high wrangler of recent years ever entered the Senate-house more imperfectly trained to produce 'paying' work than did Clerk Maxwell. But by sheer strength of intellect, though with the very minimum of knowledge how to use it to advantage under the conditions of the Examination, he obtained the position of the Second Wrangler, and was bracketed equal with the Senior Wrangler (E. J. Routh) in the higher ordeal of the Smith's Prizes."

After graduating, Maxwell remained at Cambridge and in 1855 he was elected a Fellow of Trinity College and started lecturing in the subjects of hydrostatics and optics. His scientific work at that time dealt with the measurement of mixtures of colors and the causes of colorblindness. He was also interested in electricity and published the memoir on Faraday's lines of forces. In 1856, he was elected to a professorship at Marischal College, Aberdeen, where he taught natural philosophy during the period 1856–1860. At the same time he continued the work on color sensation and made an investigation on the stability of Saturn's rings. His work on the kinetic theory of gases was also started at Aberdeen, and his first paper in this subject was read to the British Association in 1859.

In 1860, Maxwell was elected to a chair at King's College, London, and he lectured there on natural philosophy and astronomy during the period 1860–1865. It was during his stay at the college on the bank of the Thames that his important memoir on the kinetic theory of gases and his famous memoir on electricity were published. His great advances in the theory of structures, which we have already discussed (see page 202) also appeared at that time.

The teaching work at King's College occupied much of Maxwell's time, and, in order to improve conditions for his own scientific work, he decided to resign his chair and to retire to his country seat in Scotland. There he was able to throw all his energy into the preparation of the book on heat and into writing his celebrated treatise on electricity and magnetism.

In connection with the publication of G. B. Airy's paper on the strains in the interior of beams,[1] Maxwell returned to his idea of reciprocal figures and developed a general theory of stress diagrams for three-dimensional stress systems.[2] He shows that the general solution of the equations of elasticity may be expressed in terms of three functions of stress. Applying his theory to two-dimensional problems, he concludes: "The solutions of the cases treated by Mr. Airy, as given in his paper, do not exactly fulfill the conditions deduced from the theory of elasticity. In fact, the consideration of elastic strain is not explicitly introduced into the investigation." Using his theory for a rectangular simply supported beam carrying a load uniformly distributed over the upper edge, he finds that the correct solution differs from that of Airy only in the magnitude of longitudinal stresses and that the maximum value of the error is $0.314q$, where q denotes the intensity of the load.

An interesting comparison of Maxwell's scientific outlook and those of Stokes and Lord Kelvin is given in Joseph Larmor's book, "Sir George Gabriel Stokes" (p. 217): "With Maxwell the scientific imagination was

[1] *Phil. Trans.*, 1863.
[2] *Phil. Trans.*, vol. 26.

everything: Stokes carried caution to excess. Maxwell reveled in the construction and dissection of mental and material models and images of the activities of the molecules which form the basis of matter: Stokes' published investigations are mainly of the precise and formal kind, guided by the properties and symmetries of matter in bulk, in which the notion of a molecule need hardly enter. Kelvin stands perhaps halfway between them. In the main features of his activity he could rightly be described, as he has himself insisted, as a pupil of Stokes; but along with this practical quality there worked a constructive imagination which doubtless, next to Faraday, formed the main inspiration of Maxwell, though it stopped short of the bold tentative flights into the unknown which in Maxwell's work turned out to be so successful."

In 1871, Maxwell was induced to quit his retreat in Scotland and to accept a professorship at Cambridge. By that time, the university fully recognized the importance of experimental work in such branches of physics as heat, electricity, and magnetism. The chancellor, the Duke of Devonshire, generously furnished the funds for building the physics laboratory and Maxwell was elected to be its first director. During this last period of his life (1871–1879), he had to spend much of his energy in administrative work in the new laboratory. He also helped greatly in guiding the general policy of the university. He insisted that the studies in the university should maintain good contact with outside scientific bodies and that the various branches of physics should have a place in the tripos examination. He took an active part in preparing the scheme of examination introduced in 1873.

CHAPTER IX

Strength of Materials in the Period 1867–1900

59. Mechanical Testing Laboratories

As we have already noted, the introduction of iron and steel into structural and mechanical engineering made experimental study of the mechanical properties of those materials essential. Earlier experimenters had usually confined their attention to determining ultimate strengths, but it was now realized that the properties of iron and steel depend upon the technological processes involved in manufacture and that, in any case, a knowledge of ultimate strength only is insufficient when selecting a suitable material for a particular purpose. A more thorough investigation of the mechanical properties of materials became of utmost practical importance. Such tests require special equipment, and we shall see that in the period under consideration a rapid growth of material-testing laboratories took place in many countries.

Wöhler in Germany advocated the organization of government laboratories. In England, laboratories were set up by private enterprise. One of the first successful examples was that of David Kirkaldy, which was opened in 1865 in Southwark, London. In the preface of his famous book,[1] Kirkaldy states: " Messrs. Robert Napier and Sons having received orders for some high-pressure boilers and marine machinery, where lightness combined with strength was of the utmost importance, it was proposed to use *homogeneous-metal* for the one and *puddle-steel* for the other, instead of wrought-iron as ordinarily employed. As these materials, however, had then been only recently introduced, it was considered prudent, before employing them, to ascertain their relative merits; and with this view the apparatus, to be described afterwards, was erected. At the time it was only intended to test a few specimens of each, but the investigation proved both so interesting in itself and so likely to conduce to important practical results, that I was induced, under the sanction of Messrs. Napier, to extend the experiments, as leisure and opportunity offered, very considerably beyond what had been originally contemplated.

[1] D. Kirkaldy, "Results of an Experimental Inquiry into the Tensile Strength and Other Properties of Various Kinds of Wrought-iron and Steel," Glasgow, 1862.

The testing commenced on 13th April, 1858, and terminated 18th September, 1861. . . . " Figure 166 shows the machine constructed and used by Kirkaldy in his experimental work, while in Fig. 167 we see the specimens and their attachments. With this simple equipment, Kirkaldy succeeded in obtaining very important results, and his book, at the time of its appearance, contained the most complete description of mechanical properties of iron and steel then available.

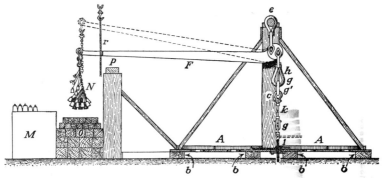

FIG. 166. Kirkaldy's testing machine.

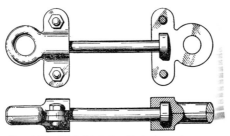

FIG. 167. Kirkaldy's tensile-test specimens.

Working with comparatively short specimens[1] and using crude measuring instruments, Kirkaldy was unable to measure elastic elongations or to calculate Young's modulus. His experiments gave the ultimate load and total elongation of specimens. In some cases, the length of the specimen was subdivided by marks $\frac{1}{2}$ in. apart and the distances between the marks were measured after the test. In this way it was shown that along most of its length the specimen stretched uniformly and remained cylindrical in form and that only those portions close to the fracture showed greater strain and a large local contraction of the cross-sectional

[1] In order to measure elastic elongations satisfactorily and to obtain Young's modulus with sufficient accuracy, Hodgkinson used rods 50 ft long.

area. From this, Kirkaldy concluded that the ratio of total elongation to initial length depended on the proportions of the specimens used. He was the first to recommend the measurement of lateral contraction and considered that the relative reduction in the cross-sectional area at the fracture was an important characteristic of materials. He considers that a mere knowledge of the magnitude of the ultimate strength is not sufficient for making decisions about the quality of a material, for "high tensile strength is not any proof whatever that the quality is good since it may be so extremely hard and brittle as not to bend and twist. When, however, we compare the breaking strain jointly with the contraction of the area at fracture, the case is vastly different, as we then have an opportunity of deciding correctly whether, on the one hand, a high breaking strain (stress) was owing to the iron being of superior quality, dense, fine, and moderately soft, or simply because it was very hard and unyielding; on the other hand, a low breaking strain (stress) might be due to looseness in the texture, or to extreme softness although very close and fine in quality." In his tables this writer gives two values for the ultimate strength; these are computed by using the initial and the fractured area of the cross section.

Kirkaldy was probably the first to recognize the effect of the shape of specimens on the ultimate strength. Comparing some of his results with those obtained for short specimens of variable cross section but of similar material, he remarks: "The amount of breaking strain borne by a specimen is thus proved to be materially affected by its shape—an important influence hitherto unascertained, and even unsuspected by any one, to the best of the writer's knowledge. This is one instance, out of many, that could be produced to show the absolute necessity (and the remark is applicable to many other subjects besides that of the present treatise) of correctly knowing the exact conditions under which any tests are made before we can equitably compare results obtained from different quarters. From the above facts it must surely be evident that we cannot judge of the merits of two pieces of iron or steel by comparing their breaking strains (stress), when one of them was of uniform diameter for some inches of the length, and the other for only a very small portion."

Kirkaldy was much interested in the fatigue strength of steel and (in his book) described the discussions of that subject at meetings of the Institution of Civil Engineers under the chairmanship of Robert Stephenson in 1850 (see page 163). He agrees with the chairman's view that vibration cannot produce recrystallization of steel. He makes his own tests and proves that the hypothesis of recrystallization is unnecessary for explaining the brittle character of fatigue fractures. He shows that a brittle fracture can be produced in soft materials by changing the shape of specimens so that lateral contraction is restrained at fracture, as in a specimen with a deep rectangular groove. He also shows that the appearance of fracture may be changed "by applying the strain so suddenly as to render the specimen more liable to snap, from having less time to stretch."

In studying the effect of cold work on mechanical properties of iron, Kirkaldy disagrees with the accepted opinion that the increase in the tensile strength is due to "the effect of consolidation." He shows by direct test that the "density of iron is decreased by the process of vise-drawing, and by the similar process of cold-rolling, instead of increased, as previously imagined." Kirkaldy finds that useful information regarding the properties of iron and steel can be found by examining the structure of these metals. He shows that the "texture of various kinds of wrought-iron is beautifully developed by immersion in dilute hydrochloric acid, which, acting on the surrounding impurities, exposes the metallic portion alone for examination." In this way he finds that "in the fibrous fractures the threads are drawn out, and are viewed externally, whilst in the crystalline fractures the threads are snapped across in clusters, and are viewed internally or sectionally. In the latter cases the fracture of the specimen is always at right angles to the length; in the former it is more or less irregular."

From this outline of Kirkaldy's work, we see that this outstanding experimenter raised and explained many important questions in the field of mechanical testing of materials and his book still is of interest to engineers occupied with this subject. Combining great practical experience with his flair for observing anything unusual, Kirkaldy was able to solve, in his private laboratory, many of the practical problems that arose in the rapidly developing structural and mechanical industries.

On the Continent, material-testing laboratories developed as government-owned concerns. They reached a particularly high level in Germany,[1] where they were run by the engineering schools and were used not only for testing for industry but also for the research work of the teaching staff and for educational purposes. Such an arrangement was beneficial both to engineering education and to industry.

The first laboratory of this kind was founded in 1871 at the Polytechnical Institute of Munich. Its first director, Johann Bauschinger (1833–1893), was the professor of mechanics at the Institute. He succeeded in getting a machine installed that had a 100-ton capacity. This device was designed by Ludwig Werder (1808–1885)[2] in 1852 for *Maschinen-*

[1] For the history of testing materials in Germany, see R. Baumann, *Beitr. Geschichte Technik u. Industrie*, vol. 4, p. 147, 1912. See also A. Leon, "Die Entwicklung und die Bestrebungen der Materialprüfung," Vienna, 1912.

[2] For a biography of Werder, see the book "Grosse Ingenieure" by Conrad Matschoss, 3d ed., 1942.

baugesellschaft Nürnberg. It proved very suitable for experiments requiring better accuracy than had been achieved before, and later most of the important European laboratories installed similar machines.

For measuring small elastic deformations, Bauschinger invented a mirror extensometer[1] which enabled him to measure, accurately, unit elongations of the order of 1×10^{-6}. With this sensitive instrument, he investigated the mechanical properties of materials with much greater accuracy than had been achieved by previous workers. Making tensile tests of iron and mild steel, he noticed that, up to a certain limit, these materials follow Hooke's law very accurately and that so long as elongations remain proportional to stress, they are elastic and no permanent set can be detected. From these tests, Bauschinger concluded that it could be assumed that the *elastic limit* of iron or steel coincided with its *proportional limit*. If he continued to increase the load on a specimen beyond the elastic limit, the elongations began to increase at a higher rate than the load and, at a certain value of the load, a sharp increase in deformation occurred which continued to grow with the time under this load when it was held constant. This limiting value of the load defines the *yield point* of the material. The yield point of mild steel is raised by loading the specimen beyond the initial yield point and the maximum value of the load gives us the new value of the yield point if the second loading is started immediately. If the second loading is made after an interval of several days, the yield point becomes somewhat higher than the maximum load in the first loading. Bauschinger also noticed that a specimen which is stretched beyond the yield point does not show perfect elasticity and has a very low elastic limit under a second loading if it follows the initial stretching at once. But, if there is an interval of several days between the loadings, the material recovers its elastic property and exhibits a proportional limit that is somewhat higher than its initial value.

Bauschinger examined the properties of mild steel when it is subjected to cycles of stress.[2] He finds that if a specimen is stretched beyond its

FIG. 168. Johann Bauschinger.

[1] For a description of this extensometer, see Bauschinger's paper in *Civiling.*, vol. 25, p. 86, 1879.

[2] See *Mitt. mech. tech. Lab. Munich*, 1886; see also *Dinglers Polytech. J.*, vol. 266, 1886.

initial elastic limit in tension, its elastic limit in compression is lowered. Its value can be raised by subjecting the specimen to compression, but if this compression proceeds beyond a certain limit, the elastic limit in tension will be lowered. By submitting the test piece to several cycles of loading, the two limits can be fixed such that the specimen behaves perfectly elastically within them. Bauschinger calls these two limits the *natural* elastic limits in tension and compression. He assumes that the *initial* elastic limits depend upon the technological processes which the material has undergone and that the *natural* limits represent "true" physical characteristics. He asserts that, so long as a specimen remains within the *natural* limits of elasticity the material can withstand an unlimited number of cycles and there is no danger of fatigue failure.

Bauschinger published his results in *Mitteilungen*, which appeared yearly, and they were read by many engineers who were engaged upon the same type of work. After this pioneering work in Munich, experimental investigations of structural materials were started in many other places. In 1871, the laboratory for testing materials was organized at the Berlin Polytechnicum[1] by A. Martens, who later published a book[2] describing the various methods of testing and the machines that were employed. The year 1873 saw the opening of a similar laboratory at the Vienna Polytechnical Institute under the direction of Prof. K. Jenny.[3] In 1879, the material-testing laboratory at the Zürich Polytechnical Institute was opened, and under the directorship of Prof. Ludwig von Tetmajer (1850–1905), it gained the reputation of being one of the most important laboratories in Europe.

C. Bach accepted a professorship at Stuttgart Polytechnical Institute in 1879 and formed a material-testing laboratory there. Bach was chiefly interested in the application of strength of materials to machine design, and he used the laboratory not only for research work in the properties of materials but also for contriving experimental solutions to various problems dealing with stress analysis in structures and in machine parts. The principal results of these investigations were collected in Bach's textbook[4] "Elasticität und Festigkeit" which became popular with mechanical engineers and consequently achieved importance in machine design in Germany.

In Sweden, the interest in mechanical testing of materials was centered about the iron and steel industry. In 1826, P. Lagerhjelm had done much

[1] For the history of development of this German laboratory, see the book "Das Königliche Materialprüfungsamt der technischen Hochschule," Berlin, 1904.

[2] A. Martens, "Handbuch der Materialienkunde für den Maschinenbau," I Teil, Berlin, 1898; II Teil with E. Heun, Berlin, 1912.

[3] For a description of that laboratory and its work, see K. Jenny, "Festigkeits— Versuche und die dabei verwendeten Maschinen und Apparate an der K. K. technischen Hochschule in Wien," Vienna, 1878.

[4] "Elasticität und Festigkeit," 1st ed., 1889; 6th ed., 1911.

work in this subject and, as we have seen (see page 102), he built a special machine for the purpose. This machine was used extensively by Knut Styffe in his work with the iron and steel used in Swedish railroad engineering.[1] Styffe was the director of the Technological Institute in Stockholm. In 1875, the Swedish steel industry set up a material-testing laboratory in Liljeholmen and installed a Werder machine of 100 tons capacity together with a hydraulic press by Amsler-Laffon. In 1896, the Werder machine was moved to the new laboratory at the Polytechnical Institute in Stockholm.

Mechanical testing was started in Russia by Lamé, who tested the Russian iron used in the construction of suspension bridges in St. Petersburg. His testing machine was placed at the Institute of Engineers of Ways of Communication where he was a professor of mechanics. Further work was done under P. I. Sobko[2] (1819–1870) who was professor of bridge engineering at the same school. The laboratory expanded greatly and became the most important laboratory in Russia under the directorship of the well-known Russian bridge engineer N. A. Belelubsky (1845–1922), who strongly advocated the establishment of international specifications in testing materials and who took an active part in the organization of the International Society for Testing Materials.

To render the experimental results of various laboratories comparable, it was necessary to have some uniformity in the technique of mechanical testing. Toward that end, Professor Bauschinger arranged a conference in Munich in 1884. It was a great success. Seventy-nine research workers from various countries attended the meeting and many took part in the discussions. A committee was elected to draw up specifications for various kinds of testing, and most of its preliminary work was approved by the second conference which took place in Dresden in 1886.

During the International Exposition in Paris in 1889, an International Congress of Applied Mechanics convened, and one section of that congress concerned itself with mechanical testing of materials. The works of Kirkaldy, Bauschinger, Barba, and others were discussed[3] and the neces-

[1] The results of the investigation were published in Jern-Kontorets Ann., 1866. A brief discussion of this important work is included in the book "Festigkeit-Proben Schwedischer Materialien," Stockholm, 1897, prepared for the International Congress for Testing Materials in Stockholm.

[2] P. I. Sobko graduated in 1840 from the Institute of Engineers of Ways of Communication and became an assistant to M. V. Ostrogradsky. He taught mechanics, edited Ostrogradsky's lectures on calculus, and published several important papers on strength of materials and on bridges. In 1857, he made a trip to the United States to gain knowledge of American bridges.

[3] The report of this section occupies the greater part of the third volume of the publication *Congrès International de Mécanique Appliquée*, Paris, 1891. It gives a good picture of the condition of mechanical testing of materials at that time.

sity of using geometrically similar test specimens was well recognized. At the concluding meeting, it was unanimously decided to approach the various governments regarding the establishment of an international society for testing materials. The first International Congress took place in Zürich with Prof. L. Tetmajer as the president. Most countries joined this organization, and the later work of the International Congresses did much to foster new methods of testing mechanical properties of materials.

60. The Work of O. Mohr (1835–1918)

Otto Mohr was born at Wesselburen in Holstein on the coast of the North Sea. When he was sixteen years old, he entered the Hannover Polytechnical Institute, and after graduating he worked as a structural engineer on the construction of railroads in Hannover and Oldenburg. During this period, Mohr designed some of the first steel trusses in Germany. He also did some theoretical work and published several important papers in the *Zeitschrift des Architekten—und Ingenieur—Vereins* of Hannover.

Mohr's paper[1] on the use of the funicular curve in finding elastic deflections of beams, his derivation of the three moments equation for a continuous beam the supports of

Fig. 169. Otto Mohr.

which are not on the same level, and also the first applications of influence lines were published.

When thirty-two years old, he was already a well-known engineer and was invited by the Stuttgart Polytechnikum to become the professor of engineering mechanics of that institute. He accepted, and during the five years 1868–1873 he offered lectures at that school in the various branches of engineering mechanics. He was especially interested at that time in devising graphic methods for solving problems in the theory of structures and broke much new ground in the science of graphical statics. Mohr was a very good professor, and his lectures aroused great interest in his students, some of whom were themselves outstanding. C. Bach and August Föppl attended his lectures at that time. In his autobiography,[2]

[1] *Z. Architek. u. Ing. Ver. Hannover*, 1868.
[2] August Föppl, "Lebenserinnerungen," Munich, 1925.

Föppl states that all the students agreed that Mohr was their finest teacher (*Lehrer von Gottes Gnaden*). The delivery of his lectures was not of the best, for sometimes words came to him very slowly and his sketches on the blackboard were poor. But the lectures were always clear and logically constructed, and he always tried to bring something fresh and interesting to the students' attention. The reason for his students' interest in his lectures stemmed from the fact that he not only knew the subject thoroughly but had himself done much in the creation of the science which he presented.

In 1873, Mohr moved to Dresden and continued teaching at the Dresden Polytechnikum up to the age of sixty-five, when he retired. He spent the last years of his life living quietly in the neighborhood of Dresden and continuing his scientific work.

Mohr's best work lies in the theory of structures, which will be discussed in the next chapter, but his contribution to strength of materials is also very important. He was the first to notice[1] that the differential equation of the funicular curve

$$\frac{d^2y}{dx^2} = -\frac{q}{H} \qquad (a)$$

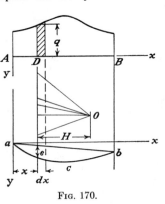

Fig. 170.

constructed for a load of variable intensity q distributed along a line AB (Fig. 170) has the same form as the differential equation

$$\frac{d^2y}{dx^2} = -\frac{M}{EI} \qquad (b)$$

of the elastic line. From this, it follows that the deflection curve can be constructed as the funicular curve for a fictitious load of intensity M by taking the pole distance equal to the flexural rigidity EI of the beam. Observing also that the ordinate e (between the funicular curve acb and the closing line ab), when multiplied by the pole distance H, gives a measure of the bending moment at the cross section D of the beam AB, Mohr concludes that the beam deflections can be calculated from the bending moments induced in a beam by a fictitious load of intensity M/EI, which greatly simplifies the problem of finding deflections. It is no longer necessary to integrate Eq. (b), and the deflections are obtained from simple

[1] See the paper in the *Z. Architek. u. Ing. Ver. Hannover*, 1868, p. 19.

Strength of Materials in the Period 1867–1900

statics. This simplification is especially great, as Mohr shows, in beams of variable cross section.

Applying his method to a simply supported beam (Fig. 171a) and considering two cross sections c and d, Mohr proves that if a load P is first applied at c and the deflection is measured at d and second the same load is applied at d and the deflection is measured at c, then the measured deflections for these two loading conditions are equal. Having this result, Mohr considers the problem in which it is required to calculate deflections at any cross section of the beam (say d) brought about by a force which may act at various positions along the beam. For this purpose he constructs the elastic line $acdb$ (Fig. 171b), corresponding to a unit load applied at d. Let y denote the deflection at any point c. Then, on the strength of the above result, he concludes that a unit load applied at c produces at d the deflection y, and any load P applied at c produces a deflection yP at d. Thus having the deflection curve $acdb$, he at once obtains the deflection at d for any position of the load. Mohr calls the curve $acdb$ the *influence line* for the deflection at d. Using the notion of superposition, Mohr shows that, if several loads $P_1, P_2, P_3 \ldots$ act on the beam and $y_1, y_2, y_3 \ldots$ are the corresponding ordinates of the influence line, then the deflection at d produced by these loads will be equal to $\Sigma P_i y_i$. This is the first use of influence lines in engineering.[1] In the same paper (of 1868) Mohr also discusses continuous beams and offers a graphical solution of the three moments equations.

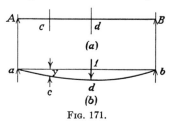

Fig. 171.

Another important piece of Mohr's work in strength of materials is his graphical representation of stress at a point.[2] In the mathematical theory of elasticity, the ellipsoid of stress is used for that purpose (see page 109), and for two-dimensional problems we have to consider the ellipse of stress. In his "Graphische Statik" (p. 226, 1866), Culmann had already shown that the stress in a two-dimensional system can be represented by a circle, (see page 195) which greatly simplifies the problem. Mohr makes a more complete study of this subject. Considering a two-dimensional case with the principal stresses σ_1 and σ_2 (Fig. 172), he shows that the normal and shearing components σ and τ of the stress acting on a plane mm defined by the angle ϕ are given by the coordinates of the point R defined by the angle 2ϕ on the circle constructed as shown

---

[1] Mohr in his "Abhandlungen," 2d ed., p. 374, 1914, notes that at about the same time and independently of him, Winkler used also the notion of influence lines. See Winkler's paper in *Mitt. Archilek. u. Ing. Ver. Böhmen*, 1868, p. 6.

[2] See *Civiling.*, 1882, p. 113.

in Fig. 173. The diameter of the circle AB is equal to $\sigma_1 - \sigma_2$ and its ends A and B define the stresses on the principal planes perpendicular to σ_1 and σ_2. Proceeding in the same way in the three-dimensional case with principal stresses σ_1, σ_2, σ_3, Mohr finds that the stress components acting on the planes passing through the principal axes may be represented by the coordinates of points on the three circles shown in Fig. 174.

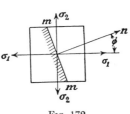

Fig. 172.

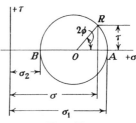

Fig. 173.

The largest of the three circles has a diameter equal to the difference $\sigma_1 - \sigma_3$ between the largest and the smallest of the principal stresses and defines the stresses on the planes passing through the axis of the intermediate principal stress σ_2. Mohr also shows how the stresses acting on planes intersecting all three principal axes can be represented by the coordinates of certain points of the diagram (Fig. 174) and proves that all those points lie in the shaded areas. If we take some value of normal stress, such as that given by OD, there exists an infinite number of planes having the same normal component of stress. The points defining the magnitudes of shearing stresses on those planes lie along the portions EF and E_1F_1 of the perpendicular through point D, and it is seen that of all planes having the same value of the normal stress component, given by the length OD, the largest value of shearing stress will correspond to planes passing through the intermediate principal axis, the stresses for which are defined by the coordinates of points F and F_1.

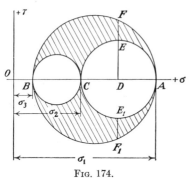

Fig. 174.

Mohr now uses this method of stress representation in developing his theory of strength.[1] At that time, most engineers working in stress analysis followed Saint-Venant and used the *maximum strain theory* as their

[1] See Z. Ver. deut. Ing., 1900, p. 1524.

criterion of failure. The dimensions of structural members were fixed in such a manner that the maximum strain at the weakest point under the most unfavorable condition of loading did not exceed the permissible unit elongation in simple tension. But for many years there had been scientists who considered that the action of shearing stress was important and must be considered. Coulomb had already assumed that failures are precipitated by shearing stresses. Vicat (see page 83) criticized the elementary beam theory in which shearing stresses were not considered, and Lüders drew attention to the lines of yielding which appear on the surface of polished specimens if stretched beyond elastic limit.[1] In Mohr's time there already existed some experimental results obtained by Bauschinger, who, by using his sensitive instruments, determined the elastic limits of steel in tension, compression, and shear. These results did not agree with the maximum strain theory. Mohr, in his paper of 1882, discusses Bauschinger's experiments and criticizes the results of the compression tests made with cubic specimens on the grounds that friction forces acting on the cube surfaces in contact with the machine plates must have a great effect upon the stress distribution so that the results are not those of a simple compression test.

Mohr uses his representation of stresses by circles (Fig. 174) to devise a strength theory which can be adapted to various stress conditions and be brought into better agreement with experiments. He assumes that of all planes having the same magnitude of normal stress the weakest one, on which failure[2] is most likely to occur, is that with the maximum shearing stress. Under such circumstances, it is necessary to consider only the largest circle in Fig. 174. Mohr calls it the *principal circle* and suggests that such circles should be constructed when experimenting for each stress condition in which failure occurs. Figure 175, for example, shows such principal circles for cast iron tested to fracture in tension, compression, and in pure shear (torsion). If there are a sufficient number of such principal circles, an envelope of those circles can be drawn, and it can be assumed with sufficient accuracy that, for any stress condition for which there is no experimental data, the corresponding limiting principal circle will also touch the envelope. For instance, when considering cast iron, Mohr suggests that the envelope be taken as the two outer tangents to the circles I and II (Fig. 175), which correspond to experimental fractures in tension and compression. The ultimate strength in shear is then found by drawing circle III, which has its center at O and is tangential to the envelope. If σ_t and σ_c are the absolute values of the ultimate strength of

---

[1] See his paper in *Dinglers Polytech. J.*, 1860. See also L. Hartmann, "Distribution des déformations dans les métaux soumis à des efforts," 1896.

[2] Mohr considers failure in a broad sense; it can be yielding of the material or fracture.

the material in tension and compression, we find from the figure that the ultimate strength in shear is

$$\tau_{ult} = \frac{\sigma_t \sigma_c}{\sigma_t + \sigma_c}$$

which agrees satisfactorily with experiments.

Mohr's theory attracted great attention from engineers and physicists. A considerable amount of experimental work was done in connection with it and this will be discussed later (see page 346).

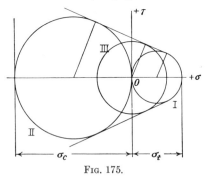

Fig. 175.

61. Strain Energy and Castigliano's Theorem

We have seen that Euler used the expression for the strain energy of bending of bars in his derivation of the differential equation of the elastic curve (see page 32). Green, in discussing the requisite number of elastic constants, assumes that the strain energy is a homogeneous function of the second degree of the strain components (see page 218). Lamé in his book on elasticity (first edition, 1852, see page 118) speaks of Clapeyron's theorem, which states that the work done by external forces on an elastic solid during deformation is equal to the strain energy accumulated in the solid.

Assuming that the displacements occurring during deformation of an elastic body are linear functions of external forces, the work T done by these forces will be

$$T = \tfrac{1}{2} \Sigma P_i r_i \tag{a}$$

where P_i denotes the force at point i and r_i is the component of the displacement of that point in the direction of the force. The strain energy V of the solid is represented by the integral

$$V = \frac{1}{2} \iiint \left\{ \frac{1}{E}(\sigma_x^2 + \sigma_y^2 + \sigma_z^2) - \frac{2\mu}{E}(\sigma_x \sigma_y + \sigma_x \sigma_z + \sigma_y \sigma_z) \right. $$
$$\left. + \frac{1}{G}(\tau_{xy}^2 + \tau_{xz}^2 + \tau_{yz}^2) \right\} dx\, dy\, dz \tag{b}$$

Strength of Materials in the Period 1867–1900

and Clapeyron's theorem states that

$$T = V \qquad (c)$$

In analyzing trusses with redundant members, the Italian military engineer L. F. Ménabréa suggested the use of the expression for the strain energy of the truss.[1] He claims that the forces $X_1, X_2, X_3, \ldots$ acting in the redundant members must be such as to make the strain energy a minimum. In this way he obtains the equations,

$$\frac{\partial V}{\partial X_1} = 0 \qquad \frac{\partial V}{\partial X_2} = 0 \qquad \frac{\partial V}{\partial X_3} = 0 \cdots \qquad (d)$$

which, together with the equations of statics, are sufficient to determine the forces in all bars of the truss. We see that Ménabréa uses the *principle of least work;* but he does not give an acceptable proof of this principle. A logical proof was supplied later by the Italian engineer Alberto Castigliano (1847–1884).

Castigliano was born in Asti, Italy. After several years of work as a teacher, he entered the Turin Polytechnical Institute in 1870 and there, as a student, he did outstanding work in theory of structures. His thesis for the engineer's degree of that school, presented at 1873, contains a statement of his famous theorem with some applications to the theory of structures. In a more extended form, the work was published by the Turin Academy of Sciences in 1875.[2] At first the importance of this paper was not appreciated by engineers, and to make it better known, Castigliano published his work together with a complete proof of his theorem and with numerous applications, in French.[3] The early death of Castigliano terminated the career of this brilliant scientist. But his theorem has become a well-recognized tenet of the theory of structures. Numerous publications of such outstanding engineers as H. Müller-Breslau in Germany and Camillo Guidi in Italy have been based upon it.

In deriving his theory, Castigliano begins with a consideration of trusses with ideal hinges. If external forces are applied at the hinges and the bars have prismatical form, the strain energy of such a system is of the form

$$V = \frac{1}{2} \sum \frac{S_j{}^2 l_j}{E A_j} \qquad (e)$$

[1] See *Compt. rend.*, vol. 46, p. 1056, 1858.
[2] *Atti reale accad. sci. Torino*, 1875.
[3] A. Castigliano, "Théorie de l' équilibre des systèmes élastiques," Turin, 1879. On the fiftieth anniversary of Castigliano's death, the book "Alberto Castigliano, Selecta" was edited by G. Colonnetti. It contains Castigliano's thesis for the engineer's degree and the principal part of his French publication. An English translation of the book was made by E. S. Andrews, London, 1919.

and, from Clapeyron's theorem, it follows that

$$\frac{1}{2}\sum P_i r_i = \frac{1}{2}\sum \frac{S_j{}^2 l_j}{EA_j} \qquad (f)$$

The summation on the left-hand side is extended over all loaded hinges, and that on the right side, over all the bars of the truss. Castigliano assumes that the system is such that the deflections r_i are linear functions of external forces. Substituting these functions into the left-hand side of Eq. (f), he can represent the strain energy as a homogeneous function of second degree of the external forces P_i. Using the same relations between the deflections and the forces, he can also represent forces as linear functions of displacements and in this way obtain the strain energy as a homogeneous function of second degree of the displacements r_i. Castigliano uses both these forms for the strain energy V in his investigation and proves two important theorems.

First he assumes that V is written down as a function of displacements r_i. If by slight changes of the forces P_i, small changes δr_i in the displacements r_i are produced, the total amount of strain energy of the truss will be changed by the small quantity

$$\delta V = \sum \frac{\partial V}{\partial r_i} \delta r_i$$

This quantity must evidently be equal to the work done by the external forces during the small changes in displacements, i.e., to

$$\Sigma P_i\, \delta r_i$$

In this way he obtains the equation

$$\sum \frac{\partial V}{\partial r_i} \delta r_i = \sum P_i\, \delta r_i \qquad (g)$$

which must hold for any arbitrarily selected values of the quantities δr_i. From this, it follows that

$$\frac{\partial V}{\partial r_i} = P_i$$

for all i. This means that the linear functions, which we obtain by differentiating V with respect to any displacement r_i, represent the corresponding forces P_i.

In the formulation of his second theorem, Castigliano assumes that V is written as a function of the external forces P_i.[1] Then, if we slightly

---

[1] Castigliano remarks that P_i are the *independent forces*. This means that reactive forces, which can be calculated by using equations of statics, are given as functions of the forces P_i in calculating the strain energy V.

change the forces, the change in the strain energy will be

$$\sum \frac{\partial V}{\partial P_i} \delta P_i$$

and, instead of Eq. (g), we obtain

$$\sum \frac{\partial V}{\partial P_i} \delta P_i = \sum P_i \, \delta r_i \qquad (h)$$

The right-hand side of this equation, representing the increment of the work $\frac{1}{2} \Sigma P_i r_i$, can be put in another form if we observe that

$$\Sigma P_i \, \delta r_i = \delta[\tfrac{1}{2} \Sigma P_i r_i] = \tfrac{1}{2} \Sigma P_i \, \delta r_i + \tfrac{1}{2} \Sigma r_i \, \delta P_i$$

from which it follows that

$$\Sigma P_i \, \delta r_i = \Sigma r_i \, \delta P_i$$

Substituting into Eq. (h), we obtain

$$\sum \frac{\partial V}{\partial P_i} \delta P_i = \sum r_i \, \delta P_i$$

Since this must hold for any arbitrarily taken values of δP_i, the writer concludes that

$$\frac{\partial V}{\partial P_i} = r_i \qquad (i)$$

Thus if the strain energy V is written as a function of the independent external forces, the derivative of that function with respect to any one of

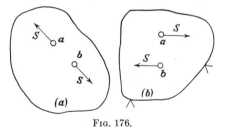

Fig. 176.

the forces gives the corresponding displacement of the point of application of the force in the direction of the force.

Having this general method of calculating deflections, Castigliano considers the two important particular cases shown in Fig. 176. He proves that if equal and opposite forces S act along line ab (Fig. 176a) of an elastic system, the derivative $\partial V / \partial S$ gives the increase of the distance ab due to deformation of the system, say of a truss which contains a and b as hinges. In the case of two forces perpendicular to the line ab in the truss and forming a couple M (Fig. 176b), the derivative $\partial V/\partial M$ gives the angle of rotation of the line ab.

Castigliano applies these results to the analysis of trusses with redundant bars and gives a proof of the principle of least work. In this analysis, he removes all redundant bars and replaces their action upon the rest of the system by forces, as shown in Fig. 177b. The system is now statically determinate and its strain energy V_1 is a function of the external forces P_i and of the unknown forces X_i acting in the redundant bars. On the strength of what was said in regard to Fig. 176a, he concludes that $-\partial V_1/\partial X_i$ represents the increase of the distance between the joints a and b. This increase is clearly equal to the elongation of the bar ab in the redundant system (Fig. 177a); thus he obtains the relationship

$$-\frac{\partial V_1}{\partial X_i} = \frac{X_i l_i}{EA_i} \qquad (k)$$

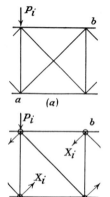

Fig. 177.

Observing that the right-hand side of this equation represents the derivative with respect to X_i of the strain energy $X_i^2 l_i/2EA_i$ of the redundant bar, and denoting the total strain energy of the system with redundant bars included by V, he put Eq. (k) into the form

$$\frac{\partial V}{\partial X_i} = 0 \qquad (l)$$

which expresses the principle of least work.

These results which were obtained initially for hinged-truss analysis were generalized by Castigliano to extend to an elastic solid of any shape. He develops expressions for the strain energy of bars under various kinds of deformation and uses these expressions in numerous applications of his theory to the analysis of statically indeterminate problems of beams and arches. Looking through all these applications, it is easy to see that little has been added to this branch of the theory of structures since Castigliano wrote his famous book.

The principal generalization of Castigliano's theory is due to F. Engesser.[1] While Castigliano always assumed that displacements of structures are linear functions of external forces, there are instances in which this assumption is not valid and his theorem cannot be applied. Engesser introduces the notion of the *complementary energy* for such cases and shows that derivatives of the complementary energy with respect to independent forces always give displacements (the relations between

[1] See his paper in Z. Architek. u. Ing. Ver. Hannover, vol. 35, pp. 733–744, 1889.

forces and displacements can be nonlinear). Let us illustrate this by the simple example of two identical bars hinged together and to the immovable supports AB (Fig. 178a). When unstressed, the bars lie along the straight line ACB. If a central force P is applied, the hinge C moves to C_1 through the distance δ and, from simple analysis, we find

$$\delta = l \sqrt[3]{\frac{P}{AE}} \qquad (m)$$

where l and A denote the length and the cross-sectional area of the bars, respectively. The curve Odb in Fig. 178b represents δ as a function of P and the area $Odbc$ gives the strain energy

$$V = \int_0^\delta P\, d\delta = \frac{lP^{\frac{4}{3}}}{4\sqrt[3]{AE}}$$

This is not a function of the second degree in the force P so that the derivative $\partial V/\partial P$ does not give the deflection δ. The complementary energy V_1, according to Engesser, is represented by the area $Odbg$, and we have

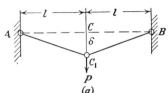

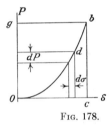

Fig. 178.

$$V_1 = \int_0^P \delta\, dP = \frac{l}{\sqrt[3]{AE}} \int_0^P P^{\frac{1}{3}}\, dP = \frac{3lP^{\frac{4}{3}}}{4\sqrt[3]{AE}}$$

The derivative of this expression with respect to P gives the value of δ represented by Eq. (m). Engesser's work has received little attention since structural analysis has been concerned mainly with systems which satisfy Castigliano's assumption.[1]

62. Elastic Stability Problems

The introduction of steel into structural engineering produced problems of elastic stability which became of vital importance. Very often engineers had to deal with slender compressed bars, thin compressed plates, and with various thin-walled structures in which failure may occur, not due to excessive direct stresses but as a result of a lack of elastic stability. The simplest problems of this kind, dealing with compressed columns, had already been investigated theoretically in sufficient detail. But the limitations under which the theoretical results could be used with confidence

[1] Recently the late Prof. H. M. Westergaard has drawn the attention of engineers to Engesser's contribution and has given examples where it can be used to great advantage. See *Proc. Am. Soc. Civ. Engrs.*, 1941, p. 199.

were not yet completely clear. In experimental work with columns, inadequate attention had been paid to the end conditions, to the accuracy with which the load was applied, and to the elastic properties of material. Hence, the results of tests did not agree with the theory, and engineers preferred to use various empirical formulas in their designs. With the development of mechanical-testing laboratories and with the improvement of measuring instruments, fresh attacks were made on experimental investigations of columns.

The first reliable tests on columns were made by Bauschinger.[1] By using conical attachments for his specimens, he ensured free rotation of the ends and central application of the load. His experiments showed that, under these conditions, the results obtained with slender bars agreed satisfactorily with Euler's formula. Shorter specimens buckled under compressive stresses exceeding the elastic limit and, for them, Euler's theory could not be applied and it was necessary to devise an empirical rule. Bauschinger made only a small number of tests—insufficient to yield a practical formula for column design.

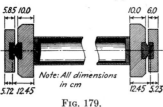

Fig. 179.

The investigation was continued by Prof. L. Tetmajer at the Zürich Polytechnical Institute.[2] Under him, a considerable number of built-up iron and steel columns were tested, and as a result it was recommended that Euler's formula should be used for structural steel in calculating critical stresses when the slenderness ratio is larger than 110. For shorter specimens, a linear formula was offered which later found wide use in Europe. Figure 179 depicts the type of end supports that were employed in Tetmajer's experiments; they allowed freedom of end rotation. Using this equipment, Tetmajer also carried out compression tests with eccentrically applied forces and verified the theoretical formulas for deflections and maximum stress. The magnitude of the compressive force at which the maximum stress reaches the elastic limit was also calculated for this case. Dividing this load by a factor of safety, the maximum safe load on the column was obtained.[3]

The state of knowledge in the field of elastic stability of structures at the end of the nineteenth century can be seen from the important book of

[1] T. Bauschinger, *Mitt. tech. mech. Lab. Munich*, Heft 15, p. 11, 1887.

[2] "Mitteilungen der Materialprüfungs—Anstalt," Heft VIII, Zürich, 1896; 2d enlarged ed., 1901.

[3] It seems that A. Ostenfeld was the first to introduce the assumed inaccuracies in column analysis and to determine the safe load on the columns as a certain proportion of the load which produces dangerous stresses. See *Z. Ver. deut. Ing.*, vol. 42, p. 1462, 1898; vol. 46, p. 1858, 1902.

F. S. Jasinsky.[1] Here we find a correlation of theoretical and experimental investigations on columns together with the author's original work. Jasinsky (1856–1899) was born in Warsaw in a Polish family. In 1872, he entered the Institute of the Engineers of Ways of Communication at St. Petersburg after passing the competitive entrance examination. After his graduation in 1877, Jasinsky started work, first on the Petersburg–Warsaw and later on the Petersburg–Moscow railroads. On these railroads, he designed and built several important structures. At the same time, he was publishing scientific papers. The papers dealing with the theory of columns were collected and published (1893) in book form to serve as a thesis for the scientific degree of Adjunct of the Institute of the Engineers of Ways of Communication. In 1894, Jasinsky became the professor of the theory of structures and the theory of elasticity at the Institute. An untimely death interrupted his brilliant professorial career, but during the five years of his teaching at the Institute, he succeeded in raising the standard of theoretical training given to Russian engineers. His textbooks on the theory of structures and theory of elasticity were widely read in Russia.

Fig. 180. F. S. Jasinsky.

Jasinsky was a great professor. In him, the school had the unusual combination of an outstanding practical engineer who was a great scientist with deep knowledge of his subject and who was also a first-class lecturer. Russian students at that time were allowed complete freedom in how they chose to study and in the distribution of their working time. Only very few attended lectures regularly; but Jasinsky's lecture room was always crowded. This was by no means due to his oratory and his manner of lecturing was always very simple. The students' interest was attracted by the clearness and logic of his presentation and by his frequent references to the new problems which arose from his great activity as a structural engineer. There were always plenty of new problems for students who desired to test their ingenuity.

[1] F. S. Jasinsky, "An Essay on the Development of the Theory of Column Buckling," St. Petersburg, 1893. See also his "Collected Papers," vol. 1, St. Petersburg, 1902, and the French translation of this work in *Ann. ponts et chaussées*, 7th series, vol. 8, p. 256, 1894.

Jasinsky's main work in the theory of columns deals with problems encountered in bridge engineering. In lattice bridges (see footnote, page 184), we have two systems of diagonals, one working in compression and the other in tension. Jasinsky was the first to investigate the stability of the compressed diagonals and to evaluate the strengthening afforded by the diagonals in tension. In his time, several failures of open bridges occurred in Western Europe and in Russia (Fig. 181). The failures were due to insufficient lateral rigidity of the upper compressed chord, and

FIG. 181. Failure of an open bridge.

no satisfactory information existed in regard to the proper selection of flexural rigidities of the chords and of the verticals of such bridges. Jasinsky made a theoretical investigation of this problem. He considered the compressed upper chord of an open bridge as a bar with simply supported ends which is compressed by the horizontal components of the tensile forces in the diagonals while lateral buckling is restrained by the verticals. To simplify the problem, the elastic resistance of the verticals was replaced by a continuously distributed reaction of an equivalent elastic medium, and the action of the diagonals was replaced by continuously distributed axial forces, the intensity of which was taken as proportional to the distance from the middle of the bar. Jasinsky found a rigorous solution for this complicated case of lateral buckling and calculated the critical values of compressive forces so that the upper chord and the verticals of an open bridge could be designed rationally.

Jasinsky also examined the solution of the exact differential equation for the lateral buckling of prismatical bars and showed that this solution gives the same value for the critical load as that obtained by Euler from the approximate equation. In this way he disproved Clebsch's remark[1] that it is only a happy accident that the approximate differential equation gives the correct value for the critical load.

Jasinsky did not content himself with the theoretical study of lateral buckling of bars for, using the experimental results of Bauschinger, Tetmajer, and Considère,[2] he prepared a table of critical compressive stresses for various slenderness ratios. This found wide application in Russia and superseded Rankine's formula. Further, he showed how, by using the "reduced" length of a bar, the table calculated for a bar with hinged ends could be applied to other types of lateral buckling.

Another engineer who contributed much to the theory of buckling in the period under consideration was Friedrich Engesser. Engesser (1848–1931) was born in the family of a music teacher in Weinheim near Manheim (in the Duchy of Baden).[3] After attending high school in Manheim, he entered the Karlsruhe Polytechnical Institute in 1865. He graduated in 1869 with the diploma of structural engineer. A rapid development of the railroad system was then taking place in Germany, and Engesser, as a young engineer, helped with the design and construction of bridges on the Schwarzwald railroad. Later, he obtained an administrative position in Baden's railroads and became occupied with the design of bridges for that system. This was a time of rapid growth of the theory of structures and Engesser took an active part in this development. He published several important articles dealing, in the main, with

FIG. 182. Friedrich Engesser.

---

[1] Clebsch, "Theorie der Elasticität fester Körper," p. 407, 1862.

[2] Considère, "Résistance des pièces comprimés," Congrès International des Procédés de Construction, vol. 3, p. 371, Paris, 1891.

[3] For a biography of F. Engesser, see Otto Steinhardt's paper "Friedrich Engesser," Karlsruhe, 1949.

statically indeterminate systems such as continuous beams, arches and bridges with redundant members and with rigid joints. He became known not only as a practical engineer but also as an expert in the theory of structures so that, in 1885, the Karlsruhe Polytechnical Institute elected him to a professorship. Engesser spent the following thirty years teaching at that school and he made immense advances in the theory of structures.

Working with the theory of lateral buckling, Engesser proposed[1] that the field of application of Euler's formula should be extended by introducing, instead of the constant modulus E, a variable quantity $E_t = d\sigma/d\epsilon$, which he called the *tangent modulus*. By determining the tangent modulus from the compression test curve for any particular case, he was able to calculate critical stresses for bars of materials which do not follow Hooke's law and for bars of structural steel beyond the elastic limit. In connection with this suggestion, a controversy with Jasinsky arose. The latter indicated[2] that the compressive stresses on the convex side decrease during buckling and that, in accordance with Bauschinger's tests, the constant modulus E instead of E_t must be used for that portion of the cross section. Later, Engesser revised his theory by introducing two different moduli for the two parts of the cross section.[3]

Engesser was the first to work with the theory of buckling of built-up columns.[4] He investigated the influence of shearing forces on the magnitude of the critical load and found that in solid columns this effect is small and can be neglected, but that in latticed struts it may be of practical importance, especially if these are made with battens alone. Engesser derived formulas for calculating in what ratio Euler's load must be diminished in each particular case in order to allow for the flexibility of lattice members.

Only in the simplest problems are rigorous solutions of the differential equation for lateral buckling known. Thus engineers are often compelled to use solutions that are only approximate. Engesser offers a method[5] for calculating critical loads by successive approximation. To get an approximate solution, he suggests the adoption of some shape for the deflection curve which satisfies the end conditions. This curve gives a bending-moment diagram, and we can calculate the corresponding deflections by using the area-moment method. Comparing this calculated deflection curve with the assumed one, an equation for determining the critical value of the load can be found. To get a better approximation, he takes

---

[1] *Z. Architek. u. Ing. Ver. Hannover*, vol. 35, p. 455, 1889.
[2] *Schweiz. Bauz.*, vol. 25, p. 172, 1895.
[3] *Schweiz. Bauz.*, vol. 26, p. 24, 1895; *Z. Ver. deut. Ing.*, vol. 42, p. 927, 1898.
[4] *Zentr. Bauw.*, 1891, p. 483.
[5] *Z. österr. Ing. u. Architek. Ver.*, 1893.

Strength of Materials in the Period 1867–1900

the calculated curve as a new approximation for that of the buckled bar and repeats the above calculation, and so on. Instead of taking an analytical expression for the initially assumed curve, the line can be given graphically and the successive approximations can be made in a graphical way.[1]

Engesser was also interested in the problem of buckling of the compressed upper chord of open bridges and derived some approximate formulas for calculating the proper lateral flexural rigidity of that chord.[2] Stability problems, however, constitute only a fraction of Engesser's work in the theory of structures. His important work on secondary stresses will be discussed in the next chapter.

While Jasinsky and Engesser were investigating various cases of lateral buckling of bars, an important paper dealing with the general theory of elastic stability was published by G. H. Bryan.[3] He shows that Kirchhoff's theorem on the uniqueness of solutions of the equations of the theory of elasticity holds only if all the dimensions of the body are of the same order. In thin bars, thin plates, and thin shells, there is the possibility of sometimes having more than one form of equilibrium for the same external forces so that the question of stability of those forms becomes of practical importance.

Later Bryan took up[4] the problem of buckling of a compressed rectangular plate with simply supported edges and gave a formula for calculating the value of the critical compressive stress. This is the first case in which stability of compressed plates is investigated theoretically. As an example for practical application of his formula, Bryan mentions the proper selection of thickness of compressed steel sheets in the hull of a ship. With the development of airplane structures, problems of buckling of thin plates have become of paramount importance and Bryan's paper gave a foundation for the consequent theory of elastic stability of thin-walled structures.

63. August Föppl (1854–1924)

We will conclude the history of strength of materials at the end of the nineteenth century by a discussion of the work of August Föppl (1854–1924). His book on this subject, which appeared in 1898, was very widely read in Germany and it was translated into Russian and French.

---

[1] A graphical method of this kind was offered by the Italian engineer L. Vianello, Z. Ver. deut. Ing., vol. 42, p. 1436, 1898. A mathematical proof of the convergency of this process was given by E. Trefftz, ZAMM, vol. 3, p. 272, 1923.

[2] See Zentr. Bauv., 1884, 1885, 1909; see also Z. Ver. deut. Ing., vol. 39, 1895.

[3] Proc. Cambridge Phil. Soc., vol. 6, p. 199, 1888.

[4] Proc. London Math. Soc., vol. 22, p. 54, 1891.

August Föppl[1] was born in the small town of Gross-Umstadt (in the Duchy of Hesse) in the family of a doctor. He received his elementary education in a public school and later went to the Darmstadt Gymnasium. At that time, the railroad through Gross-Umstadt was under construction, and the boy was so impressed that he decided to become a structural engineer and, toward that end, entered the Polytechnical Institute of Darmstadt in 1869. It seems that the standard of teaching at Darmstadt was not very high in those days, and therefore, after finishing the preparatory training, Föppl moved to Stuttgart in 1871. Mohr was teaching at the Stuttgart Polytechnicum and his lectures caused Föppl to devote most of his energy to study of the theory of structures.

FIG. 183. August Föppl.

When, in 1873, Mohr took the professorship at Dresden, Föppl also left Stuttgart and he entered the Polytechnical Institute at Karlsruhe to finish his engineering education. The engineering mechanics at that school was presented by Professor Grashof (see page 133), but apparently his lectures did not impress Föppl, for in his autobiography he severely criticizes Grashof's teaching methods. After the masterly lectures offered by Mohr, Grashof seemed too involved and lacking in originality. In 1874, Föppl graduated from the Polytechnicum with the degree of a structural engineer and decided to work as a bridge designer. But the economic situation in Germany was then at a low ebb and work on railroad construction was slowed down so that he was unable to find a satisfactory permanent position. After doing some temporary work in bridge design at Karlsruhe and after completing one year of compulsory military training, he took a post as a teacher at a trade school in 1876. First he taught in Holzminden, and later, after 1877, he was at Leipzig. Such a position did not satisfy Föppl, but it was always very difficult in Germany to get a professorship at a university or at a polytechnicum since the number of vacancies was small and the competition very great. It was necessary to publish some important work and to become well known in science in order to have some chance of capturing a professorship at a polytechnical institute. For such a position was always considered a

[1] A very interesting autobiography of Föppl was published by R. Oldenbourg, Munich and Berlin, 1925.

Strength of Materials in the Period 1867–1900

very high one in Germany, and the best engineers of the country would compete for any new vacancy. Föppl worked hard and published several important papers dealing with space structures. He also designed a market building in Leipzig. Later, he collected these papers and published them in book form.[1] It was the first book of its kind and it became very popular.

In the early 1880's, a rapid growth of the industrial applications of electricity took place in Germany, and Föppl became interested in that branch of physics. He met G. Wiedemann, the well-known professor of physics at Leipzig University and, on the latter's advice, he started to work on Maxwell's theory of electricity. This work resulted in the publication of an important book[2] on Maxwell's theory. It pioneered the use of this theory in Germany and made its author's name well known in science.

In 1893, Prof. J. Bauschinger died at Munich, and Föppl was elected during the following year to replace this outstanding worker in engineering mechanics. Thus he once again became able to devote his energy to mechanics which always held a special attraction for him. He had not only to deliver lectures in this subject but he also had to direct the work of the mechanical laboratory, which already boasted a very high reputation as a result of Bauschinger's work. Föppl's activity in both these directions was remarkably successful. He was an outstanding lecturer and knew how to hold students' interest although his classes were very large. Sometimes, he addressed as many as five hundred. To improve his teaching and to raise the standard of work, he soon started to publish his lectures in engineering mechanics. They appeared in four volumes: (1) Introduction to Mechanics, (2) Graphical Statics, (3) Strength of Materials, and (4) Dynamics. The volume on the strength of materials appeared first, in 1898. This book was at once a great success and soon became the most popular textbook in German-speaking countries. It also became known beyond the German frontiers. For instance, Jasinsky in St. Petersburg at once drew his students' attention to this outstanding work. The book was translated into Russian and was widely used by engineers interested in stress analysis. The book was also translated into French.

In his autobiography, Föppl discusses the requirements which a good textbook must satisfy. He remarks that very often writers of textbooks think more about the critics who may review their work than about the students. To satisfy the critics, authors try to present their subject in the most general terms and in as rigorous a form as possible. This makes

---

[1] August Föppl, "Das Fachwerk im Raume," Leipzig, 1892.

[2] A. Föppl, "Einführung in die Maxwell'sche Theorie der Elektricität," Leipzig, 1894.

the reading of the book difficult for beginners. Föppl presented the subject in writing in the same manner as he did by word of mouth. He usually started with simple particular cases, which a beginner could easily understand, and discussed them without introducing extraneous details. A more general discussion and a more rigorous form of presentation came later when the student had familiarized himself with the fundamentals and could appreciate the more rigorous form. There already existed very complete books on strength of materials in Germany, such as those of Grashof and Winkler. But in both of these, the authors took the mathematical theory of elasticity as the basis of their presentation of strength of materials and in this way made the subject too difficult for most students. Föppl gives all the necessary information on strength of materials in an elementary way in his book and only comes to the equations of the theory of elasticity at the end. In the later editions of his course, Föppl expanded the portion dealing with the theory of elasticity and put it into an additional volume. This new book did much to popularize this science and had a great deal to do with the adoption of more rigorous methods of stress analysis in engineering practice. This was the first book on theory of elasticity to be written specifically for engineers.

The experimental work of Föppl was also of great importance. His predecessor, Professor Bauschinger, was principally interested in the mechanical properties of materials and developed a fine technique in testing specimens of various materials. Föppl extended the field of experimental work and used the laboratory for tests relating to the various theories of failure of materials and for experimental determinations of stresses in complicated problems for which we have no theoretical solutions. When measuring the tensile strength of cement, Föppl observes that the tensile stresses are not uniformly distributed over the cross section in the standard tensile-test specimens so that the results do not give the true value of the tensile strength. Using a rubber model of the tensile-test specimen, and measuring longitudinal strain at different distances from the axis of the specimen, Föppl obtains a satisfactory picture of the stress distribution in this complicated case. Working with compression tests of cubic specimens of cement, he notes the importance of friction forces acting on the faces of the cube which are in contact with the plates of the testing machine. He investigates various methods of reducing these forces and shows that the usual tests of cubic specimens give exaggerated values for the compression strength. In all these tests, Föppl was naturally very interested in the accuracy of his testing machines. To check the tensile and compressive forces acting on specimens, he introduced a special dynamometer in the form of a heavy steel ring such as that now used in many laboratories.

In his book, Föppl follows Saint-Venant's notion and uses the maximum strain theory in deriving formulas for calculating safe dimensions of structures. But at the same time he was interested in the various other theories of strength and, to clarify the question of which should be used, he conducted some interesting experiments. By using a thick-walled cylinder of high-grade steel, he succeeded in making compressive tests of various materials under great hydrostatic pressures. He found that isotropic materials can withstand very high pressures in that condition. He designed and constructed a special device for producing compression of cubic specimens in two perpendicular directions and made a series of tests of this kind with cement specimens.

The fatigue experiments of Wöhler's type, which were started at Munich by Bauschinger, were continued by Föppl. He extended them to specimens with grooves and studied the effects of stress concentration. He also studied this question theoretically and showed that in the torsion of a shaft with two parts (of different diameters) connected by a fillet the stress concentration depends very much upon the radius of the fillet.

Föppl was the first to give a satisfactory theory of whirling of a flexible shaft rotating at high speed and he made a series of tests in his laboratory to verify his theory.

This engineer published all his experimental results in the bulletins of the laboratory. This journal was started by Bauschinger, and beginning with issue No. 24, Föppl continued to publish it until the end of his life. It was well known to engineers interested in strength of materials and, accordingly, wielded considerable influence in the growth of that science.

CHAPTER X

Theory of Structures in the Period 1867–1900

64. Statically Determinate Trusses

We discussed the various methods devised by engineers for analyzing trusses in Chap. VII. In the simple instances treated by Whipple and Jourawski, the forces in the truss members were found from the conditions of equilibrium of the joints. Later, A. Ritter and Schwedler introduced the *method of sections* while Maxwell, Taylor, and Cremona showed how to construct *reciprocal diagrams*. These methods were adequate for analyzing most of the trusses then in existence but, with the growing use of metal in structures, more complete investigations of various types of trusses became necessary.

Some theorems, fundamental in the theory of trusses, were stated by A. F. Möbius (1790–1868) who was the professor of astronomy at the University of Leipzig. In his book on statics,[1] Möbius discusses the problems of equilibrium of a system of bars hinged together and shows that if there are n hinges it is necessary to have not less than $2n - 3$ bars to form a rigid system in one plane and $3n - 6$ bars for a three-dimensional system. Möbius also indicates that there are exceptional cases in which a system with $2n - 3$ bars will not be absolutely rigid, but will admit the possibility of small relative displacements of the hinges. In studying such exceptional cases, he finds that they occur when the determinant of the system of equations of equilibrium of the truss joints vanishes. Thus he observes that, if a system has the number of bars essential for rigidity and if it be assumed that one bar, say that of length l_{ab} between the joints a and b is removed, then the system will permit motion of the members as a mechanism. As a result of this relative movement, the distance between the joints a and b may be altered. Imagine now that the configuration of the given system is such that the distance between the joints a and b is a maximum or a minimum. Then small relative motions of the system will not change the distance ab, which indicates that the insertion of the bar l_{ab} will not eliminate the possibility of small movements of the system. Such is an exceptional case.

[1] "Lehrbuch der Statik," by August Ferdinand Möbius, 2 vols., Leipzig, 1837. See vol. 2, chaps. 4 and 5.

Theory of Structures in the Period 1867–1900 305

The important work of Möbius remained unknown to engineers for many years, but when steel trusses became of practical importance and it became necessary to improve the general theory of trusses, engineers rediscovered Möbius' theorems. In this work of rediscovery, the most prominent name was that of O. Mohr.[1] He found the requirement regarding the number of bars which are necessary for forming a rigid statically determinate system. He also investigated the exceptional case of infinitesimal mobility. He demonstrated that there are statically determinate trusses which cannot be analyzed by previously suggested methods, and proposed that the principle of virtual displacements be used in studying such systems.

Figure 184a is an example of a statically determined system which cannot be treated by the old methods. In calculating the stress in any bar

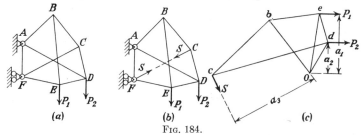

Fig. 184.

(say in the diagonal FC) by using the principle of virtual displacement, we assume that the bar is removed and its action on the rest of the system is replaced by the two equal and opposite forces, as shown in Fig. 184b. The system is no longer rigid, and we can assign virtual displacements to its hinges. The equation for calculating the forces S is now obtained by equating the work done in the virtual displacements by the given external forces P_1, P_2 and by the forces S to zero. Naturally the applicability of the method depends on the manner of taking the virtual displacement. The method used by Mohr[2] is illustrated by Fig. 184c. During small displacements of the system in Fig. 184b, the hinges B, D, and E move perpendicularly to the radii AB, AD, and FE. In constructing the diagram of displacements (Fig. 184c), we take a pole O and draw the lines Ob, Od, and Oe parallel to those displacements and the magnitude of one of these, say Ob, can be taken arbitrarily. Then the other two displacements are obtained by drawing the lines be perpendicular[3] to the bar BE and de per-

[1] *Z. Architek. u. Ing. Ver. Hannover*, 1874, p. 509; *Ziviling.*, 1885, p. 289. See also his "Abhandlungen aus dem Gebiete der technischen Mechanik," Chap. XII, 1905.

[2] See "Abhandlungen," Chapt. IV.

[3] The perpendicularity of such lines as be and BE follows from the fact that any movement of the bar in the plane of the figure can only be accomplished by rotation about the instantaneous center.

pendicular to ED. Finally, to obtain the point c on the diagram of Fig. 184c, we have only to draw the lines bc and dc perpendicular to the bars BC and CD. Having the virtual displacements of all hinges we can readily calculate the virtual work and write an equation for the determination of the unknown forces.

N. E. Joukowski (1847–1921), the well-known pioneer in the theory of aerodynamics, suggested[1] considering the displacement diagram in Fig. 184c as a rigid body hinged at the pole O and to which the forces P_1, P_2 and S (rotated by 90 deg) are applied. The expression for the virtual work is then identical with that for the moment of the forces about the pole O and the equation of equilibrium of the figure is identical with that of virtual displacements. Thus we obtain

$$P_1 a_1 + P_2 a_2 - S a_3 = 0$$

and

$$S = \frac{P_1 a_1 + P_2 a_2}{a_3} \quad (a)$$

When the force S, in diagram 184c, passes through the pole O, the distance a_3 vanishes and Eq. (a) gives an infinite value for S. This indicates that the given system fulfills the requirements of the exceptional case of infinitesimal mobility.

FIG. 185. N. E. Joukowski.

A somewhat different method of getting the virtual displacements was proposed by H. Müller-Breslau.[2] The method is illustrated in Fig. 186. Considering the same system as before and assuming that the bar FC is removed, we again obtain a nonrigid system the virtual displacements of which must be studied. Instead of making a separate diagram for these displacements, all the necessary information is now kept on the same figure. The magnitude δ of the displacement of hinge B is taken arbitrarily and the point B' is obtained by rotating this displacement by 90 deg. From the principle of the instantaneous center, we conclude that the point E' defining the rotated displacement EE' of the hinge E is obtained by drawing the line $B'E'$ parallel to BE. The point D' is obtained by drawing $E'D'$ parallel to ED and finally the point C', defining the rotated displacement of the hinge C, is obtained as the intersection

[1] In a paper presented at the Moscow Mathematic Society in 1908. See N. E. Joukowski "Collected Papers," vol. 1, p. 566, 1937.

[2] H. Müller-Breslau, *Schweiz. Bauztg.*, vol. 9, p. 121, 1887.

point of the straight lines $D'C'$ and $B'C'$ which are parallel to CD and BC. Having now the virtual displacements of all hinges, the virtual work is obtained by calculating the sum of moments about the points E', D', C', ... of the external forces and of S applied at E, D, C. ... Equating the sum to zero, the equation for calculating S is obtained. If it happens that point C' falls on the line FC, the force S becomes infinitely large; this indicates that we have the exceptional case of infinitesimal mobility.

Another general method for analyzing complicated trusses was devised by L. Henneberg.[1] The method rests upon the possibility of transforming a given complicated truss into a simpler one by removing some bars and replacing them by others, differently located. For illustration, we again take the case previously discussed (Fig. 187a). Replacing the bar FC by the bar AE, as shown in Fig. 187b, a simple truss is obtained the stresses in which can be readily found by drawing the Maxwell diagram. Let S_i' denote the force acting in the newly introduced member AE. As

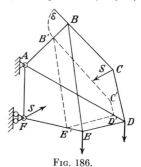

Fig. 186.

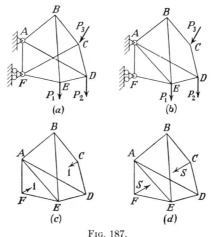

Fig. 187.

a simple auxiliary problem, we consider two equal and opposite unit forces as shown in Fig. 187c. Let s_i'' denote the force in the bar AE in this case. If, instead of unit forces, loads of magnitude S act in Fig. 187d,

[1] L. Henneberg, "Statik der starren Systeme," p. 122, Darmstadt, 1886. See also his "Graphische Statik der Starren Systeme," p. 526, 1911, and his article in "Encyklopädie der mathematischen Wissenschaften," vol. 4_1, p. 406.

then a force of Ss_i'' will be induced in the bar AE. The forces in the bars of the actual system (Fig. 187a) are now obtained by superposition of the cases 187b and 187d. The true value S will evidently be that which makes the force in the bar AE evanescent. Thus we obtain the equation

$$S_i' + Ss_i'' = 0$$

from which the force S can be calculated. If the quantity s_i'' vanishes, we have the exceptional case previously referred to. This method can be readily extended to systems in which more than one exchange of bars is required to get a simple truss and it offers a general method of analyzing complicated plane trusses. Other general methods, dealing with the same problem, were developed by C. Saviotti[1] and by F. Schur.[2]

The general theory of three-dimensional systems was also initiated by Möbius. He showed that we need $3n - 6$ bars to connect n hinges rigidly and observed that again exceptional cases of infinitesimal mobility occur. These are characterized by the vanishing of the determinant of the system of equations of equilibrium of all joints. He gave a useful practical method of deciding whether or not a given system is rigid. If for some loads we can find the forces in all members of the system without ambiguity, the above-mentioned determinant does not vanish and the system is rigid. As the simplest assumption Möbius suggests zero loads so that, if we can prove that the forces in all bars vanish for this condition, the system is rigid.

Möbius considered the very important problem of a self-contained space truss which has the form of a closed polyhedron and shows[3] that, if the plane faces of that polyhedron are triangular or are subdivided into triangles, the number of bars is just equal to the number of equations of statics and the truss is statically determinate. Figure 188 shows examples of such trusses.

Möbius' work in three-dimensional systems also remained unknown to engineers and they developed the theory of space trusses independently. This was done, in the main, by A. Föppl, who collected his work in the subject and published it in book form.[4] In this book, we find treatment of several important questions dealing with space structures for the first time. Föppl's book was recognized as an important one and it has provided a background for much of the later work in this field.

The author begins by considering types of structures such as are shown in Fig. 188. These, he calls *lattice structures* (*das Flechtwerk*). Not

---

[1] C. Saviotti, "La Statica grafica," 3 vols. Milan, 1888. See also his paper in *Atti accad. Nazl. Lincei*, (3), vol. 2, p. 148, 1875, Rome.
[2] *Z. Math. u. Physik*, vol. 40, p. 48, 1895.
[3] See his "Statik," vol. 2, p. 122.
[4] August Föppl, "Das Fachwerk im Raume," Leipzig, 1892.

knowing Möbius' work, he remarks: "A future writer of history of engineering will perhaps ask with surprise how it was possible that the general notion of lattice-structures could remain completely unknown, up to this year of 1891, despite the fact that such a large number of them has been built through the ages." Then he proves that this type of system is stat-

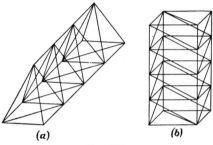

FIG. 188.

ically determinate and discusses its application as a roof truss (Fig. 189). Föppl was so interested in this system of trusses that he made a model of considerable size and tested it in his laboratory. These tests showed that the usual assumption of hinges at the joints does not give such satisfactory results in this case as it does for two-dimensional trusses. Thus, in more accurate analyses of space structures, the rigidity of joints must be taken into account.

Further on, Föppl discusses Schwedler's[1] type of cupola (Fig. 190) and he designs one of his own (Fig. 191). His was used in the construction of the large market hall in Leipzig.[2] For each configuration, Föppl outlines methods by the use of which the forces in the bars can be calculated for any kind of loading.

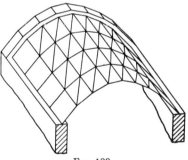

FIG. 189.

Further, he investigates the manner in which the structures must be supported so that the condition of infinitesimal mobility is eliminated. In Föppl's analysis we find use made of the method of joints and the method of sections. In more complicated structures, the method of virtual displacements and Henneberg's method were successfully applied by H. Müller-Breslau.[3]

[1] *Z. Bauwesen*, 1866.
[2] See *Schweiz. Bauzg.*, vol. 17, p. 77, 1891.
[3] *Zentr. Bauv.*, vol. 11, p. 437, 1891; vol. 12, pp. 225, 244, 1892.

This work on truss analysis was chiefly connected with the design of metal bridges and, in this, the problem of finding the most unfavorable position of a live load became of primary importance. A solution was found to be in the method of *influence lines*. In most practical structures, the principle of superposition can be applied, and therefore the stresses produced in a truss by any system of loads can be readily obtained if the effect of a unit load moving along the span is investigated. The influence lines were introduced to help to visualize the effect on stresses of a unit load when changing its position. We have already mentioned (see page 152) the pioneer work done in this field by Mohr and by Winkler in 1868.

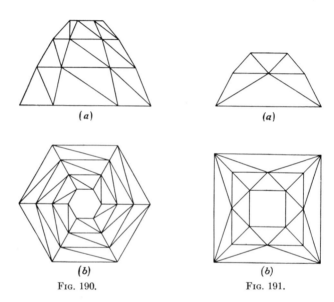

Fig. 190.   Fig. 191.

The latter used these lines in his book on theory of bridges,[1] which was one of the best books in its field and was widely read. Fränkel wrote the paper[2] containing the first systemmatic presentation of the theory of influence lines. We find a most complete description of various of their applications in the books[3] of H. Müller-Breslau, who was a professor at the Polytechnicum in Berlin and who did a great deal of work in the use of graphic methods for solving problems of the theory of structures.

[1] "Vorträge über Brückenbau," vol. 1, Theorie der Brücken, Vienna, 1872.
[2] *Civiling.*, vol. 22, p. 441, 1876.
[3] H. Müller-Breslau, "Die graphische Statik der Baukonstruktionen," 2d ed., 1887. See also his book "Die neueren Methoden der Festigkeitslehre und der Statik der Baukonstruktionen," 1st ed., 1886.

65. Deflection of Trusses

In the design of trusses, it is often required to calculate deflections. When the forces in all bars of a statically determinate truss have been found, the changes in lengths of bars can be computed by using Hooke's law so that a purely geometrical problem arises. Knowing the elongations of the component bars, how can we find the corresponding displacements of the joints of a truss? In many instances we need to know the displacements of only a few joints. Then Castigliano's theorem affords a useful method if there are forces acting on those joints. Castigliano showed also that, if there is no force applied to the joint in question, we can add a fictitious force there and, after making the calculation, assume in the final result that the added force is equal to zero. He used the notion of generalized forces and proved that, in the case of two equal and opposite forces P acting along a straight line, the derivative $\partial V/\partial P$ gives the change in the distance between the points of application of the forces, while in the case of a couple M, applied to an elastic system, the derivative $\partial V/\partial M$ gives the angle of rotation of the element of the system to which the couple is applied (see page 289).

As mentioned before (see page 205), Maxwell gave another method for calculating the displacements of joints of a truss (earlier then Castigliano). But he presented it in such an abstract form that the method remained unnoticed by engineers and proper use of it was only made after its rediscovery by Mohr.[1] Mohr, not knowing of Maxwell's publication, conceived the method by using the principle of virtual work and demonstrated its practical importance with examples. To illustrate Mohr's method, let us consider the problems previously discussed and shown in Fig. 141a (page 205). We can calculate the deflection of a joint A produced by given loads $P_1, P_2, \ldots$. In solving such a problem, Mohr considers the auxiliary case shown in Fig. 141b. To find the relation between the elongation Δ_i' of a bar i and the corresponding deflection δ_a' of the joint A, he removes the bar and replaces its action on the truss with the two equal and opposite forces s_i'. Considering the other bars as being absolutely rigid, he obtains a system with one degree of freedom upon which, the unit load, the reactions at the supports, and the forces s_i' act. Since these forces are in equilibrium, their work during any virtual displacement must vanish. Thus he obtains the equation

$$-s_i' \Delta_b' + 1 \cdot \delta_a' = 0$$

from which

$$\delta_a' = \Delta_i' \frac{s_i'}{1} \qquad (a)$$

[1] See Z. Architek. u. Ing. Ver. Hannover, 1874, p. 509; 1875, p. 17.

It is seen that, by using the principle of virtual displacement, the required relation between the elongation Δ_i' and the deflection δ_a' is established and it holds for any small value of Δ_i'. Having this relation, the deflection δ_a of the joint A under the actual loads $P_1, P_2, \ldots$ (Fig. 141a) can readily be found. Substituting the actual elongation $S_i l_i / E A_i$ of the bar i into Eq. (a), he gets the deflection of joint A due to the elongation of one bar only. Summing such deflections, over all bars, he obtains

$$\delta_a = \sum \frac{S_i s_i' l_i}{A_i E} \qquad (b)$$

This result of Mohr's coincides with that obtained previously by Maxwell (see page 206).

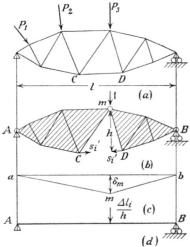

Fig. 192.

It is seen that, to calculate the deflections of several joints by the Maxwell-Mohr method, we have to solve an auxiliary problem for each joint such as shown in Fig. 141b. If the number of joints is large, the method becomes complicated. To overcome this difficulty, Mohr gave another method,[1] by which all the necessary deflections can be obtained by constructing the bending-moment diagram for a simple beam supporting some fictitious loads. As an example, let us consider the vertical deflections of the truss of Fig. 192a. The forces in all the bars of this truss and the corresponding elongations and contractions can be readily found. To calculate the vertical deflections of the joints resulting from an elongation Δl_i of one member, say bar CD, we use Eq. (a) and conclude that the deflection δ_m of the joint m (Fig. 192b) opposite the bar CD is

$$\delta_m = \Delta l_i \frac{s_i'}{1} \qquad (c)$$

where s_i' is the axial force in the bar CD due to a unit load at m. Since all bars except CD are considered as rigid, the two portions of the truss, shaded in Fig. 192b, move as rigid bodies, rotating with respect to each

---

[1] See O. Mohr, *Z. Architek. u. Ing. Ver. Hannover*, 1875, p. 17. See also his "Abhandlungen," p. 377, 1906.

other about the hinge m. Hence, the vertical deflections of all joints are evidently given by the appropriate ordinates of the diagram amb in Fig. 192c. Observing that the axial force s_i', produced in the bar CD by the unit load at m, is equal to the bending moment at m divided by the distance h, we conclude, from Eq. (c), that the deflection diagram in Fig. 192c can be considered as the bending-moment diagram for a beam AB (Fig. 192d) acted upon by the fictitious load

$$\frac{\Delta l_i}{h} \qquad (d)$$

In a similar manner the deflection resulting from the change in length of any other chord member of the truss can be found. Thus the deflection

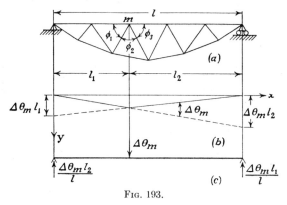

Fig. 193.

of the truss, resulting from changes in length of all chord members, can be calculated for each joint as the bending moment at the corresponding cross section of the beam AB when supporting fictitious loads that are defined for each joint by the quantity (d). These loads, as we see, are pure numbers and the corresponding bending moments have the dimension of a length, as they should. Mohr shows that the additional deflections that are due to deformation of *web members* can be evaluated in a similar manner, so that finally the calculation of deflections of joints in our truss (Fig. 192a) is reduced to the computation of bending moments in a beam induced by a properly selected system of fictitious loads.

Further progress in the determination of deflections of trusses was made by Winkler[1] who shows that the deflections of the joints of one chord of a truss can be calculated by considering the members of that chord. This calculation is especially simple if the chord is a horizontal straight line as, for example, the upper chord of the truss in Fig. 193. It is evident

[1] See his "Theorie der Brücken," II Heft, 2d ed., p. 363, 1881.

that deflections are completely defined when we know the changes in lengths of the bars of the upper chord together with their angles of rotation. Since the chord is a horizontal straight line, changes in the lengths of its bars result only in horizontal displacements of the upper joints. Thus, vertical displacements at these points depend only upon rotations of the upper chord members. Winkler[1] shows that these rotations can be readily found if we calculate the changes in the angles of each triangle of the truss due to deformation of bars. Assuming for the moment that such changes in the angles ϕ_1, ϕ_2, and ϕ_3 at an upper joint m (Fig. 193a) have been found, we obtain, by their summation, a small angle $\Delta\theta_m$ representing the angle between the two chord members at the joint m after deflection. The corresponding deflection of the upper chord, assuming that $\Delta\theta_m$ is positive, is shown in Fig. 193b. The deflection for any point of the chord to the left of joint m is equal to $\Delta\theta_m l_2 x/l$, while for any point to the right of m, it is $\Delta\theta_m l_1 (l - x)/l$. We see that these expressions for deflections are identical with those for bending moments in a simply supported beam produced by a fictitious load $\Delta\theta_m$ acting as shown in Fig. 193c. Thus, by calculating the values of $\Delta\theta_i$ for each joint i of the upper chord and using the method of superposition, the required deflections can be obtained as the bending moments for a simply supported beam subjected to fictitious loads $\Delta\theta_i$.

If the chord whose deflection is required is of polygonal form, as, for example, the lower boundary of the truss in Fig. 193a, the deflections due to changes in the angles between the bars can be calculated in exactly the same way. The deflections due to changes in the lengths of the chord members should be added to those due to rotation. These again, are obtainable as the bending moments set up in a simple beam by a properly selected set of fictitious loads.

A purely graphic method of determining the deflections of trusses was devised by M. Williot.[2] This method will be illustrated by means of the simple example of a joint A formed by two bars 1 and 2, as shown in Fig. 194, the remote ends of the bars being pinned at B and C. It is assumed that the displacements BB' and CC' of the joints B and C are known as are the changes in length of the bars 1 and 2. It is required to find the resulting displacement of joint A. We first assume that the bars are parted at A and translate them to the positions $A'B'$ and $A''C'$ parallel to their initial positions and such that BB' and CC' represent the given displacements of the joints B and C. Starting with these new positions, we keep points B' and C' fixed and give displacements $A'A_1'$ and $A''A_1''$ to the opposite ends of the bars, as shown by the heavy lines. These last

[1] See "Theorie der Brücken," 2d ed., II Heft, p. 302.
[2] M. Williot, "Notions Pratiques sur la Statique Graphique," Paris, 1877. See also *Ann. génie civil*, 2d series, 6th year, 1877.

displacements are equal to the given changes in the lengths of the bars; that is, $A'A_1'$ is the known elongation of bar 1 and $A''A_1''$ is the known contraction of bar 2. Now, to complete the construction, we have to bring the points A_1' and A_1'' together by rotating the bar $B'A_1'$ about B' and the bar $C'A_1''$ about C'. Since we are dealing with small deformations and small angles of rotation, the arcs of the circles along which the points A_1' and A_1'' travel during the course of such rotations can be replaced by the perpendiculars $A_1'A_1$ and $A_1''A_1$ and the intersection of these gives the new position A_1 of the joint A. Thus the vector AA_1 represents the required displacement of A.

Since the elongations of the bars and the displacements of the joints are very small in comparison with their lengths, it is necessary to draw

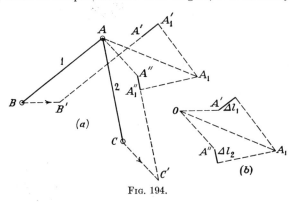

Fig. 194.

these quantities to a large scale and to make all constructions in a separate diagram, as shown in Fig. 194b. We take a pole O and lay off the given displacements OA' and OA'' of the joints B and C to a chosen scale. Then, from the points A' and A'', we draw the vectors Δl_1 and Δl_2 which are shown by heavy lines and which represent the known changes in lengths of the bars 1 and 2. Due attention must be paid to the sign of these elongations. The bar 1 supposedly increases in length so that Δl_1 is drawn in the direction from B to A. The bar 2 decreases in length; hence Δl_2 is drawn in the direction from A to C. Finally, the perpendiculars erected at the ends of the vectors Δl_1 and Δl_2 intersect at point A_1. This fixes the required displacement OA_1 of joint A. Figure 194b represents the Williot diagram for our simple structure.

The same procedure can be followed in the construction of displacement diagrams for all trusses that are formed by starting with one bar and using two new bars to form each joint. In handling such trusses, we can always assume that the displacements of two joints are known and determine the displacement of a third by using the above method. Having this displacement, we can go to the next joint and again repeat the same construction, and so on. Having all the displacements of the hinges, we can readily obtain the vertical components of the displacements of a bridge truss by projection, and construct the deflection curve as we did before by using the analytical methods of Castigliano and of Maxwell and Mohr or the semianalytical-semigraphic method of fictitious loads.

From the publications of the period under discussion, it can be seen that two different opinions were growing up regarding the relative merits of analytical and graphic methods used in theory of structures. While some engineers often enlisted the aid of graphical solutions, others thought that this was not accurate enough and preferred analytical determinations of all required quantities. Winkler discusses this question in the preface to his book in bridges.[1] He mentions that graphic methods have the advantages of being more illustrative and of making mistakes easier to discover. Graphical work is not as tiresome as analytical calculations and its solutions can usually be obtained in a shorter time than is required for numerical calculations while the results have an accuracy that is sufficient for all practical purposes. In the analysis of continuous beams a graphical solution requires only one-third of the time required for an analytical calculation, according to Winkler. But this writer also sees some advantages of numerical calculations. He indicates that numerical calculations are preferable in cases where greater accuracy is called for, especially when numerical tables are being prepared for future use. The work of Winkler and Mohr did much to popularize graphic methods in practical work in the theory of structures.

66. Statically Indeterminate Trusses

We have seen (see page 259) that Clebsch took up problems of stress analysis of trusses with redundant members. He showed that by taking the displacements of hinges as our unknowns, we can always write as many equations as there are unknowns. He considered some simple examples in which his method could be applied easily and obtained solutions for them. Further progress in treating statically indeterminate trusses was made by Maxwell. His method was rediscovered by Mohr (see page 311) and since then has been generally used by structural engineers. Another way of handling the matter, based on consideration of the strain energy, was offered by Castigliano. Let us illustrate the Maxwell-Mohr and the Castigliano methods with a simple example of a system with one redundant member, as shown in Fig. 195a. We take the diagonal AB as the redundancy. In applying the Maxwell-Mohr method, we remove the bar AB and replace its action on the rest of the system by the two equal and opposite forces X. In this way, we obtain

---

[1] "Theorie der Brücken," Vienna, 1873.

the statically determinate truss, shown in Fig. 195b, on which the given loads P_1, P_2 and the unknown forces X act. The forces in the bars of this system, produced by the given loads P_1, P_2, can be readily calculated. Let S_i denote the force acting in any bar i. To determine the force in the same bar produced by the two forces X, we solve the auxiliary problem in which the loads P_1, P_2 are removed and the forces X are replaced by two unit forces (Fig. 195c). Let s_i denote the force produced in a bar i by this pair of forces. Then the total load in any bar i produced by the loads P_1, P_2 and by forces X (Fig. 195b) will evidently be

$$S_i + s_i X \qquad (a)$$

The corresponding elongation of the bar i is $(S_i + s_i X)l_i/A_i E$. Due to such elongations, the distance between the joints A and B will be changed and the amount by which B approaches A is given by the expression[1]

$$\sum \frac{(S_i + s_i X)s_i l_i}{A_i E} \qquad (b)$$

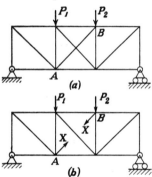

The magnitude of the forces X can now be found from the fact that, in the actual truss (Fig. 195a), the change in distance between the hinges A and B is equal to the elongation Xl/AE of the diagonal AB. This gives us the equation

$$-\sum \frac{(S_i + s_i X)s_i l_i}{A_i E} = \frac{Xl}{AE} \qquad (c)$$

FIG. 195.

From this, the unknown force X can be determined and then the forces in all bars can be calculated by using expression (a).

In using Castigliano's method, we calculate the strain energy of the system, shown in Fig. 195b. It is given by

$$V_1 = \sum \frac{(S_i + s_i X)^2 l_i}{2 A_i E}$$

The derivative $\partial V_1/\partial X$ gives us the expression (b) determining the movement of the hinge B toward hinge A. Equating this displacement (with negative sign) to the elongation of the diagonal AB, we again arrive at Eq. (c).

[1] A positive value of this summation will indicate that the distance AB decreases. A negative value will indicate an increase of the same amount.

In a similar manner, systems with several redundant members can be treated. For example, Fig. 196a shows a system with two redundant elements. We take as the redundant quantities, the two components X and Y of the reaction at support A and consider the two auxiliary problems shown in Figs. 196b and 196c. Let s_i' and s_i'' denote the forces in any bar i for these two cases, and suppose that S_i is the force brought about in bar i by the loads P_1, P_2, when the support A is removed and

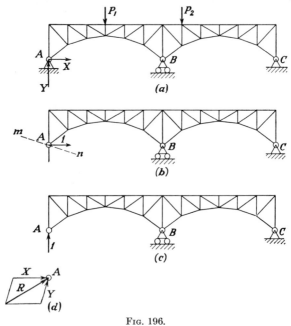

Fig. 196.

$X = Y = 0$. Then the total force in any bar i, under the actual conditions of Fig. 196a, will be

$$S_i + s_i'X + s_i''Y \qquad (d)$$

Now using Eq. (b) of the preceding article and equating the horizontal and vertical displacements of the hinge A in the actual case of Fig. 196a to zero, we obtain the following two linear equations in X and Y:

$$\sum \frac{(S_i + s_i'X + s_i''Y)s_i'l_i}{A_iE} = 0$$
$$\sum \frac{(S_i + s_i'X + s_i''Y)s_i''l_i}{A_iE} = 0 \qquad (e)$$

From these, X and Y may be calculated.

In the same way, a system with three redundant elements will give rise to three linear equations with three unknowns, and so on. We always obtain as many linear equations as there are unknowns, and the statically indeterminate quantities can be extracted from them. Practically, the problem becomes more and more complex as the number of redundancies increases, not only because the number of equations [like Eqs. (e)] becomes greater, but also because the equations are often of such a kind that all the numerical work must be done with great accuracy if the required quantities are to be found with any degree of precision.[1] Also, graphical analysis is usually not accurate enough. To get around this difficulty, it was suggested that the unknowns should be picked in such a manner as to render equations for determining the redundant quantities such that each contains only one unknown. Mohr shows[2] how this can be done in the particular case of an arch truss with three redundant bars (Fig. 197).

To explain the general method to be followed in selecting the unknowns, let us again consider the case represented in Fig. 196. By examining the corresponding Eqs. (e), it may be seen that the desired simplification is accomplished if we have

$$\sum \frac{s_i' s_i'' l_i}{A_i E} = 0 \qquad (f)$$

This equation means [as can be seen from Eq. (b) of Art. 65] that the unit load acting in X direction should produce no displacement of joint A in the Y direction. To satisfy this requirement, we first investigate in what direction the hinge A moves under the action of the horizontal unit load of Fig. 196b by constructing the Williot diagram. Let this direction be mn. Then, in calculating the two components of the reaction R at hinge A, we take the component Y perpendicular to mn, as shown in Fig. 196d. With this choice of the directions of the two components, condition (f) will be satisfied and each of Eqs. (e) will contain only one unknown.

In the arch shown in Fig. 197a, we have a system with three redundant elements. If we take the forces S_1, S_2, and S_3 in the bars 1, 2, 3 as the unknown quantities and proceed as before, we shall obtain three equations similar to Eqs. (e), each of which will contain all three unknowns S_1, S_2, and S_3. To simplify the problem by getting the three equations with one unknown each, we replace S_1, S_2, and S_3 by the statically equivalent system of three forces selected so that each of them, if acting alone, produces

---

[1] A discussion of the accuracy of the numerical calculations required in the analysis of statically indeterminate systems with many redundant members has been given by J. Pirlet, Dissertation, Technische Hochschule, Aachen, 1909; see also A. Cyran, Z. Ver. deut. Ing. vol. 54, p. 438, 1910.

[2] O. Mohr, Z. Architek. u. Ing. Ver. Hannover, vol. 27, p. 243, 1881.

no displacements corresponding to the other two. To accomplish this, we assume that the hinges A and B are attached to an absolutely rigid block AOB (Fig. 197b). Then we apply the properly selected forces X and Y and a couple M to this block. These are treated as *generalized forces*. To find the point O at which the forces X and Y must be applied, we investigate the displacement of the block ABO under the action of the couple M by using the Williot diagram. We take the point which remains stationary during this displacement for O. Since point O does not move under the action of the couple M, we conclude, from the reciprocity theorem, that a force applied at O will not produce any rotation of the block. As a result of this, one of the three equations for calculating the unknown reactions (namely, that containing M) will not involve X

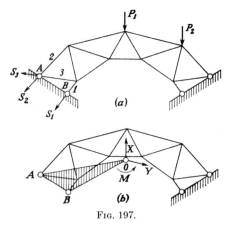

FIG. 197.

and Y and the others will contain only X and Y. By choosing the directions of the forces X and Y in the same way as was explained in the case shown by Fig. 196, we finally obtain three equations each containing only one unknown.

The methods of finding stresses in redundant systems can be readily extended to cases in which the changes in lengths of bars is due to temperature differences. Examples of this appear in Castigliano's book and also in Mohr's publications.

The use of influence lines in analyzing statically indeterminate systems was widely developed. In the drawing of these lines, the reciprocity theorem for the simple case of two forces was made by Maxwell and a general proof of the theorem was given later by E. Betti.[1] Lord Rayleigh extended the theorem to cover vibration of elastic systems[2] and proved

[1] See the Italian journal *Nuovo cimento* (2), vols. 7, 8, 1872.
[2] *Proc. London Math. Soc.*, vol. 4, pp. 357–368, 1873.

Theory of Structures in the Period 1867–1900

that if a harmonic force of given amplitude and period acts upon a system at the point P, the resulting displacement at a second point Q will be the same both in amplitude and phase as it would be at the point P were the force to act at Q. He then deduced the statical reciprocity theorem as the particular case in which the forces have an infinitely large period.[1] In his work, Lord Rayleigh uses the notions of *generalized force* and of the *corresponding generalized displacement* and considers, as particular instances, a force and a couple. He remarks: "For the benefit of those whose minds rebel against the vagueness of generalized coordinates, a more special proof of the theoretical result may here be given . . . ." Rayleigh verifies his theorem by experiments and, working with a beam, determines the influence line for the deflection of the beam at a given cross section. This is the first time that an influence line was obtained experimentally.

Lord Rayleigh's work, and especially the publication of his book "The Theory of Sound,"[2] greatly influenced the development of the theory of structures in Russia. The idea of using the reciprocity theorem, together with the notion of generalized forces, was applied by Prof. V. L. Kirpitchev (1844–1913) to the construction of influence lines for various problems of beams on two supports, of continuous beams, and of arches.[3] Later, the concepts of generalized forces and generalized coordinates were widely used by Kirpitchev in his important book on "Redundant Quantities in Theory of Structures."[4] In this way, he succeeded in considerably simplifying the presentation of various methods of analysis of statically indeterminate structures. In the preface of his book, Kirpitchev states that all engineers interested in theory of structures should study Rayleigh's "Theory of Sound." Kirpitchev's books[5] and his lectures were prominent in strength of materials in Russia at the end of the nineteenth and the beginning of the twentieth centuries.

In treating the theory of trusses it was assumed up till now that there were ideal hinges at the joints. But in actual fact the joints are usually rigid and consequently members are subjected to some bending in addition to tension or compression. This makes the stress analysis of trusses much more difficult. Only gradually were satisfactory methods developed for taking that bending into consideration. At the suggestion of Professor Asimont,[6] the presentation of a method of analyzing trusses with rigid joints was proposed by the Munich engineering school as a prize

---

[1] See *Phil. Mag.*, vol. 48, pp. 452–456, 1874; vol. 49, pp. 183–185, 1875.

[2] The first edition of that famous book appeared in 1877.

[3] See *Bull. Tech. Inst. St. Petersburg*, 1885.

[4] Kiev, 1903.

[5] In addition to the above-mentioned book, Kirpitchev published a course on strength of materials in two volumes and a course on graphical statics.

[6] Asimont, Z. *Baukunde*, 1880.

problem and as a result of this, an elaborate investigation of the problem was made by H. Manderla.[1] He shows that with such rigidity, we have to consider not only the displacements of joints, but also their rotation and he derives the necessary three equations of equilibrium for each joint. Since, in this work, the effect of axial forces on bending moments in truss members was taken into account, his set of equations was a complicated one and was not suitable for applications. For practical purposes, some approximate methods were devised in which only the chords were considered as continuous and the web members were assumed as being hinged.[2] A more accurate approximate method was offered by Mohr.[3] He assumes that the displacements of joints are not affected much by the rigidity of joints and determines those displacements by using the methods developed for trusses with ideal hinges. In this way the angles of rotation ψ_{ik} of all members (ik) can be readily calculated (say, by using the Williot diagram). As unknown quantities, he takes the angles of rotation ϕ_i of the rigid joints. The bending moment at the joint i of any member ik will then be given by the equation

$$M_{ik} = \frac{2EI_{ik}}{l_{ik}} (2\phi_i + \phi_k - 3\psi_{ik}) \qquad (g)$$

where $\phi_i - \psi_{ik}$ and $\phi_k - \psi_{ik}$ are the angles of rotation of the ends i and k of the bar ik with respect to the chord ik. From the conditions of equilibrium of joints he obtains now as many equations of the form

$$\Sigma M_{ik} = 0 \qquad (h)$$

as there are unknown rotations ϕ_i. Although the number of Eq. (h) may be considerable, Mohr shows that they can be readily solved by the method of successive approximation. Having now the angles ϕ_i, he finds the bending moments from Eqs. (g) and calculates the corresponding bending stresses (or *secondary stresses*). This method of analyzing trusses with rigid joints proved accurate enough and found wide application in practice.

Since truss analysis is based upon various simplifying assumptions, the engineers have always been interested in experimental verifications of the theoretically calculated stresses and deflections. For instance, W. Fränkel made a special extensometer for measuring the stresses in truss members. He also devised an instrument for recording the deflec-

[1] H. Manderla, Die Berechnung der Sekundärspannungen, *Allgem. Bauztg.*, vol. 45, p. 34, 1880.

[2] See E. Winkler, "Theorie der Brücken," II Heft, p. 276; see also Engesser's book "Die Zusatzkräfte und Nebenspannungen eiserner Fachwerk Brücken," Berlin, 1892.

[3] O. Mohr, *Ziviling.*, vol. 38, p. 577, 1892; see also his "Abhandlungen," 2d ed., p. 467.

tions of bridges under moving loads. These experiments[1] showed a satisfactory agreement between the measured quantities and their theoretically calculated values.

67. Arches and Retaining Walls

We have already seen (page 150) that Bresse drew up the theory of curved bars and discussed as examples a two-hinged arch and an arch with built-in ends. But in his time, engineers did not consider that an elastic theory could be applied to the design of stone arches and continued to treat these as if composed of absolutely rigid bodies. Very gradually however, after considerable experimental work by Winkler (see page 153) and de Perrodil[2] and especially after the extensive tests made by a special committee of the Society of Austrian Engineers and Architects,[3] engineers came to accept the theory of elastic curved bars as giving the proper dimensions of stone arches with sufficient accuracy. Winkler and Mohr were chiefly responsible for the introduction of this theory into practice.

In his book on strength of materials (see page 155), Winkler treated two-hinged arches and arches without hinges in great detail, while in his important paper of 1868,[4] he applied the notions of influence lines to arches. Using the principle of least work, he investigated the positions of *pressure* lines in arches[5] and formulated the principle which bears his name. This states that, from all the funicular curves which can be constructed for the acting loads, the true pressure line is that which deviates as little as possible from the center line of the arch. To arrive at this conclusion, the following reasoning can be used. Let us assume first that the strain energy of an arch may be represented with sufficient accuracy by the energy of bending alone, so that

$$V = \int_0^s \frac{M^2 \, ds}{2EI} \tag{a}$$

Denoting the distance, measured vertically, from any point of the *center line* of the arch to the corresponding point of the *pressure line* by z and calling the horizontal thrust of the arch H, we obtain

$$M = Hz$$

---

[1] W. Fränkel, *Civiling.*, vol. 27, p. 250, 1881; vol. 28, p. 191, 1882; vol. 30, p. 465, 1884.

[2] de Perrodil, *Ann. ponts et chaussées*, 6th series, vol. 4, p. 111, 1882.

[3] See *Z. österr. Ing. u. Architek. Ver.*, 1895, 1901.

[4] *Mitt. Architek. u. Ing. Ver. Böhmen*, 1868, p. 6.

[5] Die Lage der Stüzlinie im Gewölbe, *Deut. Bauztg.*, 1879, pp. 117, 127, and 130; 1880, pp. 58, 184, 210, and 243.

Then, from the principle of least work and from Eq. (a), it follows that the true pressure line is the line for which the integral

$$\int_0^s \frac{H^2 z^2}{2EI}\, ds \tag{b}$$

is a minimum. When all the loads are vertical, H is constant with respect to s and, if the arch cross section is also constant, the requirement of minimizing the integral (b) reduces to that of extremising the integral

$$\int_0^s z^2\, ds \tag{c}$$

This requirement constitutes *Winkler's principle*.

The principle also holds for an arch of variable cross section if the variation of the cross section is such that

$$I = I_0 \frac{ds}{dx}$$

where I_0 is the cross-sectional moment of inertia at the center of the span. Now, instead of the integral (c) the theory requires the minimization of the integral

$$\int_0^l z^2\, dx \tag{d}$$

It follows from Winkler's principle that, if the funicular curve constructed for the acting loads is taken as the center line of an arch, the pressure line coincides with the center line and there is no bending. Actually, we shall always have some bending due to the axial compression of the arch, which was neglected in the expression for the strain energy (a). But this bending is usually small, so that we are justified in using Winkler's principle. Thus, it has become the usual practice to take the funicular curve constructed for the dead loads for the center line of an arch.

Mohr's principal contribution to the theory of arches appears in his paper of 1870[1] in which a graphic method of arch analysis is offered. Considering a two-hinged arch (Fig. 198a), Mohr assumes first that the right-hand hinge B is free to move horizontally and calculates the horizontal displacement of this hinge produced by a vertical unit load applied at a point E. In this calculation, he uses a graphic method and determines displacements from the funicular curve constructed for certain fictitious loads as follows. Considering an element of the arch, between two cross sections at C, he observes that the angle of rotation $d\phi$ of the

[1] *Z. Architek. u. Ing. Ver. Hannover,* 1870.

portion CB due to bending of the element is $M\,ds/EI$ and that the corresponding horizontal component of the displacement BD is equal to $My\,ds/EI$. Summing these elemental displacements, he finds that the total horizontal movement of the hinge B is

$$u = \int_0^l \frac{My\,ds}{EI} = \int_0^{x_1} \frac{x(l - x_1)y\,ds}{lEI} + \int_{x_1}^l \frac{x_1(l - x)y\,ds}{lEI}$$

It is seen that this horizontal displacement can be calculated as the bending moment at the cross section E_1 of the simply supported beam A_1B_1 (Fig. 198b) which is loaded by fictitious forces of magnitude $y\,ds/EI$ applied to each element dx of the beam. Similarly, the horizontal dis-

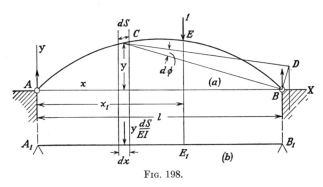

Fig. 198.

placement u_0 of the hinge B can be calculated if, instead of the vertical unit load at E, a horizontal unit load is applied at B. Since, in the actual case, the hinge B is immovable, a horizontal thrust H will be produced by the unit load at E. The magnitude of this is evidently

$$H = 1\frac{u}{u_0}$$

From this equation, Mohr concludes that for each position of the vertical unit load, the corresponding thrust is proportional to the displacement u, and the bending moment diagram constructed for the fictitious loads of the beam A_1B_1 can be taken as the influence line for the thrust H.

The method of fictitious loads together with the fundamental idea regarding the selection of the statically indeterminate quantities (as illustrated by Fig. 197) are the two main simplifications of arch analysis based on the theory of elasticity. As such, they speeded up the introduction of this analysis into practice.

In the design of retaining walls, the engineers continued to use methods based on Coulomb's assumption (see page 61) that sliding of sand occurs along an inclined plane,[1] and the principal advance lay in the development of purely graphic methods of analysis. A very useful method of this kind was proposed by G. Rebhann.[2] He considers the general case wherein the wall AB is not vertical (Fig. 199), its reaction R acts at an angle β to the horizontal, and the earth is bounded by a curved surface ACE.

Rebhann proves that if BD is the plane of natural slope, BC the plane of sliding as defined by Coulomb, and the line CD makes an angle β with the perpendicular CF to BD, then the two areas ABC and BCD must be equal. Hence, to obtain the plane of sliding, he draws the line BC which

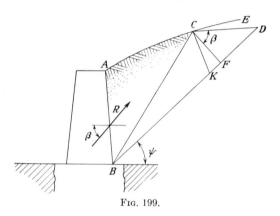

FIG. 199.

divides the area $BACD$ into halves. The writer also shows a simple way of calculating the reaction R of the retaining wall. He has only to construct the triangle KCD (Fig. 199), in which $KD = CD$. Then the area of this triangle multipled by γ, the weight of sand per unit volume, gives the reaction R per unit length of the wall.

Rankine's idea of designing retaining walls on the basis of the analysis of stresses in pulverized materials was further discussed by several engineers.[3] But the results obtained were not accepted as being more reliable than those of Coulomb's theory.

A theory for determining the necessary depth of a foundation was pro-

[1] A complete bibliography for this field can be found in the articles by Felix Auerbach and Felix Hülsenkamp, "Handbuch der physikalischen und technischen Mechanik," vol. 4_2, 1931.

[2] G. Rebhann, "Theorie des Erddruckes und der Futtermauern mit besonderer Rücksicht auf das Bauwesen," Vienna, 1871.

[3] A. Considère, *Ann. ponts et chaussées* (4), vol. 19, p. 547, 1870.

M. Lévy, *J. mathématiques* (2), vol. 18, p. 241, 1873.

E. Winkler, *Z. österr. Ing. u. Architek. Ver.*, vol. 23, p. 79, 1871.

O. Mohr, *Z. Architek. u. Ing. Ver. Hannover*, 1871, p. 344; 1872, pp. 67, and 245; see also his "Abhandlungen," p. 236, 1914.

posed by Pauker.[1] In Fig. 200, let p denote the uniformly distributed pressure transmitted from the wall (the length of the wall is perpendicular to the plane of the figure) AB to the earth. Considering now an element ab of the pulverized material under the wall, and using Eq. (a) of Rankine's theory (page 202), he concludes that, to obviate sliding of the earth, the lateral pressure σ_y must satisfy the condition

$$\sigma_y \lessgtr \frac{p(1 - \sin \phi)}{1 + \sin \phi} \tag{e}$$

where ϕ is the angle of natural slope of the material. Considering now the adjacent element bc, subjected to a pressure $h\gamma$ in the vertical direction he finds the expression

$$\sigma_y = \frac{\gamma h(1 + \sin \phi)}{1 - \sin \phi} \tag{f}$$

for the limiting value of the lateral pressure. Substituting in (e), this gives the value

$$h \lessgtr \frac{p}{\gamma} \frac{(1 - \sin \phi)^2}{(1 + \sin \phi)^2} \tag{g}$$

for the necessary depth of the foundation.

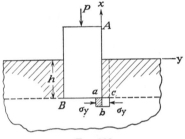

Fig. 200.

To verify this formula and to find the surfaces along which the sliding of sand occurs, some interesting experiments were conducted by V. J. Kurdjumoff in the laboratory of the Institute of the Engineers of the Ways of Communication[2] in St. Petersburg. He used a box with glass walls for his experiments and filled it up with sand. On gradually increasing the pressure P on the block AB (Fig. 200), he was able to find for each value of the depth h, the limiting value of P at which sliding of the sand impended so that the block suddenly moved downward. A photograph taken at that instant clearly shows the sliding surfaces in the sand. The experiments showed that, in order to produce this sliding, a larger pressure p than that given by Eq. (g) is required. Some elaborate experimental work with a model retaining wall was done by H. Müller-Breslau.[3] These experiments showed that the pressure on the wall may sometimes be larger than that predicted by Coulomb's theory. They also indicated clearly that the manner of piling the sand, vibrations in this material, and loading and unloading of its surface may have considerable effect on the magnitude of the sand pressure on the wall.

[1] See *J. Dept. Ways of Communication*, 1889, St. Petersburg.
[2] V. J. Kurdjumoff, *Civiling.*, vol. 38, pp. 292–311, 1892.
[3] See the book by H. Müller-Breslau, "Erddruck auf Stützmauern," Stuttgart, 1906.

CHAPTER XI

Theory of Elasticity between 1867 and 1900

68. The Work of Saint-Venant's Pupils

The most distinguished pupil of Saint-Venant was Joseph Valentin Boussinesq (1842–1929).[1] He was born in the small town of Saint-André-de-Sangonis in the province of Hérault, in the south of France. He received his preliminary education in his home town until, at the age of sixteen, he moved to Montpellier, where he studied languages and mathematics. After his graduation at the age of nineteen, Boussinesq chose the occupation of a teacher and taught mathematics at high schools in several small towns in the south of France. It seems that he was not very interested in elementary teaching, and he spent his leisure hours studying the works of such mathematicians as Fourier, Laplace, and Cauchy. He soon started with his own original work, and, in 1865, Lamé presented Boussinesq's first scientific paper (which dealt with capillarity) to the Academy.

FIG. 201. J. V. Boussinesq.

In 1867, Boussinesq obtained his doctor's degree at the University of Paris. About the same time he wrote a memoir on the theory of elasticity. The paper was sent to the Academy of Sciences where it was reviewed by Saint-Venant who immediately recognized the outstanding ability of its young author and from then on took a great interest in his work. These two men set up a correspondence and discussed the problems which occupied Boussinesq. The younger man was not a lucid writer and was often too impatient in giving logical explanations so that,

[1] For Boussinesq's biography, see Émile Picard's note in *Mém acad. sci.*, 1933.

on several occasions, Saint-Venant advised him to give clear and detailed arguments in his work.

High school teaching was not a suitable career for Boussinesq, and Saint-Venant made every effort to draw the young scientist to university work. Finally he succeeded, and Boussinesq assumed the position of professor at the University of Lille in 1873. He could now devote all his energy to science and was enabled to do important work in several branches of theoretical physics. In 1886, he was elected to membership of the Academy of Sciences and to the chair of mechanics at the University of Paris. From that time to the end of his life Boussinesq lived in Paris, withdrawn from society and politics and using all his time in the pursuit of his scientific work. He was responsible for advances in hydrodynamics, optics, thermodynamics, and the theory of elasticity. His most important contributions to our science are incorporated in the book "Application des potentiels à l'étude de l'équilibre et du mouvement des solides élastiques. . . . "[1] This is one of the most significant books in the field to have been published since Saint-Venant's famous memoirs on torsion and bending of bars.

Lamé and Kelvin had already used potential functions in their analyses of the deformation of spherical bodies but Boussinesq applies them to a variety of problems. From the point of view of practical value, the most important of his solutions are those concerning the stresses and deformations produced in a semi-infinite body by known forces applied at the boundary plane. In the simplest case, he has a force P acting perpendicularly to the horizontal boundary plane gh (Fig. 202). Taking the direction of the z axis as being positive into the solid and using polar coordinates r and θ in horizontal planes, Boussinesq obtains the following components of stress acting on the horizontal planes:

$$\sigma_z = -\frac{3P}{2\pi} z^3 (r^2 + z^2)^{-\frac{5}{2}}$$
$$\tau_{rz} = -\frac{3P}{2\pi} rz^2 (r^2 + z^2)^{-\frac{5}{2}}$$
(a)

We see that

$$\frac{\sigma_z}{\tau_{rz}} = \frac{z}{r}$$

which indicates that the direction of the resultant stress acting at any point of the horizontal plane mn (Fig. 202) is that of a line passing through the loaded point O and its magnitude is

$$S = \sqrt{\sigma_z{}^2 + \tau_{rz}{}^2} = \frac{3P}{2\pi} \frac{\cos^2 \psi}{r^2 + z^2}$$
(b)

[1] Paris, 1885.

where ψ is the angle which the direction of the resultant stress makes with the z axis. We see that the resultant stress is inversely proportional to the square of the distance from the point of application of the load P.

Considering the vertical deflections w of the boundary plane, Boussinesq finds that

$$w = \frac{P(1 - \mu^2)}{\pi E r}$$

which shows that the product wr is constant at the boundary, *i.e.*, radii drawn from the origin on the boundary surface become hyperbolas (after deformation) with asymptotes Or and Oz. At the origin the displacement becomes infinite. To overcome this difficulty, it can be assumed

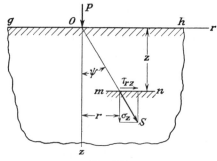

Fig. 202.

that the material near the origin is cut out by a hemispherical surface of small radius and that the concentrated force P is replaced by statically equivalent forces distributed over this surface.

Having the solution for a concentrated force, and using the principle of superposition, Boussinesq solves the problem of a load distributed over part of the boundary of a semi-infinite solid. Considering, for example, the case of a uniform load of intensity q distributed over the area of a circle of radius a, he finds that the vertical deflection at the center of the circle is

$$w_0 = \frac{2(1 - \mu^2)qa}{E}$$

while that at the perimeter is $2w_0/\pi$. In a similar way, he tackles the problems of a load distributed over the area of an ellipse and a rectangle. These problems are of interest to those who study the deflections of foundations in engineering structures.

Boussinesq also treats cases where, instead of distributed forces, the vertical displacements of a portion of the boundary plane are given. For instance, he discusses an absolutely rigid die (in the form of a circular cylinder of radius a) pressed against the plane boundary of a semi-infinite elastic solid and finds that the displacement of the die is

$$w = \frac{P(1 - \mu^2)}{2aE}$$

Thus, for a given value of the average pressure $P/\pi a^2$, the deflection is not constant but increases directly as the radius of the die. The pressure distribution under the die is obviously not uniform, and its smallest value is at the center where it is equal to half the average pressure on the circular area of contact. At the boundary of this area, the pressure becomes infinite, indicating that yielding of the material will set in along that boundary. This yielding, however, is of a local character and does not substantially affect the distribution of pressures at points at some distance from the boundary of the circle.

Boussinesq uses his solutions for proving Saint-Venant's principle. Taking a system of forces in equilibrium applied to a small portion of a body, he shows that the stresses produced are of a local nature. They die out rapidly with increase of the distance from the loaded region and become negligible at points, the distance from the loads of which is only a few times greater than the linear dimensions of the loaded region.

Several "Notes complementaires" are appended to the book "Application des potentiels. . . . " They are of great value. In one of these, the theory of vibration of bars is discussed. As a first example, Boussinesq considers a uniform thin semi-infinite bar, the axis of which is infinitely extended from the origin in the positive direction. Considering lateral vibrations, he uses the equation

$$a^2 \frac{\partial^4 w}{\partial x^4} + \frac{\partial^2 w}{\partial t^2} = 0$$

and gives solutions for various conditions at the end $x = 0$. He is especially interested in the motion set up by a force acting on the end for a very short time, as in impact, and finds that the transverse elastic waves do not retain their form, as in longitudinal vibration, but are dispersed during transmission along the bar and gradually die out. The author discusses the limiting value v of the velocity of impact of an impinging body beyond which local permanent set of the material of the bar will be produced no matter how small the mass of the striking body. He finds

$$v = \sqrt{\frac{E}{\rho}} \frac{\sigma_{up}}{E} \frac{2i}{h}$$

where $\sqrt{E/\rho}$ is the velocity of sound in the bar, h is the depth of the beam, and i is the radius of gyration of its cross section. The result, as we see, is similar to that obtained by Thomas Young for longitudinal impact (page 93).

Boussinesq also takes up longitudinal impact of bars and gives a complete solution of the problem which interested Saint-Venant (see page 239).

Boussinesq did a considerable amount of work in the theory of thin bars and the theory of thin plates.[1] He gives a new method of deriving the equations of equilibrium for thin bars which were previously obtained by Kirchhoff. In the theory of plates, he gives a new derivation of the differential equation of equilibrium and discusses the Poisson-Kirchhoff boundary conditions on the basis of a study of local perturbations produced by replacing one system of contour forces by another which is statically equivalent. In this way, he arrives at the conclusions already found by Kelvin (see page 266). This work of Boussinesq was started at the suggestion of Saint-Venant[2] and is incorporated in the "note finale du no. 73" of the latter's translation of Clebsch's book.

In conclusion, mention should be made of Boussinesq's work on the equilibrium of a pulverulent (or granulated) mass.[3] The work of his predecessors was devoted to the calculation of the limit of equilibrium of such masses, whereas Boussinesq works with the elastic deformation. He assumes that under the action of gravity forces, sufficient pressure exists between the (sand) grains for friction to prevent any sliding between the particles so that the mass deforms as an elastic body under the action of forces. He further assumes that the modulus of shear of such a solid is proportional to the average pressure $p = \frac{1}{3}(\sigma_x + \sigma_y + \sigma_z)$ at any point and, neglecting the change in volume, he establishes the necessary equations for the analysis of stresses. Limiting himself to two-dimensional cases, he succeeds in obtaining solutions which can be used in the theory of retaining walls. Saint-Venant made[4] a comparison of Boussinesq's theoretical results with experimental data and concluded that his friend's theory was an improvement upon those of Rankine and others. On his suggestion, Flamant prepared tables to facilitate retaining-wall design on Boussinesq's theory.[5] An extensive review of Boussinesq's work is given by K. Pearson[6] who concludes: "Among the many pupils of Saint-Venant, few have dealt with such a wide range of

---

[1] See *J. Mathématiques*, 2d series, vol. 16, pp. 125–274; vol. 5, pp. 163–194 and 329–344, 1879.

[2] See Saint-Venant's annotated Translation of Clebsch's book, p. 691.

[3] Essai théorique sur l'équilibre d'élasticité des massifs pulvèrulents . . . Mémoires . . . publiés par l'Académie de Belgique.

[4] *Compt. rend.*, vol. 98, p. 850, 1884.

[5] Tables numériques pour le calcul de la poussée des terres, *Ann. ponts et chaussées*, vol. 9, pp. 515–540, Paris, 1885.

[6] "History . . . ," vol. 2, part 2, p. 356.

elastic problems as Boussinesq, or contributed more useful work to elastic theory . . . he has illustrated our subject by ingenious analysis rather than by special solutions of mechanical and physical problems."

Another of Saint-Venant's pupils was Maurice Lévy (1838–1910). After graduating from the École Polytechnique (1858) and the École des Ponts et Chaussées (1861), he worked as a civil engineer in Paris while holding, at the same time, an instructorship at the École Polytechnique. Later (in 1875), he became a professor of applied mechanics at the École Central des Arts et Manufactures and 1883 saw his admission to the Academy of Sciences as a member.

Lévy was interested in a variety of problems in the theory of elasticity. He derived the differential equation of equilibrium of a rod bent in a plane by uniform lateral pressure and discusses[1] the solution wherein the center line of the bar is a circle in the unstressed state. When the rod is only slightly bent, the equation can be simplified, and from it Lévy manages to extract the critical value of pressure under which a circular ring will collapse.[2]

In his work on plates,[3] Lévy discusses Kelvin's reconciliation of Poisson's and Kirchhoff's boundary conditions, and, for a plate of finite thickness, he gives an elaborate analysis of local perturbations produced by replacing one statical system of boundary forces with another (equivalent) one. Considering bending of rectangular plates, Lévy gives a solution for the important case in which two opposite edges are simply supported while the other two may be clamped, simply supported, or entirely free.[4] This solution found wide application and a variety of particular cases were worked out by E. Estanave in his doctoral thesis.[5]

Lévy solved the two-dimensional problem of the stress distribution in a wedge subjected to pressures on its faces.[6] He suggests the use of this solution in the analysis of stresses in masonry dams. This scientist also wrote a book[7] in which we find a treatment of the graphic methods of structural analysis.

Alfred-Aimé Flamant, after graduating from the École des Ponts et Chaussées, worked on hydrotechnical constructions in the north of France. Later he became the professor of theory of structures at the École des Ponts et Chaussées and the professor of mechanics at the École Central des Arts et Manufactures. He was interested in the theory of

[1] *J. mathématiques* (3), vol. 10, 1884.
[2] Bresse was the first to obtain this solution.
[3] *J. mathématiques* (3), vol. 3, pp. 219–306, 1877.
[4] *Compt. rend.*, vol. 129, 1899.
[5] Thèse, Paris, 1900.
[6] *Compt. rend.*, vol. 127,p. 10, 1898.
[7] "La Statique Graphique," 3d ed., vols. 1–4, Paris, 1907.

granulated masses and simplified Boussinesq's theory.[1] He collaborated with Saint-Venant in translating Clebsch's book. A joint work of Saint-Venant and Flamant on longitudinal impact of bars is included in this book in the form of an appendix.

Using Boussinesq's solutions for a semi-infinite body, Flamant obtained the stress distribution produced in a semi-infinite plate of unit thickness by a force P perpendicular to its edge[2] (Fig. 203) as a particular case. The distribution of stresses in this case is particularly simple. Any element C at a distance r from the point of application of the load is subjected to a simple compression in the radial direction of magnitude

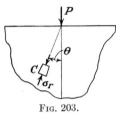

Fig. 203.

$$\sigma_r = -\frac{2P}{\pi}\frac{\cos\theta}{r} \qquad (a)$$

The stress σ_θ and the shearing stress $\tau_{r\theta}$ vanish in this case. This is the simple radial stress distribution which was used later by J. H. Michell in solving several important problems (see page 353).

69. Lord Rayleigh

Lord Rayleigh (John William Strutt)[3] was the eldest son of the second Baron Rayleigh and was born at Langford Grove near Terling on the estate of his father where he spent the greater part of his life. After studying in a private school, he entered Trinity College, Cambridge, in 1861. As an undergraduate he was coached by Routh in mathematics and mechanics. Routh had the reputation of being an outstanding teacher, and Rayleigh had a high opinion of his coach. When President of the Royal Society, he once said of Routh:[4] "I was indebted to him for mathematical instruction and stimulus at Cambridge, and I can still vividly recall the amazement with which, as a freshman, I observed the extent and precision of his knowledge, and the rapidity with which he could deal with any problem presented to him. . . . I have always been under the impression that Routh's scientific merits were underrated. It was erroneously assumed that so much devotion to tuition could leave scope for little else. . . . I think, perhaps, to the

---

[1] *Ann. ponts et chaussées*, vol. 6, pp. 477–532, 1883. See also his book "Essai théorique sur l'équilibre des massifs pulvérulents," Paris. An elementary discussion of the stability of granulated masses can be found in Flamant's textbook, "Stabilité des constructions, résistance des matériaux," Paris, 1886.

[2] See Flamant, *Compt. rend.*, vol. 114, p. 1465, 1892. An extension of the solution to cover an inclined force is due to Boussinesq, *Compt. rend.*, vol. 114, p. 1510, 1892.

[3] A biography of Lord Rayleigh, "John William Strutt, Third Baron Rayleigh," London, 1924, was written by his son, Robert John Strutt, fourth Baron Rayleigh.

[4] See the above-mentioned biography, p. 27.

credit of the mathematical system of those days, which it is perhaps now rather the fashion to abuse, and to the manner in which it was worked by Dr. Routh, that I have at no time during what I may call my subsequent scientific career had occasion to regret any of the time I spent on the mathematical course at Cambridge." However, at the same time Rayleigh noticed some "vices of the Cambridge school" and refers to the tendency to "regard a display of analytical symbols as an object in itself rather than as a tool for the solution of scientific problems."

There was no opportunity for doing experimental work at Cambridge. There were no physics courses in the early years of Rayleigh's study and not until the autumn of 1864 was he able to attend Professor Stokes' lectures on optics. He was very impressed by those lectures and especially by the experimental illustrations. There was no physics laboratory at the university and, as mentioned before, Stokes did his experimental work "with the most modest appliances."

FIG. 204. Lord Rayleigh.

Rayleigh was a very systematic worker and, as the time of the mathematical tripos examination approached, he gradually drew to the front among the pupils of Routh's class. In December, 1864, he successfully passed the examination, coming out as the senior wrangler. He remained in Cambridge after these examination successes with a view to preparing for the fellowship examination. He also wanted "to make a start in scientific investigation, but he found this extremely difficult. The facilities then existing at Cambridge were very small. In particular, he had no experimental training. . . . He had learned something from the experiments shown by Stokes, and would gladly have learned more from this source, but did not find a path open to do so. Professor Stokes was kind and civil in answering questions after the lectures, but his attempts to learn from him where similar apparatus could be procured did not lead to much. . . . He was somewhat disappointed to find that his eagerness did not meet with any response in being invited to assist, even if only in getting out and putting away the apparatus."[1]

It was easier to obtain guidance in theoretical work. Rayleigh read the more important papers of Stokes and Kelvin and studied Maxwell's

[1] See Lord Rayleigh's biography, p. 37.

great work "A Dynamical Theory of the Electro-magnetic Field." He came into correspondence with Tait and with Maxwell. The latter, at that time, was living on his estate at Glenlair, but he went to Cambridge on the occasions of the mathematical tripos examinations and then the two men were able to meet.

Fig. 205. Trinity College.

In 1866, Rayleigh was elected to a fellowship of Trinity College and made his first trip to North America in the following year. On his return in 1868, he started to do some experimenting in electricity and for that purpose purchased some simple apparatus. About the same time he read Helmholtz's famous book "Lehre von den Tonempfindung" and became interested in acoustics.[1] He made some theoretical and experimental investigations with resonators and prepared a paper on the theory

[1] See Lord Rayleigh's biography, p. 50.

of resonance,[1] which was the beginning of Rayleigh's original work on sound. Later, as a result of reading Tyndall's work, he became interested in the question of why the light of the sky is blue and published several papers on the theory of light.[2] These were applauded by Maxwell and Tyndall.

In 1871, Rayleigh was married and consequently had to vacate his fellowship at Trinity. In 1872, after a serious illness, it was recommended that he should spend the winter in a warmer climate. He took his wife and several relatives on an expedition up the Nile in a kind of sailing houseboat. During the trip, life was monotonous, but it suited Rayleigh. He began working on his famous treatise "The Theory of Sound" and "worked at it each morning in the cabin. . . . At these times it was difficult to persuade him to land, even to see the most enchanting temple."[3]

Rayleigh's father died in 1873, and the scientist had thenceforth to take a part in supervising the estate. He settled at Terling and, to allow him to continue his research, he there set up a small laboratory. He continued to work on his treatise which was finally published in 1877. Helmholtz reviewed the book in *Nature*, and on his suggestion a German translation of the book was made and printed in 1878.

In 1879, Maxwell died, and Rayleigh was elected to the chair of experimental physics at Cambridge. From his own experience he knew how important it was to give students a chance to do experimental work. "Previous to his arrival the opportunities of practical instruction for the average undergraduate had not been adequate, and had by no means reached the standard set in Germany by Helmholtz. . . . It was Rayleigh's intention to develop general elementary instruction on a much larger scale than had previously been attempted, and for this purpose considerable numbers of instruments were required. . . . "[4] To meet the deficiency, Rayleigh got up a fund to which he himself and the Duke of Devonshire contributed.[5]

In the task of administering the laboratory, Rayleigh was assisted by Glazebrook and Shaw who wrote, at that time, the book "Practical Physics" in which the early experiments of the laboratory are described. Rayleigh gave elementary courses on electrostatics and magnetism, current electricity, sound, and advanced courses on electrical measurements. But "the school of natural science at Cambridge was in those days still in

[1] *Phil. Trans.*, vol. 161, pp. 77–118, 1870.
[2] See *Phil. Mag.*, vol. 41, 1871.
[3] Lord Rayleigh's biography, p. 62.
[4] Lord Rayleigh's biography, p. 105.
[5] Much interesting information regarding the growth of the laboratory can be found in the book "A History of the Cavendish Laboratory," 1910.

its childhood and naturally the numbers attending at the Cavendish Laboratory were not great. . . . " The first course was given to a mere sixteen students, and this figure remained about the same throughout Rayleigh's tenure of the chair.[1]

In 1884, Rayleigh undertook a second trip to America and attended Lord Kelvin's lectures at Baltimore. Concerning these he discussed his experiences in later years with his son: "What an extraordinary performance that was! I often recognized that the morning's lecture was founded on the questions which had cropped up when we were talking at breakfast." When his son remarked "that when all the leading physicists in America had collected to hear him, they would have expected something more carefully prepared," Rayleigh replied, "on the contrary, they were very much impressed and he got some of them to do grinding long sums for him in the intervals."[2]

At the end of 1884, Rayleigh resigned his Cambridge professorship and settled again on his estate. There, in his "book-room" and in his laboratory, he continued his intensive scientific work to the end of his life.

He undertook also many public duties. Among these, he was professor of natural philosophy at the Royal Institution (1887–1905), president of the Royal Society (1905–1908), and chancellor of Cambridge University (1908 until his death). He won many honors during his life including the Nobel prize for physics in 1904 and the Order of Merit.

Rayleigh's principal contributions to our subject are included in his treatise "The Theory of Sound." In the first volume of this famous book he discusses vibration of strings, bars, membranes, plates, and shells. The author shows the advantages which an engineer can gain by adopting the notions of *generalized forces* and *generalized coordinates*. The introduction of these concepts and the use of the Betti-Rayleigh reciprocity theorem have brought about great simplification in handling redundant structures. The work not only covers sound proper but also covers nonsonic vibration. The writer demonstrates the benefits to be had by using *normal coordinates* and shows how, by making velocities vanish, the solutions of static problems may be extracted from vibration analysis. Thus he obtains the deflections of bars, plates, and shells expressed in terms of normal functions; these ideas have been of great importance in engineering.

In finding frequencies of vibration of complicated systems, he obtains an approximate value by assuming a suitable form for the type of motion and so transforming the problem to that of the vibration of a simple oscillator. He describes the steps which can be taken toward improving the approximation. This idea of calculating frequencies directly from an energy consideration, without solving differential equations, was

[1] Lord Rayleigh's biography, p. 106.
[2] Lord Rayleigh's biography, p. 145.

Theory of Elasticity between 1867 and 1900

later elaborated by Walter Ritz[1] and the *Rayleigh-Ritz method* is now widely used not only in studying vibration but in solving problems in elasticity, theory of structures, nonlinear mechanics, and other branches of physics. Perhaps no other single mathematical tool has led to as much research in the strength of materials and theory of elasticity.

Rayleigh uses his approximate method in several complicated problems. In the vibration of strings, he shows how an approximate solution can be found where the mass of the string is not uniformly distributed along the length and he investigates the effect, on the frequency, of a small concentrated mass attached to the string. In treating the longitudinal vibrations of bars he estimates the error involved in neglecting the inertia of lateral motion of particles not situated on the axis.[2] He shows that for a circular bar his result is in a good agreement with the results of a more elaborate theory devised by Pochhammer.[3] Again, a correction is made to the simple theory of lateral vibration of rods to allow for the rotatory inertia.[4]

An important advance in the theory of vibration of thin shells was made by Rayleigh. Two kinds of vibration have to be considered: (1) extensional modes by which the middle surface of the shell undergoes stretching and (2) flexural or inextensional modes. In the first case, the strain energy of the shell is proportional to its thickness, while in the second it is proportional to the cube. Now, on the strength of the principle that, for given displacements, the strain energy of the shell shall be as small as possible, Rayleigh concludes that "when the thickness is diminished without limit, the actual displacement will be one of pure bending, if such there be, consistent with the given conditions." Using this conclusion, he examines[5] flexural vibrations of cylindrical, conical, and spherical shells and obtains results which agree satisfactorily with experiments.

One important contribution of Lord Rayleigh's to elasticity, not to be found in his treatise, is the theory of elastic surface waves.[6] In the waves which bear his name, motion is propagated over the surface of an elastic medium in a manner similar to the rippling set up when the surface of still water is disturbed. As predicted by Rayleigh, these waves have been found to play an important part in earthquakes.

70. Theory of Elasticity in England between 1867 and 1900

The appearance of Rayleigh's treatise on sound together with that of Thomson and Tait's "Natural Philosophy" (see page 267) has had a

---
[1] W. Ritz, *Crelle's J.*, vol. 85, 1909.
[2] "The Theory of Sound," vol. I, art. 157.
[3] *J. Math. (Crelle)*, vol. 81, 1876.
[4] "The Theory of Sound," vol. I, art. 162.
[5] "The Theory of Sound," 2d ed., vol. I, p. 396.
[6] *Proc. London Math. Soc.*, vol. 17, 1887, or "Scientific Papers," vol. 2, p. 441.

great effect upon physics in general and our subject in particular. During the last quarter of the nineteenth century British elasticians have produced much fine work which will be only briefly summarized in this book primarily intended for engineers.

Sir Horace Lamb[1] (1849–1934) was born in Stockport (near Manchester) and was the son of a cotton-mill foreman. He attended the grammar school in his home town before entering Trinity College, Cambridge. He graduated in the mathematical tripos as the second wrangler and second Smith's prizeman of 1872 and remained as a lecturer at Trinity until 1875. He then became professor of mathematics in the University of Adelaide, Australia, where he remained for ten years. At the end of that period he returned to take up the chair of pure mathematics (and later applied mathematics also) in the University of Manchester. His period of tenure was extended to the year 1920 when he was elected to an honorary fellowship at Trinity College, Cambridge, and Rayleigh lecturer in mathematics, in the University of Cambridge. We are told that not only was he a first-class scientist but also a born teacher.

Fig. 206. Horace Lamb.

Lamb's principal work lay in the field of hydrodynamics, but he was also interested in the theory of elasticity and published several important papers in that field. As a result of Rayleigh's work on vibration of shells, he took up the theory of plates and shells and investigated extensional vibrations[2] of cylindrical and spherical shells, which were not considered in Rayleigh's approximate theory. Discussing boundary conditions at the edges of rectangular plates, he showed that a rectangular plate can be held in the form of an anticlastic surface by two pairs of equal forces directed normally to the plate and applied at its corners.[3] This result represents the simplest case of bending of plates and was later utilized by A. Nadai for an experimental verification of the approximate (Kirchhoff) theory of bending of plates.[4] Another interesting problem of edge

---

[1] See "Dictionary of National Biography," Oxford.
[2] *Proc. London Math. Soc.*, vol. 21, p. 119, 1891.
[3] *Proc. London Math. Soc.*, vol. 21, p. 70, 1891. The case was discussed also in Thomson and Tait, "Natural Philosophy," vol. 2, art. 656.
[4] See A. Nadai, "Elastische Platten," p. 42, Berlin, 1925.

conditions was mentioned in Thomson and Tait's "Natural Philosophy." It was stated: "Unhappily, mathematicians have not hitherto succeeded in solving, possibly not even tried to solve, the beautiful problem thus presented by the flexure of a broad, very thin band (such as a watch spring) into a circle. . . ."[1] Lamb investigated the anticlastic bending along the edge of a thin band[2] and achieved considerable progress in the solution of the beam problem.[3] Considering an infinite beam of a narrow rectangular cross section loaded at equal intervals by equal concentrated forces acting in the upward and downward directions alternately, he simplified the solution of the two-dimensional problem and obtained, for several cases, expressions for the deflection curves. It was shown in this manner that the elementary Bernoulli theory of bending is accurate enough if the depth of the beam is small in comparison with its length. It was shown also that the correction for shearing force as given by Rankine's and Grashof's elementary theory is somewhat exaggerated and should be diminished to about 0.75 of its value. Mention should be made also of Lamb's work concerning the theory of vibration of elastic spheres[4] and the propagation of elastic waves over the face of a semi-infinite solid[5] and in a solid bounded by two plane faces.[6] He took also the theory of vibration of naturally curved rods.[7] His treatment (with R. V. Southwell) of the vibration of a spinning disk is of a special interest to engineers.[8]

A. E. H. Love (1863–1940) was yet another well-known worker in elasticity to emerge from Cambridge during the period under review.[9] Born in Weston super Mare, the son of a surgeon, he went to Wolvehampton Grammar School (1874). Later he entered St. John's College, Cambridge, and graduated, as the second wrangler, in 1885. A bachelor all his life, he held a fellowship at St. John's from 1887 to 1899 and occupied the Sedleian Chair at Oxford for the remainder of his life. His work brought him many honors. He was elected a Fellow of the Royal Society in 1894 and he was intimately connected with the London Mathematical Society.

Love's chief interests lay in elasticity and geophysics but, like many of his contemporaries of the Cambridge school, he did much work in other

[1] "Natural Philosophy," vol. 2, art. 717.
[2] *Phil. Mag.* (5), vol. 31, p. 182, 1891.
[3] See *Atti congr. intern. math.*, 4th *Congr. Rome*, 1909, vol. 3, p. 12; *Phil. Mag.* (6), vol. 23, 1912.
[4] *Proc. London Math. Soc.*, vol. 13, 1883.
[5] *Trans. Roy. Soc. (London)*, (A), 1904.
[6] *Proc. Roy. Soc. (London)*, (A), 1917.
[7] *Proc. London Math. Soc.*, vol. 19, p. 365, 1888.
[8] *Proc. Roy. Soc. (London)*, (A), vol. 99, 1921.
[9] See Love's obituary notice in *J. London Math. Soc.*, vol. 16, 1941.

subjects. His principal work, "Treatise on the Mathematical Theory of Elasticity," appeared in 1892–1893 in two volumes and, in its original form, it was of a somewhat abstract character. In later editions, an effort was made by its author to change this and thus to make its contents more useful for engineers.[1] It is a standard work and ever since its appearance it has been the greatest source of information in its subject. This work contains a short historical review of the subject and numerous references to the new research in theory of elasticity.

Love started his original work in the theory of elasticity with the theory of thin shells. In connection with Rayleigh's work on the vibration of shells (mentioned above) Love made a complete investigation of bending of thin shells[2] by using Kirchhoff's method and showed that Rayleigh's assumptions regarding flexural vibration do not exactly satisfy the conditions at the edges. In the later editions of his book Love considerably extended the rigorous theory of plates by using J. H. Michell's solution of the two-dimensional problem of elasticity.[3] Love also attacked the problem of the elastic equilibrium of a solid sphere[4] and published a book "Some Problems of Geodynamics,"[5] which contains original treatments of many problems of geophysics. The last chapter contains a discussion of the propagation of seismic waves, including a correction to the theory of Rayleigh waves to allow for gravity and the important demonstration of the possibility of waves which now bear Love's name. These *Love waves* can exist in stratified media and give rise to motions transverse to the direction of propagation and in the plane of the free boundary of the transmitting solid.

FIG. 207. A. E. H. Love.

Karl Pearson (1857–1936),[6] another scientist from Cambridge, also contributed much to the development of the theory of elasticity. He was born in London and educated in Kings College, Cambridge. He grad-

[1] The book, in its second edition (1906), was translated into German and was widely used in Central Europe and Russia.
[2] *Trans. Roy. Soc. (London)*, (A), vol. 179, 1888.
[3] *Proc. London Math. Soc.*, vol. 31, p. 100, 1899.
[4] *Proc. Roy. Soc. (London)*, (A), vol. 82, p. 73, 1909.
[5] Appeared in 1911. The book was an Adams' prize essay.
[6] See "Dictionary of National Biography, 1931–1940," Oxford.

uated as third wrangler, 1879, and did some postgraduate work at Heidelberg and Berlin Universities. His first publications were literary and his original intention was to practice law. In 1884 he abandoned law for applied mathematics and became professor of applied mathematics and mechanics in University College, London. During the early years of activity in this college, Pearson contributed much to our subject. In 1891 he became a professor at Gresham College also and gave popular instruction there. He was an outstanding lecturer and could hold a large audience of students or casual hearers. In 1892 his book "The Grammar of Science" was published, which had a great influence on the younger generation. Since the end of the nineteenth century, Pearson became interested in the theory of probability and its applications in statistics and natural sciences. He became Galton professor in eugenics in 1911.

The principal contribution of Pearson to the theory of elasticity is the editing of the book the "History of the Theory of Elasticity" by Isaak Todhunter and Karl Pearson, the first volume of which appeared in 1886[1] and the second in 1893. Todhunter, who started the work, planned to review only memoirs of a mathematical nature, but Pearson, who completed it and put it into its final form, extended its scope by including reviews of papers having a physical or technical interest. In this way the usefulness of the finished volumes was considerably increased. Pearson was especially interested in Saint-Venant's work and the review of that work was extracted from the second volume of the history and published in the form of a separate volume[2] dedicated "to the disciples of Barré de Saint-Venant who have so worthily carried on his work."

At the start of his scientific career at University College, Pearson published several original papers in elasticity of which the investigation[3] "On the Flexure of Heavy Beams Subjected to Continuous Systems of Load" is of considerable interest to elasticians. In this work Pearson extends the theory of bending of beams to the cases in which volume forces, such as gravity, are acting. From the complete solution of the problem for the cases of circular and elliptical cross section Pearson concludes "that the Bernoulli-Eulerian theory, considered as an exact theory, is false in the case of beams continuously loaded. On the other hand its results approximate closely to those obtained from the true theory." Some of Pearson's papers are of interest to engineers. He investigated bending of continuous beams on elastic supports[4] and showed that in this case equations containing moments at five consecutive supports are obtained. He dis-

[1] It was dedicated to Saint-Venant.
[2] "The Elastical Researches of Barré de Saint-Venant" by Karl Pearson, Cambridge, 1889.
[3] *Quart. J. Pure Applied Math.*, no. 93, 1889.
[4] The *Messenger of Mathematics*, new series, no. 225, 1890.

cussed also the important practical problem of stresses in masonry dams.[1]

Several other British scientists were at work on elasticity during this period. J. Larmor gave an extension of the theorem of the kinetic analogue (Kirchhoff) to rods naturally curved.[2] He showed[3] also that if a shaft in torsion has a cylindrical cavity of circular section with its axis parallel to that of the shaft, the shear in the neighborhood of the cavity may be nearly twice the maximum shear that would exist if there were no cavity. Charles Chree, the well-known geophysicist, also dealt with elasticity in some of his early papers. His analysis[4] of stresses produced in a rotating sphere, a rotating ellipsoid, and rotating disks is of practical interest since it gives some refinement to the usual theory of stress distribution in rotating disks of variable thickness as in steam and gas turbines.

As we have seen, the last quarter of the nineteenth century saw an impressive number of results obtained by British scientists. The men involved did not dedicate themselves to any one branch of physics and in some cases elasticity seems to have been a distinct side line. This period of English productivity was carried over into the present century by the men that we have discussed and by the arrival on the scene of Profs. L. N. G. Filon, R. V. Southwell, G. I. Taylor, and others.

71. Theory of Elasticity in Germany between 1867 and 1900

During the last part of the nineteenth century, a considerable amount of work in the field of theory of elasticity was done by German scientists. Several books on the subject were published and the most important of them was the course of lectures by Franz Neumann, which has already been mentioned (page 251). Some of Neumann's pupils became professors of physics in German universities and they maintained the traditions of their teacher's school by organizing seminars and graduate study in physics and (incidentally) paying great attention to the theory of elasticity. The works of Kirchhoff and Clebsch, Neumann's earlier pupils, have already been discussed. Of Neumann's later pupils, the best known was Woldemar Voigt (1850–1919).[5]

W. Voigt was born in Leipzig, Saxony. There, he was given his high school education and started university work. In 1870 he was drafted into the army of Saxony and he fought in the Franco-Prussian War of

---

[1] See *Drapers' Company Research Memoirs*, Tech. Series II, London, 1904; Tech. Series V, 1907.

[2] *Proc. London Math. Soc.*, vol. 15, 1884.

[3] *Phil. Mag.* (5), vol. 33, 1892.

[4] *Proc. Cambridge Phil. Soc.*, vol. 7, pp. 201, 283, 1892; *Proc. Roy. Soc. (London)*, (A), vol. 58, 1895.

[5] A brief biography of Voigt is given in C. Runge's article Woldemar Voigt, *Nachr. Ges. Wiss. Göttingen, Math. physik. Klasse*. 1920.

1870–1871. After the war, he decided to enter the University of Königsberg where he joined Neumann's physics laboratory class. He acquired an interest in theory of elasticity and prepared (1874) his doctor's thesis on the elastic properties of rocksalt. In 1875, he became an assistant professor at Königsberg and relieved Neumann in parts of his teaching program.

In 1883, Voigt was elected to a chair at Göttingen University where he instituted courses in theoretical physics. He also set up a laboratory in which work on the theory of elasticity played an important part. He was especially interested in the elastic properties of crystals and did outstanding work in that subject. His numerous publications on this subject were later incorporated in his "Lehrbuch der Kristallphysik"[1] which is still a very important book in its field.

Voigt's work finally settled the old controversy over the rariconstant and multiconstant theories. The questions were these: "Is elastic isotropy to be defined by one or two constants? And, in the general case, is elastic aeolotropy to be defined by 15 or 21 constants?" The experiments of Wertheim and Kirchhoff could not settle the question due to imperfections in the materials with which they experimented. Voigt used thin prisms cut out from single crystals in various directions in his experiments. The elastic moduli were determined from torsional and bending tests of these prisms. In addition, the compressibility of crystals under uniform hydrostatic pressure was studied. The result he obtained definitely disproved those relations between the elastic constants which follow from the rariconstant theory. Thus the Navier-Poisson hypothesis regarding molecular forces was shown to be untenable.

Voigt became interested in Saint-Venant's theory of longitudinal impact of prismatical bars (page 239) and made a series of tests[2] with metal rods. The results did not agree with the theoretical calculations. Now Saint-Venant's theory is based upon the assumption that contact between the bars takes place at the same instant over the whole surface of the ends of the rods. This condition is difficult to realize in practice and, in order to bring his experimental results into agreement with the theory, Voigt suggested that the two rods are separated by a "layer of transition" which takes care of imperfections of the surface of contact. By proper selection of the mechanical properties of this layer, a satisfactory agreement between practice and theory can be achieved. The chief difficulty of these experiments lay in the fact that the duration of impact is extremely short. Much better agreement with Saint-Venant's theory

---

[1] Published by Teubner, Leipzig, 1910. See also Voigt's report, "L'état actuel de nos connaissances sur l'élasticité des cristaux," presented to the International Congress of Physics, Paris, 1900.

[2] See *Ann. Physik*, vol. 19, p. 44, 1883; vol. 46, p. 657, 1915.

was obtained by C. Ramsauer[1] who used helical springs instead of rods. In this way, the velocity of propagation of longitudinal waves is reduced, and the time taken by the wave to travel along the rod and back is large in comparison with the time required for flattening small irregularities of the ends. Yet another way of making end conditions more definite during impact is to use rods with spherical ends and to treat the local deformation at the ends by Hertz's theory.[2]

Engineers have been very interested in a series of experimental investigations which Voigt and his pupils made[3] to clarify ideas regarding the ultimate strength of materials. Working with specimens cut out from big crystals of rock salt, he found that tensile strength greatly depends upon the orientation of the axis of the specimen with respect to the crystal's axes. It depends also upon the surface conditions of specimens. Voigt showed that slightly etching the lateral surface of glass specimens gives rise to a considerable increase in their strength. Again, he showed that in nonhomogeneous stress distribution, the strength at a point depends not only on the magnitude of stresses at that point but also on the rate of their change from point to point. Comparing, for example, the ultimate strength in tension with that in bending, obtained for rock salt and for glass, he found that the maximum stress at fracture in bending was about twice that of failure in tension. A number of tests were made with combined stresses with a view to checking Mohr's theory. All these tests were made with brittle materials and the results obtained were not in agreement with theory. Voigt came to the conclusion that the question of strength is too complicated and that it is impossible to devise a single theory for successful application to all kinds of structural materials.

In connection with his work with crystals, Voigt did important theoretical work in elasticity. Dealing with torsion and bending of prisms cut out from crystals, he extended Saint-Venant's theory and adapted it to cover various kinds of aeolotropic materials. Concerning the deformation of plates by forces acting in their middle plane this scientist derived[4] the fourth-order partial differential equation which the stress function must satisfy in the case of an aeolotropic material. He also took up bending of aeolotropic plates, established the differential equation for lateral deflections,[5] and used it in discussing the vibration of such plates. In

[1] *Ann. Physik,* vol. 30, p. 416, 1909.
[2] Such an investigation was made by J. E. Sears, *Trans. Cambridge Phil. Soc.*, vol. 21, p. 49, 1908. See also J. E. P. Wagstaff, *Proc. Roy. Soc. (London)*, (A), vol. 105, p. 544, 1924.
[3] The principal results of these investigations are incorporated in an article by Voigt, *Ann. Physik* (4), vol. 4, p. 567, 1901.
[4] "Lehrbuch der Kristallphysik," p. 689.
[5] "Lehrbuch der Kristallphysik," p. 691.

this way a theoretical explanation of Savart's[1] experiments with nodal lines in aeolotropic circular plates was given. The analysis of nodal lines presents a very sensitive way of detecting deviations from perfect isotropy of the plate material. Finally, Voigt was the first to introduce the notions of tensor and tensor-triad into theory of elasticity. These are now widely used.

To complete our discussion of the better known elasticians amongst Neumann's pupils, mention must be made of A. Wangerin, who worked with the problems of symmetrical deformation of solids of revolution,[2] of L. Saalschütz who investigated a number of special cases of large deflections of prismatical bars,[3] and C. W. Borchardt who contributed to the theory of thermal stresses. Whereas Duhamel and Neumann investigated the cases of a circular cylinder and a sphere with temperature a function of the radial distance only, Borchardt gives solutions for both of these in the form of integrals for any temperature distribution.[4] He extended this method to the deformation produced in circular cylinders and spheres by any surface forces.[5]

A very important advance made by Heinrich Hertz[6] belongs to the period under consideration. Heinrich Rudolf Hertz (1857–1894) was born in Hamburg in the well-to-do-family of a lawyer. From his youth he showed great ability in his study and also an interest in mechanical work. During his high school years, he attended evening courses in a technical school and acquired experience in making drawings and in using measuring instruments. After graduating, he decided to study engineering and in 1877 he entered the Polytechnical Institute at Munich. But very soon he found that his interests lay more in the direction of pure science; therefore, after a year of study in Munich, he moved to Berlin where the famous physicists H. Helmholtz and G. Kirchhoff were teaching.

Fig. 208. Heinrich Hertz.

[1] *Ann. chim. et phys.*, vol. 40, 1829.
[2] *Archiv. Math.*, vol. 55, 1873.
[3] See the book by L. Saalschütz, "Der belastete Stab," Leipzig, 1880.
[4] C. W. Borchardt, "Gesammelte Werke," p. 246, Berlin, 1888.
[5] Borchardt, "Gesammelte Werke," p. 309.
[6] For a biography of this worker, see the Introduction to the "Gesammelte Werke" of H. Hertz, vol. I, and the foreword by H. Helmholtz to vol. III, Leipzig, 1894.

In October, 1878, Hertz started to work in Helmholtz's physics laboratory; owing to his previous experience at home and at Munich, he was well prepared for the work and progressed very rapidly. A prize was offered to the students by the philosophical department of the university in electrodynamics and Hertz decided to embark upon original research in that field. From his letters to his parents, we know about his first encounter with Helmholtz and about the preliminary discussion of the problem. Hertz made a very favorable impression on Helmholtz and was offered a separate room and all the necessary equipment for the work. Helmholtz became interested in the young man's experiments and visited his room every day to discuss the progress of the work with him. Hertz soon showed himself to be a first-class experimenter. During the winter of 1878–1879, the work was finished and Hertz won the gold medal of the university for it. He continued to work in electrodynamics and, in January, 1880, presented his thesis for a doctor's degree, which he obtained with the rare distinction of *summa cum laude*.

In the fall of 1880, Hertz became Helmholtz's assistant and was free to use any of the institute's equipment in his work. Working with Newton's colored rings, he became interested in the theory of compression of elastic bodies and in January, 1881, presented his famous paper[1] on that subject to the Physical Society of Berlin. He not only offered the general solution of the problem, but applied it to particular cases and prepared a numerical table to simplify practical applications. He also extended his theory to impact and derived formulas for calculating the duration of impact of two spheres and for calculating the stresses produced during that impact. The paper attracted the interest not only of physicists but also of engineers, and, at their request, he prepared another version of his paper,[2] in which a description of his experiments with compression of glass bodies and with circular cylinders is added. Covering one of the bodies with a thin layer of soot before compression, Hertz obtained, as an outline of the surface of contact, an ellipse the axes of which could be measured accurately. In this way, he was able to give an experimental proof of his theory.

Working on the compression of elastic bodies, Hertz became interested in the hardness of materials. He was dissatisfied with the methods of measuring hardness[3] and introduced his own definition of it. He advocated the use (for measuring purposes) of bodies of such form that the surface of contact be circular, and he used a sphere of definite radius

[1] *J. reine angew. Math.*, vol. 92, pp. 156–171, 1881.
[2] See "Verhandlungen des Vereins zur Beförderung des Gewerbefleisses," Berlin, November, 1882.
[3] He found the description of those methods in the book by M. F. Hugueny, "Recherches expérimentales sur la dureté des corps," Strasbourg, 1864.

which he pressed onto a flat surface of a body under investigation. As a measure of hardness, he then took that load at which permanent set was brought about in the tested material. Applying this definition in studying the hardness of glass (which remains elastic up to the instant of fracture), he took as his measure of hardness the load at which the first crack appeared along the boundary of the surface of contact. Hertz's method did not find acceptance, since in ductile materials it is very difficult to find at what load permanent set begins.[1]

At the beginning of 1883, Hertz again became interested in a problem of elasticity. Now it was the question of bending of an infinite plate floating on water and normally loaded at one point.[2] He found that the plate deflects downward under the load but that at a certain distance from the load deflections become negative. Then, at an increased distance, they again become positive, and so on. Thus the surface is wavy, and the height of the waves rapidly diminishes with increase of the distance from the load. He arrived, in this way, at the paradoxical conclusion that a plate, heavier than water, can be made to float by loading it at the center. The explanation is that due to bending the plate acquires the form of a shell and can press out more water than is equivalent to its own weight.

In the same year, Hertz solved another important problem of elasticity.[3] He considers the case of a long cylinder subjected to loads which are perpendicular to its axis and have constant intensity along its length. He finds the general solution of the problem and, as a particular instance, investigates the stress distribution in circular rollers such as are used in the construction of moving bridge supports.

In 1883, after a three-year period of assistantship in Helmholtz's laboratory, Hertz went to Kiel, becoming a lecturer at the university there. In 1885 he was elected professor of physics at the Polytechnical Institute at Karlsruhe. Whilst there, he did his famous work in electrodynamics. He discovered the propagation of electromagnetic waves through space and showed that they are similar to waves of light and heat. In this way he furnished experimental proof of Maxwell's mathematical theory. In 1889 Hertz was elected to the chair of physics at the University of Bonn. Here he worked on the discharge of electricity in rarified gases and wrote a treatise on the principles of mechanics. This was his last work, for in January, 1894, he died at Bonn. Although his investigations in theory of elasticity constituted only a few of Hertz's scientific accom-

[1] A. Föppl tried to apply Hertz's method by using, as specimens, two circular cylinders of equal diameter, the axes of which he made perpendicular to one another. See *Mitt. mech. tech. Lab.*, Munich, 1897.
[2] *Wiedemann's Ann.*, vol. 22, pp. 449–455, 1884.
[3] *Z. Math. u. Physik*, vol. 28, p. 125, 1883.

plishments, he succeeded in solving some difficult problems which were, at the same time, of great practical importance. In recent times, Hertz's theory of compression of elastic bodies has found wide application in railway engineering and in machine design.[1]

The works of L. Pochhammer (1841–1920) belong to the period under consideration. In 1876, this German scientist published an important paper on vibrations of circular cylinders.[2] Since Euler's time, the problem had been treated on the assumption that radial movements of particles during longitudinal vibrations could be neglected. Pochhammer removes this assumption and considers more general modes of vibration. In this way he is able to calculate corrections to the velocities of propagation of various modes of longitudinal vibrations. He also investigates flexural modes of vibration.

In a second paper,[3] Pochhammer discusses the bending of a beam by forces distributed over its lateral surface and shows that the neutral axis does not pass through the centroid of cross sections and that the usual elementary formula for bending stresses amounts only to a first approximation. He calculates a closer approximation for the case of a cantilever of circular cross section, when a load is uniformly distributed along its upper generator. Pochhammer extends his method to a beam in the form of a hollow cylinder and to curved bars.

71a. Solutions of Two-dimensional Problems between 1867 and 1900

During the last portion of the nineteenth century, considerable progress was made in solving two-dimensional problems of elasticity. There are two kinds of such problems. If a thin plate is submitted to the action of forces applied at the boundary in the middle plane of the plate (which we may take as the xy plane), the stress components σ_z, τ_{xz}, and τ_{yz} vanish on both faces of the plate, and it can be assumed without much error that these components vanish throughout the entire thickness of the plate. In this way, we obtain problems of *plane stress*. Another kind of two-dimensional problem arises when a long cylindrical or prismatical body is loaded by distributed forces, the intensity of which does not change along the length of the cylinder. In such a system, the portion of the body at a considerable distance from the ends has essentially a *plane deformation*, i.e., the displacements during deformation occur in

---

[1] See N. M. Belajev, "Local Stresses in Compression of Elastic Bodies" (Russian), St. Petersburg, 1924. Further development of the theory by taking not only normal but also tangential forces at the surface of contact was discussed by Fromm, *ZAMM*, vol. 7, pp. 27–58, 1927; and by L. Föppl, *Forschung Gebiete Ingenieurw.*, 1936, p. 209. See also the book by M. M. Saverin, "Surface Strength of Materials," Moscow, 1946.

[2] *J. Math. (Crelle)*, vol. 81, p. 324, 1876.

[3] See Pochhammer's book, "Untersuchungen über das Gleichgewicht des elastischen Stabes," Kiel, 1879.

planes perpendicular to the axis of the cylinder (which we may take as the z axis). In such a case, the strain components ϵ_z, γ_{xz}, and γ_{yz} vanish and we have to consider only the three strain components ϵ_x, ϵ_y, and γ_{xy}. This is known as *plane strain*. In both types of problem, we see that the number of unknowns is reduced from six to three. This naturally simplifies matters.

Clebsch was the first to discuss the problem of plane stress distribution, and he gave a solution for a circular plate (see page 257). Another case of a great practical importance was solved by H. Golovin.[1] He was interested in the deformations and stresses of circular arches of constant thickness. Considering the problem as a two-dimensional one, he succeeds in obtaining the solutions for the systems shown in Fig. 209. Under pure bending (Fig. 209a) he finds that cross sections remain plane as is usually assumed in the elementary theory of curved bars. But his stress distribution does not coincide with that of the elementary theory since the latter assumes that longitudinal fibers are subjected only to simple tension or compression stresses σ_t, while Golovin shows that there are also stresses σ_r acting in the radial direction. In the bending produced by a force P, applied at the end (Fig. 209b), not only normal stresses are produced but also shearing stresses at every cross section and the distribution of the latter does not follow the parabolic law as the elementary theory assumes. Golovin not only calculates stresses but also deflections of curved bars and, using formulas for these deflections, he is able to solve the statically indeterminate problem of an arch with built-in ends. The calculations made for usual proportions of arches show that the elementary theory is accurate enough for practical purposes. Golovin's work represents the first attempt to use the theory of elasticity in arch stress analysis.

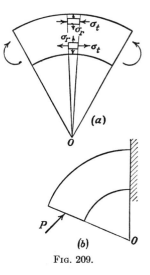

Fig. 209.

Since rigorous solutions of the theory of elasticity are known only in the simplest cases, experimental determinations of stresses have always attracted engineers. Maxwell had already showed how stresses can be measured photoelastically, but his work was not appreciated by the engineers of his time. The next attempt to use photoelasticity for measuring stresses was made, after an interval of forty years, by Carus

---

[1] *Bull. Tech. Inst. St. Petersburg*, 1880–1881.

Wilson.[1] He applies the method to the investigation of stresses in a rectangular beam loaded at the middle. He finds a considerable discrepancy between his measured stresses and the predictions of the usual beam formula. The explanation of that discrepancy was furnished by G. Stokes, who advocated calculating the stresses in the beam by superposing on the simple radial stress distribution[2] which will be produced in a semi-infinite plate (Fig. 210a), the stresses produced in a simply supported beam (Fig. 210b) by the radial tensile stress equal and opposite to compressive stresses of the infinite plate. The resultant of these tensile stresses, for each half of the beam, is balanced by the vertical reaction $P/2$ at the support and the horizontal force P/π, applied at point m (Fig. 210a). Hence, at the cross section mn of the beam, we have the bending moment

$$\frac{P}{2}l - \frac{P}{\pi}h$$

and the tensile force P/π. The corresponding normal stresses σ_x will be[3]

$$\sigma_x = \left(\frac{Pl}{2} - \frac{Ph}{\pi}\right)\frac{y}{I} + \frac{P}{2\pi h}$$

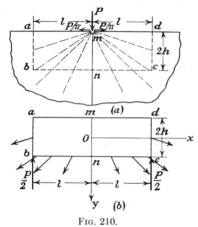

Fig. 210.

Superposing these stresses on the radial compressive stresses of the semi-infinite plate, Wilson obtained the results of his photoelastic tests with a sufficient accuracy.

Another instance of the photoelastic measurement of stresses lies in the work of Mesnager[4] who experimentally checked the radial stress distribution produced by concentrated forces acting in the middle plane of a plate. We see that, toward the end of the nineteenth century, engineers were beginning to recognize the importance of the photoelastic method. The first years of the twentieth century saw a rapid growth of its use and now photoelasticity has become one of our most powerful tools in experimental stress analysis.

The further development in the theoretical treatment of two-dimensional problems was based upon the use of the stress function. As we have seen (page 224), this function was introduced for the first time by

[1] *Phil. Mag.* (5), vol. 32, 1891.

[2] G. Stokes states that he took this stress distribution from Boussinesq's book, "Application des Potentiels . . . ," 1885. See Stokes, "Mathematical and Physical Papers," vol. V, p. 238.

[3] Assuming that the thickness of the plate is equal to unity.

[4] *Congr. intern. phys.*, vol. 1, p. 348, 1900, Paris.

Airy, who used it in his analysis of bending of rectangular beams. Airy selected his stress function so as to satisfy boundary conditions; but he overlooked the fact that it must also satisfy the compatibility equation established by Saint-Venant. Maxwell, in his work "On Reciprocal Figures, Frames, and Diagrams of Forces" corrected Airy's mistake[1] and established the differential equation for the stress function. He also showed that, if body forces are absent, the equations are identical for both kinds of two-dimensional problems and that the stress distribution is independent of the elastic constants of the material.

An important addition to the theory of two-dimensional problems was made by J. H. Michell (1863–1940).[2] Investigating two-dimensional problems,[3] he shows that the stress distribution is independent of the elastic constants of an isotropic plate if (1) body forces are absent and (2) the boundary is simply connected. If the boundary is multiconnected, as in perforated plates, for example, the stresses are independent of the moduli only if the resultant force on each boundary vanishes.

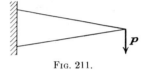

Fig. 211.

Applying the general theory to particular cases, Michell uses[4] the simple radial stress distribution that was obtained by Boussinesq and Flamant (see page 334). In this way he obtains solutions for a semi-infinite plate when the acting force is inclined to the straight edge of the plate, and for a wedge loaded at the end (Fig. 211). A conclusion regarding the accuracy of the simple beam formula can be drawn from the latter solution for beams of variable cross section. Michell also gives the solution of the problem of a force applied at an internal point of an infinite plate, and shows that, here, the stress distribution depends on the elastic constants. The stresses become infinitely large at the point of application of the force and to eliminate this difficulty it must be assumed that a small circular hole is cut out of the plate and that a distributed boundary load, statically equivalent to the given force, acts upon it. In this way, we obtain a multiconnected boundary.

Considering the stresses produced in circular disks by concentrated loads applied at the boundary, Michell obtains the same solution as was previously found by Hertz (see page 349). Then he proceeds to the solution for a heavy disk or roller on a horizontal plane. Several interesting diagrams, illustrating various systems of stress distribution in circular plates, are also given in Michell's paper.

[1] See his "Scientific Papers," vol. 2, p. 200.

[2] Michell was born in Australia. He studied at Cambridge University and received his M.A. in 1887. After several years of teaching at Cambridge, he returned to Australia and became a professor at the University of Melbourne.

[3] *Proc. London Math. Soc.*, vol. 31, pp. 100–124, 1899.

[4] *Proc. London Math. Soc.*, vol. 32, p. 35, 1900.

CHAPTER XII

Progress in Strength of Materials during the Twentieth Century

Since the end of the nineteenth century, we have seen a period of very rapid development in mechanics of materials. The International Congresses for Testing Materials attracted more and more engineers and physicists, the former interested in studying the mechanical properties of materials and the latter in studying the physics of solid bodies. Due to the efforts of Helmholtz and Werner Siemens, the Reichsanstalt was organized[1] in Berlin (1883–1887) with the purpose of bringing the research work of physicists and the requirements of industry closer together. In 1891, Oliver Lodge, in his presidential address to the British Association called attention to the work in progress at the Reichsanstalt and the importance that a similar institution would have for British industry.[2] A committee was appointed as a result of the suggestion and its members visited the Reichsanstalt and the Versuchsanstalt at Potsdam. In its report, the benefits of maintaining a close check on standardization in research were stressed. It was stated also that "it would be neither necessary nor desirable to compete with or interfere with the testing of materials of various kinds as now carried on in private or other laboratories, but there are many special and important tests and investigations into the strength and behaviour of materials, which might be conducted with great advantage at a laboratory such as is contemplated." In 1899, the plans for the National Physical Laboratory were agreed upon and this important institution came into being. In the United States, the National Bureau of Standards was founded with a large laboratory for the study of mechanical properties of materials.

Private industry started to recognize the importance of scientific research, and the setting up of industrial research laboratories progressed very rapidly. Not only was the amount of research in mechanics of materials increased as a result, but also the character of the work changed.

[1] For the history and work of this institution, see "Forschung und Prüfung 50 Jahre physikalisch-technische Reichsanstalt," by J. Stark, Leipzig, 1937.

[2] See "Early days at the National Physical Laboratory" by Sir Richard Glazebrook, its first director (1900–1919).

The new laboratories facilitated contact between research engineers and physicists and, in their work, closer attention was paid to the study of fundamental problems dealing with the structure and mechanical properties of solid bodies. After Laue's discovery (in 1912) of the interference of X rays in crystals, the possibility presented itself of studying the structure of metals in that form. A technique of producing large crystals was developed, and the study of single crystals has been of much help to our understanding of the mechanical properties of metals under various conditions.[1] The number of research workers interested in mechanical properties of materials, and the number of papers published in this field, increased immensely, and henceforth it will be possible to mention only a few of the most important of these works.

While this development in the experimental investigation of materials was progressing, the field of application of strength of materials and theory of elasticity also expanded rapidly. The use of stress analysis in structural engineering was well established during the nineteenth century. The beginning of the twentieth century saw new trends in the machine industries which made more refined stress analysis of machine parts very desirable. This new tendency found its expression in new books on machine design, of which A. Stodola's book "Die Dampfturbinen" is an especially clear example. While earlier books based the design of machine parts principally on empirical formulas and used only elementary strength of materials, Stodola in his book freely uses all the means of stress analysis which the theory of elasticity affords. He discusses stresses in plates and shells, investigates thermal stresses and stresses in rotating disks, and pays great attention to the stresses produced by various kinds of vibration. Attention is drawn to the existence of stress concentration at holes and fillets, and several methods of experimental stress analysis receive attention. This book has, to some extent, guided trends in strength of materials and is an important work on the application of scientific methods to machine design.

72. Properties of Materials within the Elastic Limit

Better accuracy in determining the modulus of elasticity was introduced by E. Grüneisen.[2] Using the interference of light in measuring small elongations, he showed that such a material as cast iron accurately follows Hooke's law for small stresses (below 140 lb per in.^{2}) and that the exponential formula $\epsilon = a\sigma^m$ proposed by G. B. Bülffinger[3] and E. Hodg-

---

[1] The most important books describing the work with single crystals are "Kristallplastizität" by E. Schmid and W. Boas, Berlin, 1935, and "The Distortion of Metal Crystals" by C. F. Elam, Oxford University Press, 1936.
[2] E. Grüneisen, *Verhandl. physik. Ges.*, vol. 4, p. 469, 1906.
[3] G. B. Bülffinger, *Comm. Acad. Petrop.*, vol. 4, p. 164, 1729.

kinson[1] and widely used by C. Bach[2] and W. Schüle[3] is completely unsatisfactory for very small deformations. A discussion of the various stress-strain formulas, proposed for materials which do not follow Hooke's law, was given by R. Mehmke.[4]

Experiments with single crystal specimens showed the moduli to have marked dependence upon orientation. Figure 212 shows,[5] for example, the variation of E and G for iron. Having this information for a single crystal, methods of calculating the average values of the moduli for polycrystalline specimens were developed[6] to such an extent that experimental results could be predicted with some accuracy.

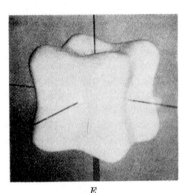 

E G

FIG. 212. The variation of the moduli E and G for iron.

By using the interference of light, small deviations from Hooke's law in "perfectly elastic materials" were investigated and it was found that *hysteresis loops* and *elastic aftereffect* in single crystals of quartz could be explained completely by the thermoelastic and piezoelectric properties of the material.[7]

Lord Kelvin had already shown[8] that if a tensile test involving very slow increase of the load is carried out, the temperature of the specimen remains equal to that of its surroundings and the stress-strain relation may be represented by a straight line OA (Fig. 213a), the slope of which gives the magnitude of the modulus E under *isothermal conditions*. If the tensile load is applied rapidly, so that there is not enough time for

[1] E. Hodgkinson, *Mem. Proc. Manchester Lit. Phil. Soc.*, vol. 4, p. 225, 1822.
[2] C. Bach, *Abhandl. u. Ber.*, Stuttgart, 1897.
[3] W. Schüle, *Dinglers Polytech. J.*, vol. 317, p. 149, 1902.
[4] R. Mehmke, *Z. Math. u. Physik.*, 1897.
[5] See the book by Schmid and Boas, p. 200.
[6] W. Boas and E. Schmid, *Helv. Phys. Acta*, vol. 7, p. 628, 1934.
[7] A. Joffe, *Ann. Physik*, 4th series, vol. 20, 1906.
[8] See Kelvin, "Elasticity and Heat," p. 18, Edinburgh, 1880.

heat exchange, the straight line OB is obtained instead of OA, and usually the modulus E under *adiabatic conditions* is larger than that obtained isothermally. Due to its sudden stretching, the specimen becomes cooler than the room temperature. If the specimen remains under constant load for a sufficient length of time, it gradually warms up and acquires the room temperature and, as a result, an additional elongation of the specimen is brought about (represented in the figure by the horizontal line BA). This is the *elastic aftereffect* and it is due to the thermoelastic property of the material. If, after complete equalization of the temperature, the specimen is suddenly unloaded, its adiabatic contraction will be represented in Fig. 213a by the line AC parallel to OB. Due to this quick shortening, the specimen warms up, and the subsequent process of cooling to room temperature will give rise to a further contraction, represented by the length CO. It is seen that, by deforming the specimen adiabatically and then allowing enough time for temperature equalization, a complete cycle, represented by the parallelogram $OBAC$, is described. The area of this represents the loss of mechanical energy per cycle. We assumed adiabatical deformation in our reasoning, whereas, practically, there is always some heat exchange during the cycle. Thus, instead of a parallelogram, a loop such as is shown in Fig. 213b is obtained. The difference between the adiabatic and the isothermal moduli is usually small[1] and the loss of mechanical energy per cycle is also very small. But if many cycles occur in succession, as in vibration, the losses in mechanical energy become important and must be considered, since they correspond to the so-called *internal friction* and are responsible for damping the motion.

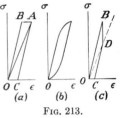

Fig. 213.

We assumed in our discussion that, after sudden stretching, the specimen remains under the action of the constant applied load. Instead, the stretched specimen could be held at constant length. Then the warming up of the specimen would result in some decrease in the initially applied force. This process of *relaxation* is represented in Fig. 213c by the vertical line BD. By suddenly releasing the specimen now, the portion DC is obtained and later, due to cooling, the closing line CO of the cycle $OBDC$.

We have considered a single crystal specimen, but we shall also have a similar phenomenon of internal friction in polycrystalline specimens. The thermoelastic effect during stretching of a crystal depends upon the crystal's orientation. Due to this fact, the temperature changes produced by stretching a polycrystalline specimen vary from grain to grain

[1] For steel it is about ⅓ of 1 per cent.

and we have to consider not only the heat exchange between the specimen and the surrounding, but also the heat flow between the individual crystals. Since the heat generated in a grain is proportional to its volume and the heat exchange depends upon the magnitude of its surface, it is evident that the temperature equalization will be facilitated and the losses of mechanical energy increased by diminution of the grain size. This is of great practical importance, since there are cases in which the damping of vibration of an elastic system depends mainly on the internal friction of the material. To increase damping in such systems, materials with small grain size should be employed.

It was assumed in our discussion that the deformation of the specimen is perfectly elastic (as can be expected with small stresses). With larger stresses, the phenomenon of internal friction becomes more complicated, since we have to consider not only the losses in mechanical energy due to heat exchange discussed above but also the losses due to plastic deformation within individual grains.[1]

73. Fracture of Brittle Materials

Considerable progress has been made in recent years in studying the fracture of brittle materials, such as glass. A tensile test of glass usually gives a small value for the ultimate strength (of the order of 10^4 lb per in.^{2}). Taking the modulus of elasticity as $E = 10^7$ lb per in.^{2}, we find that it requires only 5 lb-in. of work per cubic inch to produce fracture of this material. At the same time, calculations based on the forces necessary to disrupt the molecules give values about 30,000 times larger, for the energy.

To explain this discrepancy, A. A. Griffith proposed[2] the theory by which it is assumed that this large reduction in the strength of glass is due to the presence of microscopic cracks which act as *stress raisers*. Considering a crack as a narrow elliptical hole, and using the well-known solution for stress distribution around an elliptical hole in a plate that is uniformly stretched in one direction (Fig. 214), he finds that due to the

[1] A considerable amount of research work in measuring the damping capacity of various materials was done by Rowett, *Proc. Roy. Soc. (London)*, vol. 89, p. 528, 1913, and, more recently, by Otto Föppl and his collaborators at the Wöhler Institute, Brunswick. See *VDI*, vol. 70, p. 1291, 1926; vol. 72, p. 1293, 1928; vol. 73, p. 766, 1929. The importance of the thermoelastic cause of internal friction was shown by the investigations of C. Zener and his associates. See *Phys. Rev.*, vol. 52, p. 230, 1937; vol. 53, p. 90, 1938; vol. 60, p. 455, 1941. The results of these investigations are presented in the book, "Elasticity and Anelasticity of Metals," by C. Zener, 1948.

[2] A. A. Griffith, *Trans. Roy. Soc. (London)*, (A), vol. 221, p. 163, 1920. See also *Proc. Intern. Congr. Applied Mechanics*, Delft, 1924, p. 55. See also A. Smekal's article in the "Handbuch der physikalischen und technischen Mechanik," vol. 4_2, p. 1, 1931.

hole, the strain energy of the plate (of unit thickness) is reduced by the amount

$$\frac{\pi l^2 \sigma_0^2}{4E} \tag{a}$$

where l denotes the length of the crack and the thickness of the plate is taken as unity. Considering the magnitude of stress σ_0 at which the crack begins to spread across the plate and thus produce fracture, Griffith observes that such a spreading out becomes possible without any extra work only if the increase in *surface energy* due to the increment dl of the crack length is compensated by a corresponding reduction in strain energy of the plate. This gives the equation

$$\frac{d}{dl}\left(\frac{\pi l^2 \sigma_0^2}{4E}\right) dl = 2S\, dl \tag{b}$$

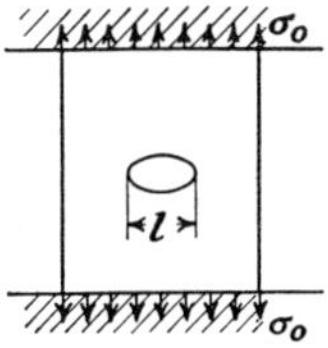

where S denotes the *surface tension*. From this equation, he finds

$$l = \frac{4SE}{\pi \sigma_0^2} \tag{c}$$

Fig. 214.

Thus he can obtain the length of the crack if the ultimate tensile strength σ_0 of glass is found by tensile tests. Taking the value[1]

$$S = 5.6 \times 10^{-4} \text{ kg per cm}$$

for the surface tension and substituting $E = 7 \times 10^5$ kg per cm², $\sigma_0 = 700$ kg per cm², he obtains

$$l \sim 1 \times 10^{-3} \text{ cm}$$

To verify his theory, Griffith made experiments with thin glass tubes subjected to internal pressure. Making artificial cracks (with a diamond) parallel to the cylinder's axis and of different lengths, he found the critical values of σ_0 experimentally. These were in satisfactory agreement with the theoretical predictions of Eq. (c). Griffith conducted further experiments with thin glass fibers and found the ultimate tensile strength to be equal to 3.5×10^4 kg per cm² for a fiber of 3.3×10^{-3} mm in diameter. This is fifty times higher than the ultimate tensile strength previously mentioned. This relatively enormous strength of thin fibers can be explained, on the basis of Griffith's theory, if it be observed that in the process of drawing the thin fiber the cracks that were initially perpendicular to the length of the fiber are eliminated. Griffith noticed that, with the elapse of time, the fibers lost part of their strength and, experimenting with fibers immediately after they were drawn, he succeeded in

---

[1] Griffith obtained the value of S by extrapolating the results of tests made on melted glass at various temperatures.

obtaining the immense ultimate strength of 6.3 × 10⁴ kg per cm² for a diameter of 0.5 mm. This is about one-half of the theoretical strength predicted by consideration of molecular forces. All these experiments clearly showed that in brittle materials the tensile strength is greatly reduced by such imperfections as internal microscopic cracks.

A further investigation of the same question was made with single crystal specimens under tension. Working with specimens of rock salt, A. Joffe found[1] that the ultimate tensile strength of that material is only 45 kg per cm², if tested in air of room temperature. If the same kind of specimen is tested in hot water, it reaches its yield point at a stress of 80 kg per cm² and then stretches plastically, finally fracturing at a stress of 16,000 kg per cm², which is not far removed from the theoretical strength of 20,000 kg per cm² as calculated by F. Zwicky.[2] These experiments demonstrated that the smoothing effect, upon the surface of the test piece, has a great effect on the tensile strength.[3]

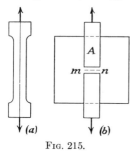

Fig. 215.

Very interesting experiments, concerning the tensile strength of sheets of mica, were conducted by E. Orowan.[4] He showed that if instead of a tensile-test specimen, such as that depicted in Fig. 215a (cut out from a mica sheet), he used sheets of mica for the tests and produced tension over the area mn by clamps A, as shown in Fig. 215b, then the ultimate strength is increased about ten times. This again showed that imperfections along the longitudinal edges of specimen (a) reduce the strength considerably and that by eliminating their effects by means of arrangement (b), higher values of the ultimate strength can be obtained.

If the strength of brittle materials is affected so much by the presence of imperfections, it seems logical to expect that the value of the ultimate strength will depend upon the size of specimens and become smaller with increase of dimensions, since then the probability of having weak spots is increased. The size effect was noticed in cases of brittle fracture such as those produced by impact[5] or by fatigue tests,[6] but an explanation of this fact on a statistical basis was furnished later by W. Weibull.[7] He

[1] See *Z. Physik*, p. 286, 1924. See also *Proc. Intern. Congr. Applied Mechanics*, Delft, 1924, p. 64.

[2] F. Zwicky, *Physik. Z.*, vol. 24, p. 131, 1923.

[3] Further discussion of the *Joffe effect* is given in the book by E. Schmid and W. Boas, p. 271.

[4] *Z. Physik*, vol. 82, p. 235, 1933.

[5] M. Charpy, *Assoc. intern. essai matériaux*, 6th *Congr. New York*, 1912, vol. 4, p. 5.

[6] R. E. Peterson, *J. Applied Mechanics*, vol. 1, p. 79, 1933.

[7] W. Weibull, *Roy. Swed. Inst. Eng. Research, Proc.*, nr. 151, 1939, Stockholm.

showed that if, for a given material, we make two series of tensile tests with geometrically similar specimens of two different volumes, V_1 and V_2, the corresponding values of ultimate strength will be in the ratio

$$\frac{(\sigma_{ult})_1}{(\sigma_{ult})_2} = \left(\frac{V_2}{V_1}\right)^{1/m} \tag{a}$$

where m is a constant of the material. Determining m from Eq. (a), we can predict the value of σ_{ult} for any size of specimen. Such experiments were made by N. Davidenkov.[1] Using steel with a large content of phosphorus and experimenting with specimens of two diameters d (10 mm and 4 mm) and length l (50 mm and 20 mm), he found two values (57.6 kg per mm^2 and 65.0 kg per mm^2) for σ_{ult}. Equation (a) then gave $m = 23.5$. Using this value, the ultimate strength of specimens with $d = 1$ mm, $l = 5$ mm was predicted as 77.7 kg per mm^2. Experiments gave the value 75.0 kg per mm^2, so that the theory was in a reasonable agreement with the experiments.

Using the same statistical method Weibull calculated (in the previously mentioned paper) the ultimate strength for pure bending and found the ratio

$$\frac{(\sigma_{ult})_{\text{bending}}}{(\sigma_{ult})_{\text{tension}}} = (2m + 2)^{1/m} \tag{b}$$

for a rectangular bar. This formula was also verified by Davidenkov's experiments. From tests, he found the value 1.40 for this ratio, while the theoretical figure (calculated for $m = 24$) was 1.41.

Several other practical types of stress distribution were discussed in Weibull's paper. Also several experimental investigations are mentioned, the results of which show satisfactory agreement with the statistical theory. Of course this theory is applicable only to brittle fracture. If a material shows considerable plastic deformation before rupture, this alleviates all local stress concentrations at the "points of imperfection" and makes the average stresses responsible for the failure.

In technical testing of brittle materials, the specimens are usually subjected to compression, but it is difficult to obtain a uniform distribution of compressive stresses in these tests. Owing to friction at the surfaces of contact between the specimen and the plates of the testing machine, lateral expansion is prevented and the material in these regions is in a condition of three-dimensional compression. As a result, the material near the surfaces of contact remains intact while that at the sides of the specimen is crushed out. To eliminate friction, A. Föppl covered the surfaces of contact with paraffin[2] and the type of failure under this condi-

[1] For an English translation of his paper, see *J. Applied Mechanics*, vol. 14, p. 63, 1947.
[2] See *Mitt. mech. tech. Lab. Munich*, nr. 27, 1900.

tion was quite different from any previously found. A cubic specimen failed by subdividing into plates parallel to the lateral sides. If a cylindrical specimen (the height of which is two to three times larger than its diameter) is used in compression tests, we obtain an approximately uniform stress distribution in the middle portion and, in this way, the end effect is practically eliminated. Another method of producing uniformly distributed compression was proposed by Siebel and Pomp[1] and is shown in Fig. 216. The cylindrical specimen is compressed between two conical surfaces with the angle α equal to the angle of friction.

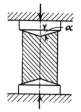

FIG. 216.

The resultant pressure in such a case is parallel to the axis of the cylindrical specimen.

Experiments with brittle materials subjected to high hydrostatic pressures showed that under such conditions the brittle material may behave as a plastic[2] one. Using cylindrical specimens of marble and sandstone and producing axial compression combined with lateral pressure, Theodore von Kármán succeeded[3] in obtaining barreled forms of compression such as are usually produced in plastic materials such as copper.

74. Testing of Ductile Materials

Recent research on the deformation of single crystal specimens has given us much new knowledge concerning the straining of ductile materials. These investigations[4] have revealed that plastic deformation during the stretching of a single crystal consists of sliding of material on certain crystallographic planes in a definite direction. For example, if we are testing a single aluminum crystal specimen having a face-centered cubic lattice structure (Fig. 217), sliding will occur parallel to one of the octahedral planes (such as abc) and the direction of sliding will be parallel to one of the sides of the triangle. In a tensile-test specimen sliding will occur, not on 45 deg planes on which the maximum shearing stress acts, but on the most unfavorably oriented octahedral plane. This explains why, in testing single crystal specimens, we obtain scattered results for the value of the tensile load at which yielding begins. These values depend, as we see, not only on the mechanical properties of the material

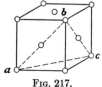

FIG. 217.

---

[1] Mitt. Kaiser-Wilhelm Inst. Eisenforsch. Düsseldorf, vol. 9, p. 157, 1927.

[2] See F. Kick, VDI, vol. 36, pp. 278, 919, 1892; F. D. Adams and J. T. Nicolson, Phil. Trans., vol. 195, p. 363, 1901.

[3] VDI, vol. 55, p. 1749, 1911.

[4] See the previously mentioned books by C. F. Elam and by E. Schmid and W. Boas.

but also on the orientation of the crystal axes with respect to the axis of the specimen.

Assume, for example, that, in our (single crystal) tensile-test specimen (Fig. 218) of cross-sectional area A, the direction of the most unfavorably oriented octahedral plane of sliding is defined by the normal n and the direction of sliding by the line pq. The shearing stress, acting on the plane of sliding is

$$\tau = \frac{P}{A} \cos \alpha \sin \alpha \qquad (a)$$

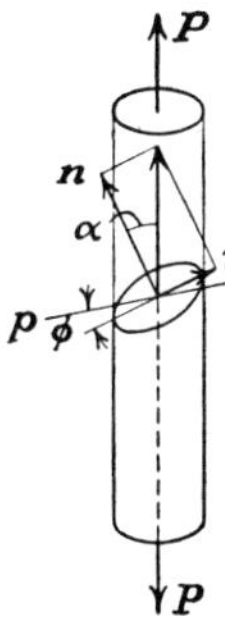

Experiments show that the onset of sliding is brought about, not by the magnitude of τ, but on the magnitude of the component $\tau \cos \phi$ of this stress in the direction of sliding pq. The sliding begins when this component reaches the definite value $\tau \cos \phi = \tau_{cr}$. Substituting into Eq. (a), we obtain

$$\tau_{cr} = \frac{P}{A} \cos \alpha \sin \alpha \cos \phi \qquad (b)$$

FIG. 218.

It is seen that, while τ_{cr} is constant for a given material, the load P at which yielding begins depends on the values of the angles α and ϕ.

The process of stretching, due to sliding, is shown in Fig. 219. We can assume that it is produced in two steps: (1) translatory motion along the sliding planes (Fig. 219b) and, (2) then the rotation of the specimen by the angle β to its original direction (Fig. 219c). From this mechanism of stretching it is clear that (1) the angle between the direction of the tensile force P and the planes of sliding changes during the distortion and (2) the initially circular cross section of the specimen becomes elliptical with the ratio of its principal axes equal to $1 : \cos \beta$.

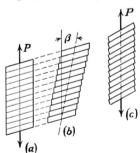

FIG. 219.

Numerous experiments with single crystals have yielded results that uphold the above conclusions. For instance, Fig. 220 represents a stretched copper-aluminum single crystal[1] specimen. The same experiments also show that the magnitude of τ_{cr} at which sliding begins is usually very small. Figure 221 gives the relation between the shearing strain and shearing stress obtained[2] in testing single crystal aluminum specimens at various temperatures. It is seen that the value of τ_{cr} is very

[1] See C. F. Elam, "The Distortion of Metal Crystals," p. 182, 1935.
[2] See W. Boas and E. Schmid, Z. *Physik*, vol. 71, p. 703, 1931.

small but the magnitude of τ required for continuation of sliding gradually increases with deformation. This is the result of *strain hardening* of the material.

Calculations of τ_{cr}, based on the consideration of molecular forces, give enormous values. This indicated that the process of slipping does not

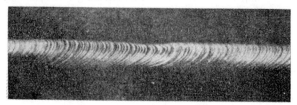

FIG. 220. A stretched copper-aluminum single crystal specimen.

consist solely of the rigid translatory motion of atomic planes relative to one another; and we must assume the existence of some local imperfections from which slip may start, under the action of a small force, and spread over the entire plane of sliding. A model illustrating the possibility of such sliding (which starts at a point of imperfection) was described by L. Prandtl.[1] Another mechanical model was proposed by G. I. Taylor.[2] Assuming a local disturbance in the atomic distribution (a "dislocation") he showed that this disturbance would move along the sliding plane in the crystal under the influence of a small stress τ_{cr} and would cause a relative displacement of one portion of the crystal with respect to the other. Using his model, Taylor was able not only to explain the onset of sliding at very small values of τ_{cr} but also the phenomenon of strain hardening shown by the curves in Fig. 221.

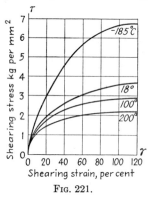

FIG. 221.

The results of tests of ordinary polycrystalline test pieces can be better understood with information regarding the properties of single crystal specimens. J. A. Ewing and W. Rosenhain[3] did very interesting experiments with polished iron tensile-test specimens. Microscopic observation of the surface of the metal showed that, even at comparatively low

---

[1] The model was discussed by Prandtl in a seminar on the theory of elasticity in 1909 (Göttingen) and its description was published in *ZAMM*, vol. 8, pp. 85–106, 1928.

[2] *Proc. Roy. Soc. (London)*, vol. 145, pp. 362–404, 1934.

[3] *Trans. Roy. Soc. (London)*, (A), vol. 193, p. 353, 1900.

tensile loads, "slipbands" appeared on the surface of some grains. The bands indicated sliding along definite crystallographic planes in those grains. Since the elastic properties of a single crystal may be very different in different directions, and since the crystals are distributed at random, the stresses in a tensile-test piece are not uniformly distributed and sliding may occur in the most unfavorably oriented individual crystals before the average tensile stress reaches the yield point value. If such a specimen is unloaded, the crystals which suffered sliding cannot return to their initial shape completely and, as a result, some residual stresses will be left in the unloaded specimen. Some *aftereffect* in the specimen can be ascribed to the action of these stresses. The yielding of the individual crystals also contributes to the losses of energy during loading and unloading and will increase the area of the hysteresis loop discussed on page 357. If the specimen is subjected to a tensile test for a second time, the grains in which sliding occurred will not yield before the tensile load reaches the value applied in the first loading. Only when the load exceeds that value will the sliding start again. If the specimen is submitted to compression after a previous loading in tension, the applied compressive stresses, combined with the residual stresses (induced by the preceding tensile test), will produce yielding in the most unfavorably oriented crystals before the average compressive stress reaches the value at which slip bands are produced in a specimen in its virgin state. Thus, the tensile-test cycle raises the elastic limit in tension, but the elastic limit in compression is lowered. It is seen that consideration of the sliding in individual crystals and of residual stresses produced by that sliding furnishes an explanation of the *Bauschinger effect*.[1]

In studying the tensile strength of structural steel, engineers have been particularly interested in the phenomenon of sudden extension at the yield point. The fact that, at a certain value of tensile stress, a sudden drop in the tensile load occurs and that after this the metal stretches considerably at a somewhat lower stress is well known. C. Bach introduced the terms of *upper* and *lower* yield point[2] for these two values of stress. Further experiments showed that the lower yield point is less influenced by the shape of the specimen than the upper. Thus more practical significance is attributed to it. Experiments with bending and torsion showed that the characteristic lines of yielding (Lüders' lines) appear at much higher stresses than in the case of a homogeneous stress distribution, so that the beginning of yielding depends not only upon the magnitude of the maximum stress but also on the stress gradient. Recently, important experiments dealing with steel at its yield point have been conducted

[1] An interesting model illustrating hysteresis loops and the Bauschinger effect, was made by C. F. Jenkin. See *Engineering*, vol. 114, p. 603, 1922.
[2] C. Bach, *VDI*, vol. 58, p. 1040, 1904; vol. 59, p. 615, 1905.

under the direction of A. Nadai. They showed that the beginning of yielding depends very much on the rate of strain.[1] The curves in Fig. 222 show the results obtained for mild steel for a wide range of rates of strain ($u = d\epsilon/dt = 9.5 \times 10^{-7}$ per sec to $u = 300$ per sec). It is seen that not only the yield point, but also the ultimate strength and the total elongation, depend greatly upon the rate of strain.

To explain the sudden stretching of steel at its yield point, it has been suggested[2] that the boundaries of grains are built up of a brittle material

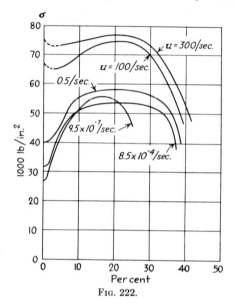

FIG. 222.

and form a rigid skeleton which prevents the plastic deformation of grains at low stress. Without such a skeleton the tensile-test diagram would be like that indicated in Fig. 223 by the dotted line. Owing to the presence of the rigid skeleton, the material remains perfectly elastic and follows Hooke's law up to the point A, at which it breaks. Then the plastic grain material suddenly obtains the permanent strain AB, after which the material follows the usual plastic material curve BC. This theory explains the condition of instability of the material at the upper yield point.

It also accounts for the fact that materials with small grains usually show higher values for yield-point stress and, as a result, a larger stretch

[1] M. J. Manjoine, *J. Applied Mechanics*, vol. 11, p. 211, 1944.
[2] See P. Ludwik and Scheu, *Werkstoffanschuss VDI, Ber.*, no. 70, 1925; W. Köster, *Archiv Eisenhüttenw.*, no. 47, 1929.

at yielding (defined by the length of the horizontal line AB in Fig. 223). Again, it explains the fact that, in high-speed tests, the increase in yield-point stress is accompanied by increase of the amount of stretching at this stress (as may be seen from the curves in Fig. 222).

In studying plastic deformation of tensile-test specimens beyond the yield point, some progress has been made[1] by introducing the notions of *true stress* and *natural strain* in construction of tensile-test diagrams. The true stress is obtained by dividing the load by the actual cross-sectional area of the deformed specimen, corresponding to that load. In calculating natural strains,[2] the actual length l is used instead of the initial length l_0. Then the increments of strain are given by the equation $d\epsilon = dl/l$, and the natural elongation is defined by the integral

$$\epsilon = \int_{l_0}^{l} \frac{dl}{l} = \log_e \frac{l}{l_0}$$

When the tensile load on a specimen reaches its maximum value, local contraction (or "necking") takes place. The stresses do not remain uniformly distributed over the cross section of the neck and a three-dimensional stress distribution obtains instead of simple tension. The distribution of stresses over the minimum cross section at the neck was investigated by N. N. Davidenkov[3] for the case of circular test bars. He found that the maximum tensile stress acts at the center of the cross section and that the minimum occurs at the boundary. The values of these stresses are given approximately by

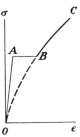

Fig. 223.

$$\sigma_{\max} = \sigma_a \frac{R + 0.50a}{R + 0.25a} \qquad \sigma_{\min} = \frac{\sigma_a R}{R + 0.25a}$$

where σ_a denotes the average tensile stress, a is the radius of the minimum cross section, and R is the radius of curvature of the profile of the neck. In addition to stresses in the direction of the axis of the specimen, stresses σ_r and σ_θ in the radial and the tangential directions will also be present. Davidenkov showed that these two stresses are equal and are given by the equation

$$\sigma_r = \sigma_\theta = \frac{\sigma_a R}{R + 0.25a} \frac{a^2 - r^2}{2Ra}$$

---

[1] A discussion of various aspects of tensile-test experiments and a corresponding bibliography are given in the book "Der Zugversuch" by G. Sachs, Leipzig, 1926. See also the paper on the same subject by C. W. MacGregor, *Proc. ASTM*, vol. 40, p. 508, 1940, and his article in the "Handbook of Experimental Stress Analysis," 1950.

[2] Mesnager suggested this notion in a paper presented to the International Congress of Physics; see *Congr. intern. phys.*, Paris, 1900, vol. 1, p. 348.

[3] For an English translation of this paper, see *Proc. ASTM*, vol. 46, 1946.

where r denotes the radial distance of a point from the axis of the specimen.

In studying the phenomemon of fracture, L. Prandtl suggested[1] that two types of fracture are identified: (1) cohesive or *brittle fracture* perpendicular to the tensile force and (2) *shear fracture*. In testing cylindrical specimens of structural steel, we obtain the so-called *cup-and-cone* fracture. In the center of this, the surface is perpendicular to the axis of the specimen and is of the brittle type. The outer portion of the fracture forms a conical surface inclined at approximately 45 deg to the direction of tension; this represents shear fracture. P. Ludwik noticed[2] that fracture starts at the center of the minimum cross section of the neck and spreads, as a brittle fracture, over the middle portion of the cross section, while the outer material continues to stretch plastically. A microscopic study[3] of mild-steel fractures shows that some grains are split along the cubic planes while others are ruptured along the octahedral planes. The crystals which were drawn out during plastic deformation give a fibrous appearance to the surface of the break.

75. *Strength Theories*

Most of our information regarding the mechanical properties of ductile materials is obtained from tensile tests, while that of brittle materials is usually found from compression tests. To have some basis for selecting working stresses for the various cases of combined stresses which are encountered in practice, various strength theories have been advanced.[4] Such scientists as Lamé and Rankine assumed as the criterion of strength the maximum principal stress, but later on the *maximum strain theory* was generally accepted, principally under the influence of such authorities as Poncelet and Saint-Venant. Thus it was assumed that failure and yield-point stretching occur under any kind of combined stress if the maximum strain reaches the appropriate critical value (which is found from tensile tests).

Maxwell suggested the use of the expression for strain energy in determining the critical values of combined stresses. He showed that the total strain energy per unit volume can be resolved into two parts: (1) the strain energy of uniform tension or compression and (2) the strain energy of distortion. Considering now the strain energy of distortion, Maxwell makes the statement: "I have strong reasons for believing that

---

[1] "Verhandlungen deutscher Naturforscher und Ärzte," Dresden, 1907.

[2] *Schweiz. Verband Materialprüfung. Tech., Ber.,* no. 13, 1928.

[3] C. F. Tipper, *Metallurgia*, vol. 39, p. 133, 1949.

[4] A complete bibliography on this subject can be found in the article by Hans Fromm; see "Handbuch der physikalischen und technischen Mechanik," vol. 4_1, p. 359, 1931.

when [the strain energy of distortion] reaches a certain limit then the element will begin to give way." Further on he states: "This is the first time that I have put pen to paper on this subject. I have never seen any investigation of the question, *Given the mechanical strain in three directions on an element, when will it give way?*" We see that Maxwell already had the theory of yielding which we now call *the maximum distortion energy theory*. But he never came back again to this question and his ideas became known only after publication of Maxwell's letters.[1] It took engineers a considerable time before they finally developed the theory identical with that of Maxwell.

Duguet developed[2] a theory for tension similar to that proposed by Coulomb for compression (page 51). The maximum shear theory for dealing with various cases of combined stresses was proposed by Guest[3] for mild steel. This theory is a particular instance of that of O. Mohr, previously discussed (see page 286). Experiments with brittle materials such as sandstone showed[4] that Mohr's theory cannot be brought into agreement with test results from brittle fractures.

Beltrami[5] suggested that, in determining the critical values of combined stresses, the amount of strain energy stored per unit volume of the material should be adopted as the criterion of failure. This theory does not agree with experiments, however, since a large amount of strain energy may be stored in a material under uniform hydrostatic pressure without the onset of fracture or yielding.

An improvement upon this theory of strength was proposed by M. T. Huber,[6] who considers only the energy of distortion in determining the critical state of combined stress. The expression for the distortion energy is

$$U = \frac{1}{12G} [(\sigma_1 - \sigma_2)^2 + (\sigma_1 - \sigma_3)^2 + (\sigma_2 - \sigma_3)^2] \qquad (a)$$

where σ_1, σ_2, and σ_3 are the three principal stresses. In simple tension, the distortion energy at the yield-point stress is obtained by substituting $\sigma_1 = \sigma_{yp}$, $\sigma_2 = \sigma_3 = 0$ into expression (a). This gives

---

[1] The letters from James Clerk Maxwell to his friend William Thomson were originally issued in *Proc. Cambridge Phil. Soc.*, part 5, vol. 32. Later on they were published in book form by the Cambridge University Press, New York and Cambridge, 1937.

[2] Duguet, "Déformation des corps solides," vol. 2, p. 28, 1885.

[3] J. J. Guest, *Phil. Mag.*, vol. 50, p. 69, 1900.

[4] Theodore von Kármán, *VDI*, vol. 55, p. 1749, 1911; R. Böker, *Forsch. Gebiete Ingenieurw.*, nr. 175–176, 1915.

[5] *Rendiconti*, p. 704, 1885; *Math. Ann.*, p. 94, 1903.

[6] M. T. Huber, *Czasopismo tech.*, vol. 15, 1904, Lwów. See also A. Föppl and L. Föppl, "Drang und Zwang," 2d ed., vol. 1, p. 50. The same idea was had independently by R. von Mises; see *Göttinger Nachr.*, p. 582, 1913.

$$U = \frac{\sigma_{yp}^2}{6G} \qquad (b)$$

Equating expression (a) to expression (b), we find that the condition of yielding for any three-dimensional stress system is

$$(\sigma_1 - \sigma_2)^2 + (\sigma_1 - \sigma_3)^2 + (\sigma_2 - \sigma_3)^2 = 2\sigma_{yp}^2 \qquad (c)$$

For a plane stress distribution (taking $\sigma_3 = 0$), we obtain the condition of yielding for two-dimensional cases from Eq. (c), viz.,

$$\sigma_1^2 - \sigma_1\sigma_2 + \sigma_2^2 = \sigma_{yp}^2 \qquad (d)$$

For instance, for pure shear we have $\sigma_1 = -\sigma_2 = \tau$, and Eq. (d) gives the value

$$\tau_{yp} = \frac{\sigma_{yp}}{\sqrt{3}} = 0.5774\sigma_{yp} \qquad (e)$$

for the yield-point stress in shear. This is in good agreement with test results.

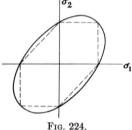

FIG. 224.

Taking σ_1 and σ_2 as rectangular coordinates, the condition of yielding (d) can be represented by the ellipse shown in Fig. 224. The irregular hexagon in the same figure represents the condition of yielding, which follows from the maximum shear theory.

From the principal stresses σ_1, σ_2, and σ_3, we can calculate the shear stresses τ_{oct} acting on the octahedral planes,[1] i.e., the planes equally inclined to the principal axes. We find

$$\tau_{oct} = \tfrac{1}{3}\sqrt{(\sigma_1 - \sigma_2)^2 + (\sigma_2 - \sigma_3)^2 + (\sigma_3 - \sigma_1)^2} \qquad (f)$$

Substituting into Eq. (c), the condition of yielding, we obtain

$$\tau_{oct}^2 = \tfrac{2}{9}\sigma_{yp}^2$$

That is, the *distortion energy theory* of yielding can be expressed in the following simple manner: in any case of combined stress, yielding begins when the octahedral shearing stress reaches the value $\tau_{oct} = \sigma_{yp}\sqrt{2}/3$.

A considerable amount of experimental work has been done to investigate the accuracy of the distortion energy theory. At the suggestion of A. Nadai, tests with thin-walled tubes of iron, copper, and nickel subjected to axial tension and internal hydrostatic pressure were made by W. Lode.[2] In this way, a two-dimensional stress condition with various

---

[1] This use of the term has nothing to do with that illustrated by Fig. 217.
[2] *Z. Physik*, vol. 36, p. 913, 1926.

values of the ratio $\sigma_1:\sigma_2$ can be produced (the stress in the radial direction can be neglected). It was found that the experimentally obtained points occupied positions along the limiting ellipse shown in Fig. 224 with some accuracy. Similar experiments were also made by M. Roś and A. Eichinger[1] at the material testing laboratory of the Zürich Polytechnical Institute. The results again upheld the distortion energy theory. G. I. Taylor and H. Quinney[2] produced various kinds of two-dimensional stress conditions by using combined tension and torsion of thin-walled tubes of aluminum, copper, and mild steel. With the first two of these metals, they found especially good agreement between their test results and those predicted by the distortion energy theory.

G. Sachs[3] arrived at relation (e) (between the yield stress in shear and the yield stress in tension) in a completely different way. From Eq. (b), Art. 74, we know that the yield-point load on a single crystal specimen depends on the orientation of the crystal. Now, considering polycrystalline specimens as systems of crystals distributed at random, neglecting the effect of crystal boundaries and assuming that all crystals start to yield simultaneously, Sachs calculated the relation between σ_{yp} and the critical shearing stress τ_{cr} by approximate averaging method.[4] For crystals with face-centered cubic lattice structure (aluminum, copper, nickel), he found

$$\sigma_{yp} = 2.238\,\tau_{cr}$$

Similarly, for pure shear

$$\tau_{yp} = 1.293\tau_{cr}$$

and finally,

$$\tau_{yp} = \sim 0.577\sigma_{yp}$$

Within the limits of accuracy of his computations, this coincides with result (e) of the distortion energy theory. Sachs assumes that approximately the same result will be obtained for crystals with body centered cubic lattice structure (as in iron). We see that his work gives some physical basis for the result obtained before by assuming that yielding of a material begins when the amount of distortion energy stored per unit volume reaches a certain value for each material.[5]

[1] See *Proc. Intern. Congr. Applied Mechanics*, Zürich, 1926.
[2] *Trans. Roy. Soc. (London)*, (A), vol. 230, pp. 323–362, 1931.
[3] G. Sachs, *VDI*, vol. 72, p. 734, 1928.
[4] A suggestion of making such a calculation was made by Prandtl in a discussion during the meeting of the Society of Applied Mathematics and Mechanics, Dresden, 1925. See a footnote in the paper by W. Lode, *Z. Physik*, vol. 36, p. 934, 1926.
[5] A discussion of various strength theories with a complete bibliography of the subject is given in "Die Bruchgefahrfester Körper" by M. Roś and A. Eichinger, Bericht Nr. 172 der EMPA, Zürich, 1949.

76. Creep of Metals at Elevated Temperatures

During the nineteenth century some investigations on the strength of materials at high temperatures were made, but the tests were of the same kind as those at room temperature. The purpose was to find the ultimate strengths of materials and their moduli of elasticity at elevated temperatures. During the last thirty years, problems concerning the mechanical properties of materials at high temperatures have become of great practical importance. Economical use of fuels has demanded the use of ever-increasing temperatures in all types of steam-driven plants. From about 650°F in 1920, temperatures have now risen to 1000°F in steam power stations. Thus knowledge of properties of metals at high temperatures has become essential in steam-turbine design. Engineers encounter the same problem in the design of diesel and gas engines. The development of gas turbines and jet-propulsion engines has become possible only with introduction of metals which can stand high temperatures. New problems in the oil and chemical industries also present problems in which the mechanical properties of metals at high temperatures are of importance. To satisfy the requirements of industry, high-temperature creep tests[1] of metals were started in several places. P. Chevenard started this work in France.[2] J. H. S. Dickenson conducted experiments on the flow of steels at dull red heat in England.[3]

It was found that, at elevated temperatures, steel specimens *creep* under the action of a constant tensile force in the same way as lead creeps at room temperature. Attempts to establish a "limiting creep stress," below which no creep would occur, revealed that no such limit exists and that, as more and more sensitive measuring instruments were used, smaller and smaller stresses were found capable of causing creep in loaded specimens. It became clear that the usual methods of selecting proper dimensions of structural members (on the basis of some specified working stresses) are not applicable for work at elevated temperatures. The designer has to consider deformations of his structure due to creep and select dimensions accordingly. They must be chosen in such a way that deformations during the service life of the structure, say twenty to thirty years in the case of a power-generating station, will not exceed certain permissible limits.

To obtain the necessary information, several laboratories started testing tensile-test pieces under constant load. The results of such tests are

---

[1] Information about this kind of work and a complete bibliography on the subject can be found in the books, "Creep of Metals" by H. J. Tapsel, Oxford, 1931, and "Properties of Metals at Elevated Temperatures" by George V. Smith, New York, 1950.

[2] *Compt. rend.*, vol. 169, pp. 712–715, 1919.

[3] *J. Iron Steel Inst. (London)*, vol. 106, p. 103, 1922.

usually represented by curves such as those shown in Fig. 225, in which the elongations are represented as a function of time. These curves are obtained with different values of the tensile load. For example, considering the curve $OABCD$, we see that, at the instant of applying the load, the specimen stretches by the amount OA and then creeps with a gradually diminishing creep rate. This is indicated by the slope of the curve AB. Along the portion BC, the slope remains practically constant and the specimen stretches at a constant creep rate. Finally, along the portion CD, the creep rate increases with time; this is mainly due to the perpetual diminution of the cross section which causes the creep to continue at an ever-increasing tensile stress. The portion BC of the curve is the most important for engineers, and they usually try to work within this range of "minimum creep rate." It is assumed[1] that during the period of time corresponding to the portion BC of the curve, the strain hardening produced by creep is continuously overcome by the annealing effect of the high temperature and that, due to this balance, creep takes place at a constant rate.

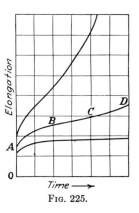

Fig. 225.

The duration of laboratory tests usually does not exceed a few thousand hours so that prediction of the creep deformation during the lifetime of a structure demands some extrapolation of the test results. Experiments with various steels have shown that, in the portion AB of the curve (Fig. 225), the excess of creep rate above the minimum creep rate decreases geometrically as time increases arithmetically. On the strength of this, the following relation between the creep rate and time is usually applied[2]

$$\frac{d\epsilon}{dt} = v_0 + ce^{-\alpha t} \quad (a)$$

The quantity v_0 denotes the minimum creep rate which, together with the constants c and α, has to be determined from the experimental creep-time curve obtained for the temperature and the stress for which the extrapolation must be made. By integration we have

$$\epsilon = \epsilon_0 + v_0 t - \frac{c}{\alpha} e^{-\alpha t} \quad (b)$$

---

[1] This idea was introduced by R. W. Bailey, who contributed much to the study of creep. See *J. Inst. Metals*, vol. 35, p. 27, 1926.

[2] This method of extrapolation was proposed by P. G. McVetty. See *Proc. ASTM*, part II, vol. 34, pp. 105–129, 1934.

The constant ϵ_0 can again be obtained from tests, so that the stretch of the specimen at any time t can be readily computed from Eq. (b). For large values of t, the last term in Eq. (b) becomes negligible, and instead of the creep-time curve AC (Fig. 226), we can use the asymptote BD. In this way experiments made with a metal at 850°F under various values of stress are used to draw the curves shown in Fig. 227.[1] Using such curves, the stretch for any assumed values of stress σ and service time can be obtained.

For the calculation of stresses in structures at elevated temperatures, it is desirable to have an expression for that part of the total creep which

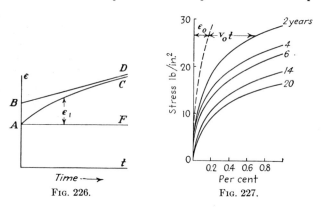

Fig. 226.   Fig. 227.

occurs after the initial stretching (shown by ϵ_1 in Fig. 226). It is usually assumed that this quantity can be represented by an equation of the form

$$\epsilon_1 = \phi(\sigma)\psi(t)$$

where ϕ is a function of σ only and ψ is a function of t only. In practice, ϕ is often taken as a power function of σ and the experimental curve AC in Fig. 226 is taken for ψ. In this way, we obtain the equation

$$\epsilon_1 = a\sigma^m \psi(t) \qquad (c)$$

where a and m are two constants which have to be selected so as to fit experimental curves. Using this equation, we can solve some practical problems encountered in structural design for elevated temperatures.

As an example, let us consider pure bending of a rectangular beam of depth h and width b. We assume that cross sections remain plane during bending and that the material has the same properties in tension as it has in compression so that the neutral axis passes through the centroid of the

---

[1] These results are taken from McVetty's paper, *Proc. ASTM*, vol. 34, 1938.

cross section. Let $\sigma_{\max}$ denote the maximum stress and σ the stress at a distance y from the neutral axis. Then, using Eq. (c), we have

from which
$$\epsilon_{\max} = \frac{h}{2\rho} = a\sigma^m{}_{\max}\psi$$
$$\epsilon = \frac{2y}{h}\epsilon_{\max} = a\sigma^m\psi \qquad (d)$$

$$\sigma = \sigma_{\max}\left(\frac{2y}{h}\right)^{1/m} \qquad (e)$$

Substituting this into the expression for bending moment, we find

$$M = 2\int_0^{h/2} b\sigma y\, dy = \sigma_{\max}\frac{bh^2}{6}\frac{3m}{2m+1}$$

so that $\sigma_{\max}$ can be obtained and the curvature $1/\rho$ can be calculated from Eq. (d) for any time t. The curvature produced at the instant of applying the moment M is neglected in this derivation.

Few experiments have been conducted concerning creep under combined stresses. At room temperature, R. W. Bailey[1] tested lead pipes subjected to internal pressure and to combined internal pressure and axial loading. He also investigated the properties of steel pipes at 900 and 1020°F under axial tension and torsion. F. L. Everett[2] tested steel pipes at high temperatures, in torsion. Not having enough experimental information regarding creep under combined stress, it is necessary to adapt the results of simple tension creep tests to the solution of these more complicated problems. This is usually accomplished by assuming that (1) during plastic deformation, the directions of the principal stresses (σ_1, σ_2, and σ_3) coincide with the directions of principal strains (ϵ_1, ϵ_2, and ϵ_3,) (2) the volume of the material remains constant so that, for small deformations

$$\epsilon_1 + \epsilon_2 + \epsilon_3 = 0 \qquad (f)$$

and (3) the maximum shearing stresses are proportional to the corresponding shearing strains, i.e.,

$$\frac{\epsilon_1 - \epsilon_2}{\sigma_1 - \sigma_2} = \frac{\epsilon_2 - \epsilon_3}{\sigma_2 - \sigma_3} = \frac{\epsilon_3 - \epsilon_1}{\sigma_3 - \sigma_1} = \theta \qquad (g)$$

where θ denotes a certain function of σ_1, σ_2, σ_3 to be determined from experiments. From Eqs. (f) and (g), it follows that

[1] World Power Conference, Tokyo, 1929; *Engineering*, vol. 129, pp. 265–266, 327–329, and 772, 1930.
[2] *Trans. ASME*, vol. 53, 1931; *Proc. ASTM*, vol. 39, pp. 215–224, 1939.

$$\epsilon_1 = \frac{2\theta}{3}\left[\sigma_1 - \frac{1}{2}(\sigma_2 + \sigma_3)\right]$$
$$\epsilon_2 = \frac{2\theta}{3}\left[\sigma_2 - \frac{1}{2}(\sigma_1 + \sigma_3)\right] \qquad (h)$$
$$\epsilon_3 = \frac{2\theta}{3}\left[\sigma_3 - \frac{1}{2}(\sigma_1 + \sigma_2)\right]$$

These equations are comparable with the stress-strain relations of Hooke's law, but they differ in two respects. The quantity $2\theta/3$ appears instead of the constant $1/E$ and a factor $\frac{1}{2}$ replaces Poisson's ratio.

To adapt Eqs. (h) to creep at a constant rate, we can divide these equations by the time t. Using the notation $\dot{\epsilon}_1$, $\dot{\epsilon}_2$, and $\dot{\epsilon}_3$ for the principal creep rates and denoting the factor before the brackets by κ, we arrive at

$$\dot{\epsilon}_1 = \kappa[\sigma_1 - \tfrac{1}{2}(\sigma_2 + \sigma_3)]$$
$$\dot{\epsilon}_2 = \kappa[\sigma_2 - \tfrac{1}{2}(\sigma_1 + \sigma_3)] \qquad (k)$$
$$\dot{\epsilon}_3 = \kappa[\sigma_3 - \tfrac{1}{2}(\sigma_1 + \sigma_2)]$$

Applying these equations to simple tension when $\sigma_1 = \sigma$ and $\sigma_2 = \sigma_3 = 0$, we have

$$\dot{\epsilon} = \kappa \cdot \sigma \qquad (l)$$

We have already stated that experimental results for the constant creep rate in simple tension can be represented satisfactorily by a power function

$$\dot{\epsilon} = b\sigma^m \qquad (m)$$

where b and m are two constants of the material. To create agreement between Eqs. (l) and (m), we must take

$$\kappa = b\sigma^{m-1} \qquad (n)$$

To establish the form of the function κ for the general case represented by Eqs. (k), we use Eq. (c), Art. 75, which shows the relation between the yielding condition for a three-dimensional stress system and for simple tension. Thus we substitute the value

$$\sigma_e = \frac{1}{\sqrt{2}}\sqrt{(\sigma_1 - \sigma_2)^2 + (\sigma_2 - \sigma_3)^2 + (\sigma_3 - \sigma_1)^2} \qquad (p)$$

into Eq. (n) as an equivalent tensile stress. For simple tension, we again obtain Eq. (m) in this way and, for the general case, Eqs. (k) give

$$\dot{\epsilon}_1 = b\sigma_e^{m-1}[\sigma_1 - \tfrac{1}{2}(\sigma_2 + \sigma_3)]$$
$$\dot{\epsilon}_2 = b\sigma_e^{m-1}[\sigma_2 - \tfrac{1}{2}(\sigma_3 + \sigma_1)] \qquad (q)$$
$$\dot{\epsilon}_3 = b\sigma_e^{m-1}[\sigma_3 - \tfrac{1}{2}(\sigma_1 + \sigma_2)]$$

Substituting values found from simple tension creep tests for b and m, we get formulas for calculating the creep rates $\dot\epsilon_1$, $\dot\epsilon_2$, $\dot\epsilon_3$ in the general case. Similar formulas have been used by several authors[1] in solving such important problems as creep in thick-walled cylinders subjected to internal pressure and in rotating disks. The stresses σ_1, σ_2, σ_3 cannot be obtained from statics in these cases and step-by-step integration must be employed in solving the problems.

77. Fatigue of Metals

Modern developments in the machine industry have brought the hazard of fatigue of metals under cycles of stress into prominence. The main cause of mechanical failure in service is the fatigue of machine parts and consequently the study of this phenomemon has been one of the most important fields of research in material-testing laboratories[2] in the twentieth century. The notions of an *endurance limit* and a *range of stress* have been well established since the time of Wöhler and Bauschinger. Having the endurance limit for complete reversal of stress, the fatigue properties for other kinds of cycle were usually based on Gerber's assumption[3] that the range of stress R and the mean stress σ_a are related by a parabolic law:

$$R = R_{max}\left(1 - \frac{\sigma_a{}^2}{\sigma_{ult}{}^2}\right)$$

Determination of the endurance limit in the laboratory using slow-speed testing machines demanded much time and expense; therefore many attempts were made to find out if any relations exist between the endurance limit and the other mechanical properties which can be investigated by static tests. These efforts have met with little success, although it was found that the endurance limit of ferrous metals subjected to complete reversals of stress is approximately given by 0.40 to 0.55 × ultimate strength.

A considerable amount of work was done to find what relation (if any) exists between the phenomenon of hysteresis and fatigue. Bauschinger had already introduced the notion of the *natural proportional limits* which are fixed after submitting the material to cycles of stress; these limits

[1] F. K. G. Odqvist, Plasticitetsteori . . . , *Proc. Roy. Swed. Inst. Eng. Research*, 1934; R. W. Bailey, *J. Inst. Mech. Engrs.*, vol. 131, p. 131, 1935; C. R. Soderberg, *Trans. ASME*, vol. 58, pp. 733–743, 1936.

[2] A description of this work can be found in the following books: "The Fatigue of Metals" by H. J. Gough, London, 1924; "The Fatigue of Metal" by H. F. Moore and J. B. Kommers, New York, 1927; the mimeographed lectures of H. J. Gough given at the Massachusetts Institute of Technology during the summer school of June–July, 1937; and also in the publication by M. Roś and A. Eichinger, Bericht Nr. 173 der EMPA, Zürich, 1950.

[3] W. Gerber, *Z. bayer. Architek. u. Ing. Ver.*, 1874.

were supposed to define the *safe range* in fatigue tests. This idea was developed further by L. Bairstow.[1] Using a mirror extensometer in conjunction with a slow loading and unloading machine, he measured the widths of the hysteresis loops and found that, when these widths were plotted against the maximum of the applied cyclical stress, the results of his tests gave an approximately straight line. Bairstow suggests using the intersection of this line with the stress axis for determining the *safe range of stress*. Subsequent endurance tests verified this assumption. Since that time, several methods have been devised for rapidly determining fatigue ranges from hysteresis loops. Instead of measuring the widths of hysteresis loops, Hopkinson and Williams[2] made calorimetric measurements of the energy dissipated per cycle and suggested a quick method for determining endurance limits. In more recent times, O. Föppl and his collaborators at the Wöhler Institute at Brunswick have done extensive work in measuring the *damping capacity of metals* and its relation to fatigue strength.[3]

Another rapid method of determining endurance limits was offered by Ewing and Humfrey.[4] They used specimens of Swedish iron with polished surfaces and examined them with a microscope after submitting them to cycles of reversed stress. They found that if stresses above a certain limit were applied, *slipbands* appeared after a while on the surface of some of the crystals. Some of these slipbands seemed to broaden as the number of cycles increased and finally a crack started following the outline of one of the widened slipbands. It was assumed that stresses which produce slipbands are outside the safe range. If such cycles of stress are continued, a continual sliding along the surfaces takes place. This is accompanied by friction, similar to that between sliding surfaces of rigid bodies. As a result of this friction, the material gradually wears along the surfaces of sliding (according to this theory) and a crack results. Further investigation by Gough and Hanson[5] has revealed that slipbands may be formed at lower stresses than the endurance limit of the material and that they may develop and broaden without leading to the formation of a crack.

To get a deeper insight into the mechanism of failure in fatigue tests, Gough[6] applied a new method of attack by using a precision X-ray tech-

[1] *Trans. Roy. Soc. (London)*, (A), vol. 210, p. 35, 1911.
[2] B. Hopkinson and G. T. Williams, *Proc. Roy. Soc. (London)*, (A), vol. 87, 1912.
[3] English summaries of this work have been published by G. S. Heydekamp, *Proc. ASTM*, vol. 31, 1931, and by O. Föppl, *J. Iron and Steel Inst. (London)*, vol. 134, 1936.
[4] J. A. Ewing and J. C. W. Humfrey, *Trans. Roy. Soc. (London)*, (A), vol. 200, p. 241, 1903.
[5] H. J. Gough and D. Hanson, *Proc. Roy. Soc. (London)*, (A), vol. 104, 1923.
[6] A summary of this work was presented to the Royal Aeronautical Society. See *J. Roy. Aeronaut. Soc.*, August, 1936.

nique. Starting with single-crystal specimens, he showed that the mechanism of deformation of ductile metallic crystals under cycles of stress is the same as under static conditions; i.e., the sliding occurs along certain crystallographic planes in definite directions and is controlled by the magnitude of the shearing stress component in the direction of sliding. The X-ray analysis showed[1] that if the cycles were outside the safe range, "the crystallographic planes became distorted in such a manner that, while their average curvature was inappreciable . . . yet severe local curvatures existed. It was suggested that large local strains—also, possibly, actual lattice discontinuities—must occur in such distorted planes, which, under the application of a sufficiently great range of external stress or strain, would lead to the formation of a progressive crack; under a lower magnitude of cyclic straining, the condition might be stable."

In experimenting with crystalline materials such as mild steel, the preliminary static tests showed that no permanent change occurs in perfect grains if they are kept within the elastic range. Between the elastic limit and the yield point, a few perfect grains break up, forming a small proportion of smaller grains and crystallites (the limiting small size of these crystal fragments was 10^{-4} to 10^{-5} cm). After yield, every perfect grain breaks up, forming smaller grains and a large number of crystallites. By the application of cycles of stress, it was found "that if the applied range of stress exceeds the safe range, progressive deterioration and break-up into crystallites sets in leading to fracture exactly as in a static test to destruction." Thus the experiments showed that the fracture of metals under statical and fatigue stressing is accompanied by the same changes in structure.

A considerable amount of work has been done in studying the factors which affect the endurance limit. Work on machines which could apply stress cycles of differing frequencies showed that, up to a frequency of 5,000 cycles per minute, no appreciable frequency effect was observable. By increasing the frequency to a value in excess of 1,000,000 cycles per minute, Jenkin[2] found an increase in the endurance limit of more than 30 per cent for Armco iron and aluminum.

Fatigue tests made with steel specimens initially stretched beyond their yield point showed that a moderate degree of stretching produces some increase in the endurance limit. With further increase in the cold working, a point may be reached at which a drop in the endurance limit may occur due to overwork.[3] Experiments in which cycles of stress above the endurance limit were applied a number of times before starting the normal endurance test, showed that there is a limiting number of cycles of over-

[1] See the above-mentioned paper by H. J. Gough.
[2] C. F. Jenkin, *Proc. Roy. Soc. (London)*, (A), vol. 109, p. 119, 1925; vol. 125, 1929.
[3] H. F. Moore and J. B. Kommers, *Univ. Illinois Eng. Exp. Sta. Bull.* 124, 1921.

stress (depending on the magnitude of the overstress) below which the endurance limit is not affected. But above this limit, the number of cycles withstood was observed to decrease. By plotting the maximum stress of the cycles of overstress against the limiting numbers of these cycles, a *damage curve* for the tested material was obtained.[1] The area below this curve defines all those degrees of overstressing which do not cause damage. The damage curve can be used in dealing with machine parts working below the endurance limit, but subjected from time to time to cycles of overstress. A formula was established for calculating the number of cycles of overstress of various intensities which a machine part can withstand before failure.[2] In airplane structures, a statistical analysis of service stresses in various parts is sometimes made[3] and fatigue tests so planned as to adjust the cycles applied in the laboratory to the service condition of the part under investigation.

A considerable amount of research work has been done in connection with the behavior of metals subjected to cycles of stresses while exposed to the action of an oxidizing agent. Haigh observed[4] some lowering of the endurance limit when specimens of brass were subjected to the action of salt water, ammonia, or hydrochloric acid while under alternating stress. He also pointed out that the damaging effect of ammonia on brass was not produced unless the corrosive substance and the alternating stresses were applied simultaneously. Further progress in *corrosion fatigue* was made by McAdam[5] who investigated the combined effect of corrosion and fatigue on various metals and alloys. These tests proved that in most cases severe corrosion prior to an endurance test is much less damaging than slight corrosion which takes place at the same time. The tests also showed that, while the endurance limit of steel increases approximately in the same proportion as the ultimate strength when tested in air, the results obtained when testing in fresh water are quite different. It was found that the corrosion fatigue limit of steel having more than 0.25 per cent carbon cannot be increased but may be decreased by heat-treatment. Experiments conducted *in vacuo* have shown[6] that the endurance limit of steel is about the same as that obtained from tests in air, while those with copper and brass specimens revealed an increase of endurance limit of no less than 14 per cent and 16 per cent, respectively. All these results are of great practical importance since there are many cases of failure in service which can be attributed to corrosion fatigue.

[1] H. J. French, *ASST*, vol. 21, p. 899, 1933.
[2] B. F. Langer, *J. Applied Mechanics*, vol. 4, p. 160, 1937.
[3] H. W. Kaul, *Jahrb. deut. Luftfahrt-Forsch.*, 1938, pp. 307–313.
[4] *J. Inst. Metals*, vol. 18, 1917.
[5] D. J. McAdam, *Proc. Intern. Cong. Testing Materials*, Amsterdam, 1928, vol. 1, p. 305.
[6] H. J. Gough and D. G. Sopwich, *J. Inst. Metals*, vol. 49, p. 93, 1932.

Progress in Strength of Materials during the Twentieth Century 381

Most of our information on fatigue is obtained from bending and direct tension-compression tests and it is of great importance to establish rules on how to use it in cases of combined stresses. To obtain the necessary experimental data, a combination of reversed bending and reversed torsion acting in phase was used by H. J. Gough and H. V. Pollard.[1] By varying the ratio of the maximum bending moment to the maximum torsion moment it was found that, in mild carbon steel and nickel-chromium steel ($3\frac{1}{2}$ per cent Ni), the limiting values of the bending stress and of the shearing stress may be found from the equation

$$\frac{\sigma^2}{\sigma_e^2} + \frac{\tau^2}{\tau_e^2} = 1 \qquad (a)$$

where σ_e is the endurance limit for bending and τ_e is the endurance limit for torsion.

We come to the same result by using the maximum distortion energy theory. Writing Eq. (c), page 370, in the form

$$(\sigma_1 - \sigma_2)^2 + (\sigma_1 - \sigma_3)^2 + (\sigma_2 - \sigma_3)^2 = 2\sigma_e^2 \qquad (b)$$

and substituting for σ_1 and σ_3 the values

$$\frac{1}{2}(\sigma \pm \sqrt{\sigma^2 + 4\tau^2}), \qquad \sigma_2 = 0, \qquad \tau_e = \frac{\sigma_e}{\sqrt{3}}$$

into this equation, we obtain Eq. (a). The tests made by R. E. Peterson and A. M. Wahl[2] also uphold the use of Eq. (b) for ductile materials.

Bitter experience has taught us that most fatigue cracks start at points of stress concentration; these are always present in actual machine parts around fillets, grooves, holes, keyways, etc. Wöhler did the first experiments on the effect of stress concentration on fatigue (see page 169). Bauschinger, who continued Wöhler's work, did some tests[3] with grooved specimens. Using grooves of 0.1 and 0.5 mm he found only small reductions in the endurance limit of mild steel. A more systematic study of stress concentration was made by A. Föppl.[4] He submitted rectangular specimens with circular holes to tension-compression tests and found that the influence of holes is much smaller than the theoretical formula predicts. This discrepancy was explained by the fact that high stresses were used in these tests to reduce the number of cycles necessary to produce fracture. With smaller stresses, the influence of the holes would be more pronounced. Föppl also made tension-compression and bending tests with circular grooved specimens and again came to the above con-

---
[1] *Proc. Inst. Mech. Engrs.*, vol. 131.
[2] See paper presented before the Society for the Promotion of Engineering Education, June 23, 1941, Ann Arbor, Mich.
[3] See *Mitt. mech. tech. Lab. Technischen Hochschule, Munich*, nr. 25, 1897.
[4] *Mitt. . . . Munich*, nr. 31, 1909.

clusion. Finally, he conducted torsional fatigue tests on circular specimens ($d = 20$ mm) with fillets of 1 and 4 mm radius. Again, in order to obtain fracture after a comparatively small number of cycles ($n = 10^5$), high stresses were employed. Once more the effect of the fillets was found to be much smaller than was expected from Föppl's theoretical investigation.[1]

A further examination of the influence of stress concentration on fatigue was made by R. E. Peterson.[2] He arrived at the following conclusions from the results of his experiments with specimens of various diameters: (1) In some cases fatigue results are quite close to theoretical stress concentration values. (2) Fatigue results for alloy steels and quenched carbon steels are usually closer to theoretical values than are the corresponding fatigue results for carbon steels not quenched. (3) With a decrease in the size of specimens, the reduction in fatigue strength due to a fillet or hole becomes somewhat less and, for very small fillets or holes, the reduction in fatigue strength is comparatively small. On the basis of these results, the theoretical values of the stress concentration factor were recommended in the design of larger machine parts and in the use of the finer grained steels, such as alloy steels and heat-treated carbon steels.

Various theories have been advanced to explain the *size effect* found in fatigue tests involving stress concentrations. Peterson[3] shows that this effect can be explained by applying the statistical method recommended by W. Wiebull (for static test results obtained with brittle materials) to the analysis of experimental results (see page 360). Whether or not this method will yield a satisfactory explanation of the size effect in fatigue tests depends upon our having an adequate quantity of experimental data covering specimens of a large variety of sizes. Fatigue tests with large specimens made on a Timken machine[4] and those made with a marine shaft machine at Stavely, England,[5] show a considerable reduction of endurance limit for large specimens.

The problem of reducing the damaging effect of stress concentration is of primary importance in machine design. Some relief can be gained by judicious changes in design such as eliminating sharp reentrant corners

[1] A. Föppl was probably the first to give an approximate formula for the factor of stress concentration at the fillet of a shaft in torsion. See *VDI*, 1906, p. 1032. A more accurate solution of the same problem was given by Willers, *Z. Math. u. Physik.*, vol. 55, p. 225, 1907. An experimental determination of the stress concentration factor by using glass specimens was made by A. Leon, *Kerbgrösse und Kerbwirkung*, Vienna, 1902.

[2] *J. Applied Mechanics*, vol. 1, pp. 79, 157, 1933; vol. 3, p. 15, 1936.

[3] R. E. Peterson, "Properties of Metals in Materials Engineering," published by the American Society for Metals, Cleveland, Ohio, 1948.

[4] O. J. Horger and H. R. Neifert, *Proc. ASTM*, vol. 39, p. 723, 1939.

[5] S. F. Dorey, *Engineering*, 1948.

Progress in Strength of Materials during the Twentieth Century 383

and introducing fillets having generous radii, by designing fillets of correct profile, by introducing relieving grooves, etc. But sometimes these measures cannot be adopted and recourse should then be made to some means of improving the material at the dangerous spot. An improvement can sometimes be brought about by suitable surface heat-treatment of the material. Remarkable improvement has been obtained by cold rolling; the idea of using surface cold work for improving the fatigue property of materials was advanced by O. Föppl who made a number of investigations on small-size specimens.[1] T. V. Buckwalter and O. J. Horger[2] applied this method to full-scale axles and developed a machine capable of testing axles up to 14 in. in diameter. This illustrates the modern trend to carry out fatigue tests and also, in certain cases, static tests to destruction on full-size members. Wöhler started his famous work on fatigue by testing full-size railway axles (see page 168) and gradually turned to laboratory testing of small specimens. Now, to test design improvements and to study the size effect, recourse is again being had to full-scale specimens.

78. Experimental Stress Analysis

Modern work in strength of materials has involved the introduction of new measuring devices into experimental research. These have enabled us to add considerably to our knowledge of stress distributions in engineering structures. There are several reasons for this development of experimental stress analysis. Theoretical formulas of strength of materials and theory of elasticity are derived under the assumptions that the materials involved are homogeneous, perfectly elastic, and that they follow Hooke's law. In point of fact, materials sometimes deviate considerably from perfect homogeneity and perfect elasticity, and it becomes of practical importance to verify formulas derived for ideal materials. Only in the simplest cases can theory furnish us with the complete solution of the stress distribution problem. In the majority of instances engineers have to use approximate solutions, the accuracy of which must be checked by direct test. The modern trend in engineering design is toward utmost economy in weight of material, and this results in increases of allowable stresses and the reduction of factors of safety. Such a tendency can be permitted with safety only if designers have accurate information regarding the properties of materials and are in possession of accurate methods of stress analysis. For such analysis, an accurate knowledge of service conditions of the structure is necessary, especially with regard to the external forces acting. Often the forces acting on a structure are only

[1] See his papers in *Mitt. Wöhler Inst.*, Brunswick; see also *Maschinenbau*, vol. 8, pp. 752–755, 1929.
[2] T. V. Buckwalter and O. J. Horger, Trans. *ASM*, vol. 25, pp. 229–244, 1937. See also S. Timoshenko, *Proc. Inst. Mech. Engrs. (London)*, vol. 156, p. 345, 1947.

known approximately so that, in order to improve our knowledge, recourse has to be taken to the investigation of stresses in actual structures under service conditions. All these considerations explain the significance of the recent advances made in experimental stress analysis.[1]

The new progress in this phase of elasticity has involved the production of new measuring instruments and new methods of approach in determining stresses. For the direct measurement of strain, various kinds of strain gauge have been developed; these can be used not only in laboratory work, but also in the field for taking measurements on actual structures in service. The application of wire resistance strain gauges has proved particularly useful. These strain gauges have not only simplified the measurement of stresses under static conditions but have offered the possibility of recording cycles of stresses when the structure withstands variable forces or when a machine part is in motion. Such records are of great help in the proper analysis of vibration problems and have greatly improved our knowledge of the dynamic forces which may act on a structure. Another kind of strain gauge, the *electric-inductance gauge*, has been widely used in measuring stresses produced in rails under rolling locomotive wheels and in measuring the angles of twist of rotating shafts.[2]

The photoelastic method of stress analysis, inititated by Maxwell (see page 271), has found very wide use in the twentieth century. Mesnager used the method for checking Flamant's theory of stress distribution around the point of application of a concentrated force.[3] He also used it for solving a practical problem of stresses in an arch bridge.[4] The photoelastic method gives us the difference of the two principal stresses. Mesnager suggested that the sum of the principal stresses should be obtained from measurement of the change in thickness of the plate model at the point under consideration. This idea was utilized by E. G. Coker, who devised a special lateral extensometer for measurements of these changes in thickness. He also introduced the use of celluloid, which has greatly simplified the preparation of models for photoelastic tests. Coker's work[5] did much to popularize the method. Many young research

[1] The Society of Experimental Stress Analysis was organized in the United States; its activities and publications have attracted the interest of a large group of engineers. The new publication of this society, "Handbook of Experimental Stress Analysis," edited by Prof. M. Hetényi, will contribute much to the improvement of our knowledge in experimental research in strength of materials and theory of elasticity.

[2] A complete discussion of this type of strain gauge is given in an article by B. F. Langer; see "Handbook for Experimental Stress Analysis," pp. 238–272.

[3] *Ann. ponts et chaussées*, vol. 4, pp. 128–190, 1901.

[4] *Ann. ponts et chaussées*, vol. 16, pp. 133–186, 1913.

[5] E. G. Coker's experimental investigations, together with the theoretical works of L. N. G. Filon, are collected in the book "A Treatise on Photo-elasticity," Cambridge, 1931.

workers in photoelasticity obtained their initial experience by working in Coker's laboratory at University College, London. Further progress in this aspect of elasticity was made by Z. Tuzi,[1] who introduced the so-called *fringe method* of stress measurement and photographic recording, by the use of which results are obtainable immediately after loading, so that errors due to time effect are minimized. He also utilized a new material with greater optical sensitivity for his models, phenolite. A purely optical method in photoelasticity was developed by H. Favre.[2] The photoelastic method has become very popular amongst engineers, and we find all the equipment necessary for photoelasticity, not only in scientific research institutions and university laboratories, but also in many industrial organizations. The techniques of preparing models and making the measurements have been much improved, and we now have sufficiently accurate experimental solutions of many two-dimensional problems for which no theoretical treatments exist.[3]

In recent times, some attempts have been made to develop three-dimensional photoelastic stress analysis by using the method of *freezing the stress* in the model. This phenomenon was first observed by Maxwell. But, as yet, three-dimensional photoelasticity is in its infancy.[4]

For the approximate determination of stresses at the weakest points of complicated structures (to which theoretical stress analysis cannot be applied easily), the method of *brittle coatings* offers many advantages.[5] If the surface of a model or specimen is coated with a brittle lacquer before application of forces and then gradually loaded, the first crack in the lacquer will reveal the spot of maximum strain and the direction of the crack indicates the direction of the maximum tensile stress; it is perpendicular to the crack. The magnitude of the stress can also be determined if a preliminary tensile test of a calibration bar, coated with the same kind of lacquer, is made. For better accuracy, sensitive strain gauges can now be used at the weak spots in the direction of the crack and perpendicular to it. In this way, Dietrich and Lehr[6] have analyzed stresses in crankshafts, connecting rods, and other machine parts.

---

[1] Z. Tuzi, *Sci. Papers Inst. Phys. Chem. Researches (Tokyo)*, vol. 7, pp. 79–96, 1927. See also *Proc. 3d Intern. Congr. Applied Mechanics*, Stockholm, 1931, vol. 2, pp. 176–180.

[2] See *Schweiz. Bauz.*, v. 20, 1927.

[3] See M. M. Frocht, "Photoelasticity," New York, 1941.

[4] The history of this new development can be found in the paper by M. Hetényi, *J. Applied Mechanics*, December, 1938. See also the article on three-dimensional photoelasticity in the "Handbook of Experimental Stress Analysis," pp. 924–965, 1950.

[5] A description of this method is given in Hetényi's article in the "Handbook for Experimental Stress Analysis," pp. 636–662.

[6] *VDI*, vol. 76, pp. 973–982, 1932.

Considerable research has been done in studying the residual stresses usually produced in machine parts by various technological processes. For example, in drawing a copper bar through a die, the stresses are not uniformly distributed over the cross section. The metal near the surface is stressed more than that at the center, as shown in Fig. 228a. When the tensile force P is removed, the bar contracts uniformly and a uniform stress is subtracted from the stresses shown by the curve mn. Thus the residual stresses represented by the shaded area in Fig. 228b finally obtain. The outer fibers are in tension, while the inner ones are in compression. If the bar is now cut lengthwise, the residual stresses will be partially released, the two portions assuming the curved shapes shown in Fig. 228c. Similar residual stresses are produced by cold rolling, as shown in Fig. 228d. If extensometers are attached to the outer surfaces of the bar, before the lengthwise cut is made, the peak residual stress can be measured[1] during the bending shown in Fig. 228c.

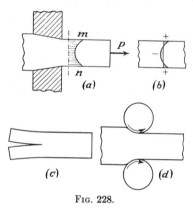

FIG. 228.

Residual stresses are of great practical importance. They cause undesirable warping in the process of machining and *season cracking* must also be attributed to them. This occurs in cold-drawn tubes of several copper alloys which, after cold work, have not been properly annealed. To measure the residual stresses which exist in a structural member, we have to cut the member into thin layers. By making measurements of deformation occurring in this process, we obtain information from which residual stresses can be calculated. The first experiments of this kind were made by N. Kalakoutzky,[2] who studied residual stresses in guns. To investigate residual stresses in rectangular bars (Fig. 229), we start by machining off a thin layer of thickness Δ. If that layer had a tensile residual stress σ, the machining produces the same effect on the deformation of the bar as the application of the two eccentric tensile forces $\sigma b\Delta$ (Fig. 229a). These two forces will produce a longitudinal strain ϵ given by

$$\epsilon = \frac{\sigma b \Delta}{Ebh} \qquad (a)$$

[1] Some experiments of this kind were made by the writer at the Research Laboratory of the Westinghouse Electric Corporation in 1924.

[2] N. Kalakoutzky, "The Study of Internal Stresses in Cast Iron and Steel," St. Petersburg, 1887, English translation, London, 1888.

and curvature $1/\rho$ according to the formula:

$$\frac{1}{\rho} = \frac{\sigma b h \Delta}{2EI} \qquad (b)$$

By measuring the change in curvature after planing off each layer, we obtain sufficient information for calculating the initial residual stresses by using Eqs. (a) and (b). The method can be improved by using a chemical process for taking off thin layers of material instead of machining. This method was devised by N. N. Davidenkov.[1] It was also applied in investigating residual stresses in cold-drawn tubes.

Residual stresses in solid cylindrical bars can be investigated by a boring-out process. By machining off the material in thin layers from inside and taking measurements of the corresponding deformations of

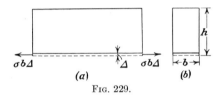

Fig. 229.

the cylinder in the longitudinal and tangential directions, we obtain sufficient information for calculating the residual stresses[2] distributed symmetrically with respect to the axis of the cylinder. This method has been used in studying residual stresses produced during various heat-treatment processes and for examining the plastic deformation caused by nonuniform cooling.[3] The same method is also used in connection with heat-treatment. Flame hardening and nitriding are now often used to improve the fatigue strength of machine parts at points of high stress concentration. In this way, high compressive residual stresses are created at the surface which, in combination with the cycles of stress produced in service, result in a more favorable stress fluctuation.

Residual stresses have great importance in welded structures. Due to high local temperatures set up in the process of welding and to the subsequent cooling, high residual stresses usually appear. The magnitude of these stresses in welded steel plate structures can be investigated by

---

[1] *J. Tech. Phys.*, (*U.S.S.R.*), vol. 1, 1931.

[2] This method of calculation was developed by G. Sachs, *Z. Metallkunde*, vol. 19, pp. 352–357, 1927; *Trans. ASME*, p. 821, 1939.

[3] H. Buchholtz and H. Bühler, "*Stahl u. Eisen*," vol. 52, pp. 490–492, 1932. See also the papers by S. Fuchs, *Mitt. Forsch.- Inst.*, vol. 3, pp. 199–234, 1933, and H. Bühler, H. Buchholtz, and F. H. Schultz *Archiv Eisenhüttenw.*, vol. 5, pp. 413–418, 1942.

using the *rosette strain gauge*[1] which measures the strain in the plane of a plate in three directions. Having such measurements, the magnitudes and directions of the principal strains and the corresponding principal stresses can be calculated. To obtain the residual stresses set up in a plate, two measurements are necessary, one before cutting out a small portion of the plate, and the other after. The differences of the two readings give the strains which result from the isolation of the element and which define the magnitude of residual stresses.

Methods of experimental stress analysis are continuing to develop in various directions, and we can expect further progress in that field and consequent improvement of our knowledge of strength of materials. Over a long period of time, analytical methods have prevailed in strength of materials. These methods are useful, but not always sufficient for the solution of engineering problems. The wide introduction of experimental methods has brought new life to research in strength of materials.

[1] A description of various kinds of rosette strain gauge, their theory and applications, can be found in the article by J. H. Meier in the "Handbook of Experimental Stress Analysis," pp. 390–437.

CHAPTER XIII

Theory of Elasticity during the Period 1900–1950

79. Felix Klein

Ever since the time of Gauss, Göttingen has been an important center of mathematics. Gauss was not only a great mathematician but he was also interested in the applications of that subject to the solutions of problems of astronomy, physics, and geodetic surveying. As a result of his work, a tradition of contact between mathematics and its applications was established at Göttingen. His successors, the outstanding mathematicians Dirichlet, Riemann, and Clebsch upheld this tradition and usually gave lectures in various branches of theoretical physics as well as in pure mathematics. At the end of the nineteenth and the beginning of the twentieth century, this idea of linking these subjects gained new impetus under the influence of Felix Klein.

Felix Klein (1849–1925) was born in Düsseldorf.[1] After graduating from the Gymnasium, he entered Bonn University where he soon became an assistant of the professor of physics, Professor Plucker. Klein

Fig. 230. Felix Klein.

was awarded his doctor's degree before he was twenty (in 1868) and moved to Göttingen, where Clebsch was teaching. In 1871, he became a lecturer

---

[1] See Klein's autobiographical notes added to several parts of his "Gesammelte Mathematische Abhandlungen," Berlin, 1921. See also *Universitätsbund Göttingen E.V. Jahrgang* 26, *Mitteilungen*, 1949, Heft 1; R. Mises, *ZAMM*, vol. 4, p. 86, 1924. For the history of mathematics at Göttingen, see "Der Mathematische Unterricht . . ." by Klein und Schimmack, Teil I, p. 158, 1907.

at that university, but he left Göttingen in the following year when he was elected to a chair at the University of Erlangen (before the age of twenty-three). There he drew up his famous program lecture "das Erlanger Programm" in which the importance of associating mathematics with its various fields of application was stressed. In 1875, Klein became professor of mathematics at the Munich Polytechnical Institute. At that school he not only offered the usual courses of calculus but also special mathematical lectures pitched at a higher level He thought that it would be to the advantage of engineering education if some engineering students took more mathematics and thus prepared themselves for research work in applied science as well as for teaching. From Munich, Klein moved, in 1880, to Leipzig University, and, in 1886, he became a professor at the University of Göttingen. He constantly stressed the importance of pure mathematics in technical physics and applied mechanics. These subjects had not been presented before in German universities, but due to his influence, the three new chairs of applied mathematics, technical physics (electrical engineering), and applied mechanics were established at Göttingen.

Klein made a trip to the United States in 1893 and visited the Chicago International Exposition. He participated in the Mathematical Congress of that year and gave a series of lectures at Evanston to American mathematicians. After that time, Klein kept in contact with American scientists and often had pupils from the United States in his classes. While he was in the United States, he studied the organization of those universities which had schools of engineering and which gave mathematical training not only to students electing that subject as their speciality but also to future engineers. He was impressed by the fact that many universities and scientific institutions were supported by private endowments and were independent of the government.

Upon his return to Göttingen, Klein decided to try American methods and established contact with representatives of German industry. He explained to these people how beneficial it would be for industry to organize, at Göttingen, institutes of applied mathematics, applied mechanics, and applied physics, in which young research scientists could be trained. He described how the activities of those institutions would influence methods of teaching mathematics and how they would help in training teachers of mathematics with some knowledge of engineering sciences. In 1898 the *Göttinger Vereinigung zur Förderung der angewandten Physik und Mathematik* was founded and, with the financial support of this society, several research institutions were brought into being at Göttingen university. Klein succeeded in finding the financial support for the new institutions and also in attracting as their directors outstanding workers in the applied sciences. In 1904, when the work of organization was completed, C. Runge[1] was in charge of applied mathematics, L. Prandtl was the director of the institute of applied mechanics, and H. T. Simon headed the department of electrical engineering. To maintain contact between these new institutions and the school of pure mathematics, a seminar was organized under the chairmanship of Klein with the collaboration of the professors of applied sciences. Strength of materials, theory of structures, theory of elasticity, electrotechnics, etc., were taken as the subjects at the seminar in consecutive semesters. For each two-hour session, one of the students had to prepare a paper, with direction from the appropriate professor, on a topic suggested in advance. After the presentation of the papers, discussion by the professors and students followed. Klein's remarks were usually of special interest, for, with his great knowledge and his outstanding ability as a teacher, he would see some points of more general interest in each specific problem and with some suitable remarks would show the connection of the work presented with problems in related fields. This kind of seminar, in which the representatives of pure mathematics and applied sciences collaborated, was a novelty and it attracted the interest of many young engineers not only in Germany but also from other countries, so that the gathering usually had a cosmopolitan air. This new movement had great success, and many students selected Göttingen for their graduate work. Many doctor's theses in the applied sciences were written there, and the university supplied professors of applied mechanics for many engineering schools. These former pupils of F. Klein, C. Runge, and L. Prandtl have contributed much to engineering mechanics in Germany since that time.

Klein's activity was not limited to this work of organization at Göttingen. For he applied his energy and his broad knowledge to some scientific projects of international importance. He led in the compilation and publication of the mathematical encyclopedia ["Encyklopädie der mathematischen Wissenschaften" (1901–1914)]. Being in touch with mathematicians of foreign countries, he succeeded in getting the collaboration of outstanding scientists and made the work on the encyclopedia an international enterprise. Klein himself edited the portion dealing with mechanics. In the four volumes, prepared under his leadership, an up-to-date account of the various branches of theoretical and applied mechanics was presented and a very complete bibliography drawn up. Historical information was given also. The publication of those volumes was a great event in the development of mechanics in our time. It has considerably helped the work of individual scientists and has done much

[1] For the biography of C. Runge, see the book, "Carl Runge," by Iris Runge, in which a vivid picture of academic life at Göttingen in the beginning of the twentieth century is also given.

to bring engineering mechanics to a higher level by establishing closer contact between the pure and applied sciences.

Another way in which Klein improved relations between the pure and applied sciences was through his work in revising the programs of mathematical education in high schools and universities. He advocated the setting up of institutes of applied sciences in the German universities. He also wanted engineering schools to participate to a greater extent in the training of teachers of mathematics by raising their standards in the fundamental sciences. These ideas had a great effect upon mathematical education in Germany and later, through the International Commission of the International Congresses of Mathematics, they influenced mathematical teaching methods in other countries.

Fig. 231. Ludwig Prandtl.

The tendency in strength of materials in the twentieth century has been toward greater use of mathematical tools, and the principal contributions to our science have been made by men who have combined engineering education with a substantial preparation in the fundamental sciences. This progress in the engineering sciences has been in the direction foreseen by Klein.

80. *Ludwig Prandtl*

L. Prandtl was born in 1875 in Freising near Munich in the family of a professor of the school of agriculture. He received his engineering education at the Munich Polytechnical Institute. After graduating, he remained at the school as an assistant of Prof. A. Föppl who, at that time, had just started teaching there and was reorganizing the program of engineering mechanics and that of the material-testing laboratory. Prandtl helped him in that work and, at the very beginning of his career, he completed an important piece of research.

Föppl was interested at that time in the theory of bending of curved bars and had made a considerable number of tests on the strength of railway couplings. He thought that the simple straight-beam formula could be applied with sufficient accuracy to the calculation of the maximum bending stresses in hooks. Professor C. Bach at the Stuttgart Polytechnical Institute was of a different opinion and used Winkler's theory of bending of curved bars, which is derived on the assumption that cross sections of a curved bar remain plane during bending. Prandtl obtained the rigorous solution for the pure bending of a curved bar of narrow rectangular cross section. This showed that cross sections remain plane during pure bending and that the stress distribution is very nearly that given by Winkler's theory.[1]

Working on bending of circular plates, Prandtl noticed that deflections are proportional to the load for small deflections only and that for larger deflections the plate exhibits greater rigidity than the theory predicts. This was perhaps the first experimental proof of this limitation on the application of the usual plate theory. In 1899 he wrote, as his doctor's thesis, a paper[2] on lateral buckling (*kipperscheinung*) of beams of narrow rectangular cross section bent in the plane of maximum flexural rigidity. He obtained solutions for several particular cases[3] of practical importance, calculated the table of Bessel's function which is required in this problem, and made a series of experiments, the results of which showed good agreement with his theory. In making these tests, Prandtl developed a technique of accurately determining the critical load which was later used by many investigators in elastic stability. This paper on lateral buckling initiated numerous investigations on the lateral stability of beams[4] and curved bars.[5]

After finishing his doctor's thesis, Prandtl worked in industry. However, he soon returned to academic life and, as early as 1900, he accepted the chair of engineering mechanics at the Hannover Polytechnical Institute. Whilst there, he published his important paper on the *membrane analogy* of the torsion problem.[6] In this paper he shows that, by using a soap film, all the information on stress distribution in torsion can be obtained experimentally. Some work on this was later done by his pupil H. Anthes.[7] The practical importance of the analogy was recognized by A. A. Griffith and G. I. Taylor, who used[8] the soap-film method for deter-

---

[1] The rigorous solution for bending of circular bars was obtained earlier by Golovin, but his paper was published in Russian and remained unknown in Western Europe. Prandtl never published his work; reference to it was made in A. Timpe's paper, *Z. Math. u. Physik*, vol. 52, p. 348, 1905.

[2] L. Prandtl, "Kipperscheinungen," Dissertation, Munich, 1899.

[3] The simplest case of buckling in the case of pure bending was solved independently by A. G. M. Michell, *Phil. Mag.* (5), vol. 48, p. 298, 1899.

[4] The writer investigated lateral buckling of I beams; see S. Timoshenko, *Bull. Polytech. Inst. St. Petersburg*, vol. 4, 1905; vol. 5, 1906. A further investigation of the same problem was made by E. Chwalla, *Bauing.*, vol. 17, 1936. See also *Sitzber. Akad. Wiss. Wien Math.-Naturw. Klasse*, Abt IIa, vol. 153, 1944, and O. Pettersson, Bulletin Nr. 10, Royal Institute of Technology, Stockholm, 1952.

[5] See S. Timoshenko, *Bull. Polytech. Inst. Kiev*, 1910.

[6] *Physik. Z.*, vol. 4, 1903.

[7] H. Anthes, *Dinglers Polytech. J.*, 1906, p. 342.

[8] *Tech. Rept. Advisory Comm. Aeronautics*, London, vol. 3, pp. 910, 938, 1917–1918.

mining torsional rigidities of bars of various complicated cross sections.

Prandtl's accomplishments in engineering mechanics became known to Klein and, on the strength of Föppl's recommendation, he decided to bring Prandtl to Göttingen and to entrust him with the important work of developing engineering mechanics at that university. This unusual action[1] of bringing a young engineer to a school of mathematics and of giving him such responsibility once more showed Klein's foresight. Prandtl's subsequent activity at Göttingen added much to the great reputation of that institution.

FIG. 232. Institute of Applied Mathematics and Mechanics at the University of Göttingen.

At the meeting of the International Mathematical Congress in Heidelberg in 1904, Prandtl presented his famous paper "Über Flüssigkeitsbewegung bei sehr kleiner Reibung" in which the boundary layer was discussed for the first time. Klein, who heard Prandtl's presentation, at once understood the great importance of the paper to future work in hydrodynamics and aerodynamics and told his young colleague that his paper was the best in the Congress.

Prandtl started his work at Göttingen in the fall of 1904. In the same year, Prof. C. Runge, a close friend, joined him to teach applied mathematics. In complete harmony, they both worked to improve the Institute of Applied Mathematics and Mechanics[2] and, due to their activity,

[1] Some of the professors were not in favor of teaching the applied sciences at the university, but Prandtl's work proved that full harmony could be maintained with the high ideals of university education. See W. Voigt, "Festrede," Göttingen, 1912.

[2] On the wall of the building the plate can be seen commemorating Thomas Young, who, as a student, worked on physics in that building at the end of the eighteenth century (see page 90).

it soon became an important center attracting young men interested in the applications of mathematics to engineering.[1] At that time, Prandtl's graduate students worked mainly on strength of materials. H. Hort[2] wrote a thesis on temperature changes during tensile tests of steel, and S. Berliner[3] made an investigation of hysteresis cycles in tensile tests of cast iron, and the writer started the work on lateral buckling of I beams which has already been mentioned. Following the failure in construction of the Quebec Bridge, Prandtl became interested in buckling of built-up columns and showed[4] that the diagonals and battens, which connected the heavy chords of the compressed members of the bridge, had inadequate cross-sectional dimensions.

About the same time, Theodore von Kármán came to Göttingen and started his work on buckling of columns in the plastic region[5] (his doctor's thesis) under Prandtl's direction. After getting his degree, von Kármán stayed at Göttingen for several years as Prandtl's assistant and made an investigation on bending of curved tubes[6] (a problem which Prandtl was interested in). He also did experimental work on the compressive strength of stone under combined longitudinal and lateral pressure.[7] Prandtl himself drew up his theory of the two types of fracture of solids and also devised his model for explaining hysteresis loops (see page 364).

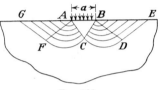

Fig. 233.

He was interested in the plastic deformation of beams and a thesis on the subject was written by H. Herbert[8] under his direction. Several problems of plastic deformation had already been solved by Saint-Venant (see page 242). Prandtl made further progress in this field and solved[9] the more complicated two dimensional problem of a semi-infinite body under a uniform pressure p distributed over a strip of width a (Fig. 233). He showed that, at a certain critical value p_{cr} of pressure, the triangular prism ABC moves downward while the triangular prisms BDE and AFG move up under the action of pressures transmitted by the sectors BCD and ACF. The sliding occurs along the surfaces of maximum

---

[1] See A. Sommerfeld's paper for Prandtl's sixtieth anniversary in *ZAMM*, vol. 15, p. 1, 1935.
[2] H. Hort, "Über die Wärmevorgänge beim Zerreissversuch," Dissertation, 1906.
[3] S. Berliner, *Ann. Physik*, vol. 20, p. 527, 1906.
[4] *VDI*, vol. 55, 1907.
[5] *Forschungsarb.*, no. 81, 1910, Berlin.
[6] *VDI*, vol. 55, p. 1889, 1911.
[7] *VDI*, vol. 55, p. 1749, 1911; *Forschungsarb.*, no. 118, 1912, Berlin.
[8] H. Herbert, Dissertation, Göttingen, 1909.
[9] See *Z. angew. Math. u. Mechanik*, vol. 1, p. 15, 1921.

shearing stress shown in the figure. Assuming that, during yielding, the maximum shearing stress is equal to one-half of the yield-point stress in tension, the critical pressure is

$$p_{cr} = \sigma_{yp}\left(1 + \frac{\pi}{2}\right)$$

Rapid progress in the theory of plasticity began with the appearance of this paper. The subject now constitutes an important branch in the mechanics of materials and is treated in the books of A. Nadai,[1] W. W. Sokolovsky,[2] and R. Hill.[3]

Prandtl's paper has historical significance in another respect. It appeared in the first issue of the *Zeitschrift für angewandte Mathematik und Mechanik*. This journal has acquired a great influence under the editorship of R. von Mises not only in Germany but also elsewhere. Several countries have followed this example and started publishing journals of applied mechanics. These have meant greatly improved facilities for bringing new work in the subject before workers in the field of mechanics of materials.

Prandtl's papers on the strength of materials represent only a small portion of his contribution to engineering mechanics. His principal accomplishments lie in aerodynamics, but, as this does not belong to the subject of our discussion, we shall only mention that, since 1906 when the *Motorluftschiff—Studiengesellschaft* was organized, Prandtl has had to devote more and more of his time to aerodynamics. In 1908, the first small wind tunnel was constructed in Göttingen to Prandtl's plans. The research work in this tunnel and the techniques of making measurements and of presenting results were so successful that, very soon, experimenters in aerodynamics all over the world accepted Prandtl's methods. During the First World War, a second (larger) wind tunnel was built at Göttingen, and Prandtl published his outstanding papers on "Tragflügeltheorie," which showed how airplanes should be designed on the basis of wind-tunnel experiments with small models. His ideas again won general acceptance and now airplane designers all over the world use them in their work.

The principal characteristic of Prandtl's papers is their originality. His publications usually opened new fields for research, and he always had enough pupils who were ready to follow in the directions indicated by his pioneer works. Many of these pupils are now teachers both in Germany and in other countries.[4] In appraising Prandtl's contribution

[1] A. Nadai, "Theory of Flow and Fracture of Solids," 2d ed., 1950.
[2] W. W. Sokolovsky, "Theory of Plasticity" (Russian), Moscow, 1946.
[3] R. Hill, "The Mathematical Theory of Plasticity," Oxford, 1950.
[4] Several of Prandtl's pupils and former assistants are at present active in the United

to engineering mechanics, we have to consider not only his publications but also the great school of mechanics which he developed at Göttingen.

81. Approximate Methods of Solving Elasticity Problems

New problems of machine design and theory of structures in the twentieth century have demanded greater accuracy of stress analysis than was previously the case. The elementary formulas of strength of materials are often not accurate enough, and the use of the theory of elasticity in solving practical problems has rapidly become more common. Thus the theory of elasticity, which was previously taught in few schools and was usually of a theoretical character, has now become a very important applied science that is included in the programs of many engineering schools. Textbooks in this subject have also changed their form, and authors often discuss problems not only of theoretical but also of practical interest.[1]

Only in simple cases are rigorous mathematical solutions of elasticity problems available, and the modern trend in the theory is toward the use of various approximate methods. One of these methods is based on the use of analogies.[2] We have already mentioned the membrane analogy which was suggested by Prandtl and which has proved very useful in studying the torsion problem. This analogy was extended by Vening Meinesz[3] to the theory of bending. Using the equation of a nonuniformly loaded membrane, the writer simplified[4] the analysis of bending of beams. Use of the membrane analogy in two-dimensional photoelastic work was proposed by J. Den Hartog[5] for determining the sum of two principal stresses. This was successfully followed up[6] by E. E. Weibel in studying stress concentration at fillets.

States. We may mention Theodore von Kármán, who has done much work in aerodynamics in this country, A. Nadai who did important work on plasticity at the Westinghouse research laboratories, and W. Prager who is leading the new researches in plasticity at Brown University. Professor and Mrs. Flügge are working in the division of engineering mechanics at Stanford University. Professor J. Den Hartog and the writer have also spent some time in graduate work at Göttingen and consider themselves pupils of Prandtl.

[1] The first book on theory of elasticity that was written specifically for engineers was A. Föppl's "Die wichtigsten Lehren der höheren Elastizitätstheorie, technische Mechanik," vol. 5, 1907, Leipzig. It was expanded later by its writer in collaboration with L. Föppl into the three-volume publication "Drang und Zwang," 1920–1947.

[2] The use of analogies in stress analysis is described in the "Handbook of Experimental Stress Analysis." See also the book "Technische Dynamik," by C. B. Biezeno and R. Grammel, Berlin, 1939.

[3] *Ingenieur (Utrecht)*, 1911, p. 108.

[4] See *Bull. Inst. Engrs. Ways of Communication*, St. Petersburg, 1913; see also *Proc. London Math. Soc.* (2), vol. 20, p. 398, 1922.

[5] *Z. angew. Math. Mech.*, vol. 11, p. 156, 1931.

[6] *Trans. ASME*, vol. 56, p. 601, 1934.

Electrical analogies also are used in solving elasticity problems. One of this kind was demonstrated by L. S. Jacobsen in the torsion of a shaft of variable diameter. He showed[1] that it is possible, by proper variation of the thickness of a plate having the same boundary as an axial section of the shaft, to make the differential equation of the potential function coincide with that for the stress function of the shaft. Using this analogy, it was found possible to investigate the important question of stress concentration at a fillet connecting two shafts of different diameters. A further advance in this field was made by A. Thum and W. Bautz[2] who used, instead of a plate of variable thickness, an electrolytic tank of variable depth.

Another interesting analogy was pointed out by Muschelišvili.[3] It related stresses produced by a two-dimensional steady-state temperature distribution and the stresses produced by dislocations. By means of this, steady-state thermal stresses in circular and rectangular tubes were investigated photoelastically by E. E. Weibel.[4]

An analogy exists that is based on the fact that the biharmonic differential equation for the Airy stress function coincides with the equation for lateral deflections of a plate bent by forces and couples distributed along the edge. It has been used[5] in the solution of two-dimensional elasticity problems.

A very useful method of getting approximate solutions has been proposed by C. Runge.[6] He uses finite difference equations instead of differential equations. Considering torsional problems, which require integration of the equation

$$\frac{\partial^2 \phi}{\partial x^2} + \frac{\partial^2 \phi}{\partial y^2} = C$$

where ϕ is the stress function, he replaces this relation by the finite difference equation

$$\phi_1 + \phi_2 + \phi_3 + \phi_4 - 4\phi_0 = C_1 \qquad (a)$$

This must be satisfied at every point (such as O in Fig. 234) of a square net which subdivides the cross section of the twisted bar. Writing Eq.

[1] *Trans. ASME*, vol. 47, p. 619, 1925.
[2] *VDI*, vol. 78, p. 17, 1934.
[3] *Bull. Elecrotech. Inst. St. Petersburg*, vol. 13, pp. 23–37, 1916; see also *Rend. accad. Nazl. Lincei*, 5th series, vol. 31, pp. 548–551, 1922; M. A. Biot, *Phil. Mag.*, vol. 19, p. 540, 1935.
[4] *Proc. 5th Intern. Congr. Applied Mechanics*, Cambridge, Mass., p. 213, 1938.
[5] K. Wieghardt, *Mitt. Forschungsarb.*, vol. 49, pp. 15–30, 1908. See also H. Cranz, *Ing.-Arch.*, vol. 10, pp. 159–166, 1939.
[6] *Z Math. u. Physik*, vol. 50, p. 225, 1908.

(a) for each inner point of the net (Fig. 234) and observing that ϕ vanishes at the boundary, we obtain a system of linear equations which is sufficient for calculating all values of ϕ. The torsional rigidity and maximum shearing stress can be calculated from these values. In the case of symmetry, shown in Figure 234, we have only to write three equations, *viz.*, those for points 0, 1, and 5. Thus direct solution of these does not present any difficulty. To achieve better accuracy, it is necessary to use a finer net, and it naturally gives rise to a larger number of equations.

If the number of equations is large, an iteration process[1] can be used in solving them. We assume some arbitrary initial values for ϕ. Using them in equation (a) for ϕ_1, ϕ_2, ϕ_3, ϕ_4, we find a new value of ϕ_0 for all points of the net, and, with these, calculations must be repeated. After several repetitions of this process, sufficiently accurate values for ϕ are obtained.

Finite difference equations were successfully used by L. F. Richardson[2] in two-dimensional problems of elasticity where the stress function has to satisfy a differential equation of fourth order. Further development of the method is due to R. V. Southwell who has improved it and applied it systematically to a large number of physics problems, including many in elasticity.[3]

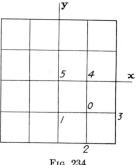

Fig. 234.

The Rayleigh-Ritz method has proved to be very useful for getting approximate solutions of elasticity problems. For finding the frequency of the fundamental mode of vibration of a complicated system, Rayleigh suggested assuming some shape for that mode, deriving an expression for the resulting frequency, and then selecting the parameters, defining the assumed shape, in such a way as to make the expression for the frequency a minimum. W. Ritz,[4] working with bending of rectangular plates, considers the expression for the potential energy:

$$I = \iint \left\{ \frac{D}{2} \left[\left(\frac{\partial^2 w}{\partial x^2} + \frac{\partial^2 w}{\partial y^2} \right)^2 - 2(1 - \mu) \left(\frac{\partial^2 w}{\partial x^2} \frac{\partial^2 w}{\partial y^2} - \left(\frac{\partial^2 w}{\partial x \, \partial y} \right)^2 \right) \right] - wq \right\} dx \, dy \quad (b)$$

[1] See H. Liebmann, *Sitz. ber. math.-naturw. Abt. bayer. Akad Wiss. München*, 1918. p. 385.

[2] *Phil. Trans.*, (A), vol. 210, pp. 307–357, 1910.

[3] See R. V. Southwell, "Relaxation Methods in Engineering Science," Oxford, 1940, and "Relaxation Methods in Theoretical Physics," Oxford, 1946.

[4] For a biography of this outstanding scientist, see his "Collected Papers," Paris, 1911.

where w denotes deflections, q is the intensity of lateral loading, and D is the flexural rigidity. He observes that the true solution for the deflections w must make this energy a minimum and takes that solution in the form of a series

$$w = a_1\phi_1(x,y) + a_2\phi_2(x,y) + a_3\phi_3(x,y) + \cdots \qquad (c)$$

where each of the functions ϕ satisfies the boundary conditions of the plate and the coefficients a_1, a_2, a_3, ... have to be calculated from the system of linear equations

$$\frac{\partial I}{\partial a_1} = 0, \qquad \frac{\partial I}{\partial a_2} = 0, \qquad \frac{\partial I}{\partial a_3} = 0 \cdots \qquad (d)$$

These express the conditions for minimizing the integral I when the approximate solution (c) obtains.[1] Experience shows that usually a small number of terms in series (c) is sufficient to give good results.

By using this method in the bending of bars under combined lateral and axial forces, approximate formulas have been obtained. Bars with some initial curvature and circular rings were also treated in a similar manner.[2] By applying the Ritz method to calculations of the deflection of a membrane, in the membrane analogies, simple formulas for calculating torsional and bending stresses in bars of various cross sections have been derived.[3] The same method has yielded useful results in studying the vibration of bars of variable cross section and of rectangular plates with various edge conditions.

The Ritz method has also been used[4] in conjunction with the principle of least work. This states that, if we have a body with given forces acting on its boundary and if we consider such changes of stress components as do not affect the equations of equilibrium and the boundary conditions, the true stress components are those making the variation of strain energy vanish. For example, in a two-dimensional problem with a stress function ϕ, the strain energy is given by

$$V = \frac{1}{2E} \iint \left[\left(\frac{\partial^2 \phi}{\partial x^2}\right)^2 + 2\left(\frac{\partial^2 \phi}{\partial x\, \partial y}\right)^2 + \left(\frac{\partial^2 \phi}{\partial y^2}\right)^2 \right] dx\, dy \qquad (e)$$

[1] In this way the upper limit for the minimum value of I is obtained. E. Trefftz showed how the lower limit for the same quantity can be obtained; see *Math. Ann.*, vol. 100, p. 503, 1928.

[2] See S. Timoshenko's paper in the "Festschrift zum siebzigsten Geburtstage, August Föppl," 1923.

[3] See S. Timoshenko, *Bull. Inst. Engrs. Ways of Communications*, St. Petersburg, 1913; see also *Proc. London Math. Soc.*, (2), vol. 20, p. 398, 1922.

[4] See S. Timoshenko, "Theory of Elasticity," St. Petersburg, 1914; also S. Timoshenko, *Phil. Mag.*, vol. 47, p. 1095, 1924.

Taking the stress function in the form of a series

$$\phi = a_1\phi_1(x,y) + a_2\phi_2(x,y) + a_3\phi_3(x,y) + \cdots \quad (f)$$

in which each term satisfies the boundary conditions of the problem, and substituting it in expression (e), we can calculate the values of the coefficients $a_1, a_2, a_3 \ldots$ from the equations

$$\frac{\partial V}{\partial a_1} = 0, \quad \frac{\partial V}{\partial a_2} = 0, \quad \frac{\partial V}{\partial a_3} = 0 \cdots$$

Substituting these values into expression (f), we obtain the stress function which can be used for an approximate calculation of stresses. By using this method, the problem of stress distribution in wide flanges of I beams under flexure (effective width of flanges) has been treated.[1] Also, *shear lag* in thin-walled structures has been treated[2] by this means.

82. Three-dimensional Problems of Elasticity[3]

Study of Saint-Venant's problem on torsion and the bending of a cantilever beam has been continued in this century and rigorous solutions for several new shapes of cross section have been found.[4] In the bending case, nonsymmetrical sections have been taken up and the point of application of the bending load fixed, at which no torsion is produced.[5] It was shown that, in semicircular and isosceles triangular cross sections, only a small displacement of the load from the centroid is necessary to eliminate torsion. In thin-walled sections, this displacement may become large and of great practical importance. This question was cleared up by R. Maillart,[6] who introduced the notion of the *shear center* and showed how this point can be found.

In Saint-Venant's solution of the torsion problem, it is assumed that the torque is applied by means of shearing stresses distributed over the ends of a bar in the same way as at any intermediate cross section. If

---

[1] Theodore von Kármán, "Festschrift . . . August Föppl," p. 114, 1923.

[2] Eric Reissner, *Quart. Applied Math.*, vol. 4, p. 268, 1946.

[3] Only a few papers having practical importance are mentioned in this article. For a complete bibliography in this field, see the articles by A. Timpe and O. Tedone in *Encyklopädie Math. Wiss.*, vol. 4₄, 1914; and the articles by E. Trefftz and J. W. Geckeler in "Handbuch der Physik," vol. 6, 1928.

[4] See E. Trefftz, *Math. Ann.*, vol. 82, p. 97, 1921; *ZAMM*, vol. 2, p. 263, 1922.

[5] See S. Timoshenko, *Bull. Inst. Engrs. Ways of Communication*, St. Petersburg, 1913. See also the paper by M. Seegar and K. Pearson, *Proc. Roy. Soc. (London)*, (A), vol. 96, p. 211, 1920.

[6] R. Maillart, *Schweiz. Bauztg.*, vol. 77, p. 195, 1921; vol. 79, p. 254, 1922. See also A. Eggenschwyler, Diss. E. Tech. Hochschule, Zürich, 1921. A further study of the subject was made by C. Weber, *ZAMM*, vol. 4, p. 334, 1924. See also E. Trefftz, *ZAMM*, vol. 15, p. 220, 1935.

the distribution of stresses at the ends differs from this, a local disturbance in the stresses results, and Saint-Venant's solution is only valid in regions at some distance from the ends. These local disturbances have been studied by several investigators.[1] In thin-walled open sections, application of Saint-Venant's principle is not as easy and the angle of twist may depend considerably on the manner in which the twisting forces at the ends are applied. The writer investigated[2] the problem of twisting in an I beam with a built-in end and found that, to get a satisfactory value for the angle of twist, not only Saint-Venant's torsional stresses but also bending stresses in the flanges must be considered. The differential equation for the angle of twist per unit length produced by the torque T then becomes

$$C\theta - C_1\theta'' = T \qquad (a)$$

C is the Saint-Venant torsional rigidity and C_1 is a constant depending on the flexural rigidity of the flanges. Experiments showed satisfactory agreement with the theoretical results of Eq. (a) and indicated that there exists some limitation on the applicability of Saint-Venant's principle, since bending of the flanges is not of a local character. C. Weber[3] reached the same result in studying twisting of channel sections as did V. Z. Vlasov in his general discussion of torsion of thin-walled open sections.[4]

The use of functions of a complex variable in solving Saint-Venant's problem has been demonstrated by N. I. Muschelisvili.[5] He found solutions for several new shapes of cross section and examined[6] the torsion of bars built up of two different materials, as in reinforced concrete.

The problem of combined torque and axial load has been further studied. Thomas Young (see page 92), considering the torsion of a circular shaft, found that the expression for torque consists of two terms, one proportional to the angle of twist per unit length θ and the other proportional to θ^3. Young showed that to obtain the second term it is necessary, in discussing deformation, to take small quantities of higher order into consideration. These are usually omitted in the analysis of torsion. Considering a rectangular element $abcd$ on a cylindrical surface coaxial

[1] F. Purser, *Proc. Roy. Irish Acad.*, (A), vol. 26, p. 54, 1906; A. Timpe, *Math. Ann.*, vol. 71, p. 480, 1912; K. Wolf, *Sitzsber. Akad. Wiss. Wien, Math.-Naturw. Klasse*, vol. 125, p. 1149, 1916. The case of a rectangular bar with built-in ends was discussed by S. Timoshenko, *Proc. London Math. Soc.*, vol. 20, p. 389, 1921. A proof of Saint-Venant's principle was given by J. N. Goodier, *Phil. Mag.* (7), vol. 24, p. 325, 1937.
[2] *Bull. Polytech. Inst. St. Petersburg*, 1905–1906.
[3] *ZAMM*, vol. 6, p. 85, 1926.
[4] See V. Z. Vlasov's book, "Thin Walled Elastic Bars," Moscow, 1940.
[5] *Rend. accad. nazl. Lincei*, 6th series, vol. 9, pp. 295–300, 1929.
[6] *Compt. rend.*, vol. 194, p. 1435, 1932.

with the shaft and of radius r, we conclude that, due to the twist, it takes on the form of a parallelepiped ab_1c_1d (Fig. 235) and that the shearing strain is $r\theta$. The corresponding shearing stresses $Gr\theta$ give the well-known formula for torque

$$T_1 = \int_0^a Gr\theta \cdot 2\pi r^2\, dr = I_p G\theta \qquad (b)$$

But, if we assume that the distance between the two cross sections of the shaft remains unchanged during torsion, we see that such linear elements as ab and cd are elongating and their unit elongation is

$$\epsilon = \sqrt{1 + r^2\theta^2} - 1 \sim \tfrac{1}{2} r^2 \theta^2 \qquad (c)$$

The corresponding tensile stresses $Er^2\theta^2/2$ give the additional torque about the axis of the shaft:

$$T_2 = \frac{1}{2}\int_0^a Er^2\theta^2 \cdot r\theta \cdot 2\pi r^2\, dr = \frac{\pi a^6}{6}\theta^3 E \qquad (d)$$

FIG. 235.

which, as we see, is proportional to θ^3. Young was the first to consider the small quantities of higher order which give rise to Eq. (c). The correction (d) to the usual formula is very small in a steel shaft, but it becomes important for such materials as rubber, which may assume very large deformations within the elastic range. It is also of importance in thin-walled open sections.[1] The theory of large deformation has been developed, in recent times, by several investigators,[2] and Biot[3] has showed that Thomas Young's method, as explained above, gives the correct answer for circular shafts. Cases of torsion of thin-walled open sections combined with axial force and with pure bending have been examined by J. N. Goodier,[4] who applied large deformation theory, and again it was shown that elementary derivations, based on Thomas Young's assumption, agree with the results of more elaborate investigation.

Some progress has been made with the torsional problem for a circular shaft of variable diameter. J. H. Michell[5] and, independently, A. Föppl[6] showed that the distribution of stresses is defined by one stress function,

[1] See the paper by Buckley in *Phil. Mag.*, vol. 28, pp. 778–787, 1914; also the paper by C. Weber in "Festschrift . . . A. Föppl," Berlin, 1924.

[2] See R. V. Southwell, *Trans. Roy. Soc. (London)*, (A), vol. 213, p. 187, 1913; C. B. Biezeno and H. Hencky, *Proc. Acad. Sci. Amsterdam*, vol. 31, pp. 569–578 and 579–592, 1928; E. Trefftz, *ZAMM*, vol. 12, pp. 160–165, 1933.

[3] M. A. Biot, *Phil. Mag.*, vol. 27, pp. 468–489, 1939.

[4] *J. Applied Mechanics*, vol. 17, pp. 383–387, 1950. See also A. E. Green and R. T. Shield, *Trans. Roy. Soc. (London)*, A, vol. 244, p. 47, 1952.

[5] *Proc. London Math. Soc.*, vol. 31, p. 140, 1900.

[6] *Sitzsber. Math.-Naturw. Abt. bayer. Akad. Wiss. Münich*, vol. 35, pp. 249, 504, 1905.

and gave that function for a conical shaft. Cases of shafts in the form of an ellipsoid, hyperboloid, and paraboloid of revolution were solved by the same method. C. Runge[1] suggested an approximate method for calculating local stresses at a circular fillet joining two cylindrical shafts of different diameters.

Several problems of axially symmetrical stress distribution in a solid of revolution have received treatment. J. H. Michell[2] and A. E. H. Love[3] showed that, in these cases, all the stress components can be represented in terms of a single stress function. The relation between this stress function and that of two-dimensional problems has been investigated by C. Weber.[4] By taking the function for axially symmetrical stress distribution in the form of polynomials, the rigorous solutions for bending of circular plates under several symmetrical types of load distribution were obtained.[5]

The extension of a cylinder by surface shearing stresses applied near the ends and the compression of a cylinder between two rigid plates have been examined by L. N. G. Filon,[6] who gave the solution in terms of Bessel's functions. The stress distribution produced in a long cylinder by a uniform normal pressure acting on a narrow circumferential band is of practical importance, since it corresponds approximately to that induced when a short collar is shrunk onto a much longer shaft. It has been discussed by several authors,[7] and we now have sufficient information concerning this problem.

General solutions of the differential equations of elasticity have recently attracted the interest of engineers. Boussinesq showed that the three components of displacement u, v, w can be defined by three biharmonic functions.[8] P. F. Papkovich[9] has shown that Boussinesq's solution can be simplified and presented in the following form:

$$u = B\frac{\partial \omega}{\partial x} + \phi_1 \qquad v = B\frac{\partial \omega}{\partial y} + \phi_2 \qquad w = B\frac{\partial \omega}{\partial z} + \phi_3$$

---

[1] See F. A. Willers, *Z. Math. u. Physik*, vol. 55, p. 225, 1907. See also L. Föppl, *Sitzsber. Math.-naturw. Abt. bayer. Akad. Wiss. München*, vol. 51, p. 61, 1921.

[2] *Proc. London Math. Soc.*, vol. 31, p. 144, 1900.

[3] "Mathematical Theory of Elasticity," 4th ed., p. 274, 1927.

[4] *ZAMM*, vol. 5, 1925.

[5] See the paper by A. Korobov, *Bull. Polytech. Inst. Kiev*, 1913. A concentrated force applied at the center of the plate was investigated by A. Nadai; see his book "Elastische Platten," p. 315, 1925.

[6] *Phil. Trans.*, (A), vol. 198, p. 147, 1902.

[7] A. Föppl and L. Föppl, "Drang und Zwang," 2d ed., vol. 2, p. 141, 1928. M. V. Barton, *J. Applied Mechanics*, vol. 8, p. 97, 1941; A. W. Rankin, *J. Applied Mechanics*, vol. 11, p. 77, 1944; C. J. Tranter and J. W. Craggs, *Phil. Mag.*, vol. 38, p. 214, 1947.

[8] "Application des Potentiels," p. 281, Paris, 1885.

[9] *Compt. rend.*, 1932, p. 513.

where B is a constant,

$$\omega = \omega_0 + \tfrac{1}{2}(\phi_1 x + \phi_2 y + \phi_3 z)$$

and the functions ϕ_1, ϕ_2, ϕ_3, and ω_0 satisfy the Laplacian equation

$$\frac{\partial^2 \phi}{\partial x^2} + \frac{\partial^2 \phi}{\partial y^2} + \frac{\partial^2 \phi}{\partial z^2} = 0$$

This general solution was independently obtained by H. Neuber,[1] who used it in the solution of practical problems dealing with stress concentration in circular shafts having grooves or ellipsoidal holes.[2] Neuber adopted his solution also to two-dimensional stress distributions and derived formulas for the stress concentration factors in grooves and elliptic holes and in a plate subjected to tension, shear, or bending in its plane. The numerous tables and charts given in Neuber's book[3] are very useful for analyzing stress concentrations in machine parts.

83. Two-dimensional Problems of Elasticity

Further progress in handling two-dimensional problems of elasticity has been made in this century, and the use of the rigorous solutions in practical stress analysis has become very common. A. Mesnager[4] solved two-dimensional problems by using stress functions in the form of polynomials and applied his results to several problems in bending of beams of narrow rectangular cross section. He showed that the elementary formulas of strength of materials give correct values for normal and shearing stresses in a cantilever loaded at its free end. He also showed that the rigorous solution for a uniformly loaded beam gives some small corrections to the elementary formulas, which can be neglected for practical purposes.

M. C. Ribière[5] used Fourier series in discussing the bending of rectangular beams. Further study of the same problem was made by L. N. G. Filon[6] who applied the general solution to particular cases of practical interest. H. Lamb[7] considered an infinite rectangular strip loaded at constant intervals by equal concentrated forces acting in the upward and downward directions alternately. In this way, he studied the deflections

[1] *ZAMM*, vol. 14, p. 203, 1934. See also *Ing. Arch.*, vol. 5, pp. 238–244, 1934; vol. 6, pp. 325–334, 1935.

[2] The case of a spherical cavity was investigated by R. V. Southwell, *Phil. Mag.*, 1926, and by J. N. Goodier, *Trans. ASME*, vol. 55, p. 39, 1933. The general case of ellipsoidal holes was discussed by M. A. Sadowsky and E. Sternberg, *J. Applied Mechanics*, vol. 16, pp. 149–157, 1949.

[3] H. Neuber, "Kerbspannungslehre," Berlin, 1937.

[4] *Compt. rend.*, vol. 132, p. 1475, 1901.

[5] See his thesis "Sur divers cas de la flexion des prismes rectangles," Bordeaux, 1889.

[6] *Phil. Trans.*, (A), vol. 201, p. 63, 1903.

[7] H. Lamb, *Atti congr. intern. matemat.*, IV*th* *Congr.*, Rome, 1909, vol. 3, p. 12.

produced by a concentrated load. The same question was taken up again by Th. von Kármán,[1] who derived an accurate formula for the deflection produced in a simply supported beam by a concentrated force. Fourier series were employed by Clebsch in studying circular disks (see page 257). The same method was used for a circular ring by O. Venske.[2] A. Timpe investigated[3] several particular cases and obtained the solutions of Golovin for bending of a portion of the ring by couples and forces applied at the ends. The circular ring constitutes the simplest case of a multiply connected region and the general solution for it contains multi-valued terms. Timpe gives the physical explanation of the multiply valued solutions by considering the residual stresses which can be set up by cutting the ring, displacing one end with respect to the other and then joining them by some means. As mentioned before (see page 353), a general discussion of the solutions of two-dimensional problems for multiply connected boundaries was made by J. H. Michell,[4] who shows that the stress distribution does not depend on the elastic constants of the material if there are no body forces and the surface forces are such that their resultant at each boundary vanishes. This conclusion is of a great practical importance in cases where the photoelastic method of stress analysis is used. The case of a circular disk subject to concentrated forces at any point was discussed by R. D. Mindlin.[5] As a particular example of a stressed circular ring, the problem of compression by two equal and opposite forces acting along a diameter was discussed by the writer.[6] It was shown that, at a cross section situated at some distance from the points of application of the loads, the hyperbolic stress distribution given by Winkler's elementary theory is accurate enough for practical purposes. Other examples of deformation of circular rings have been discussed by L. N. G. Filon[7] and by H. Reissner.[8] The problem of two equal and opposite forces acting in the radial direction and applied at the outer and the inner boundaries of the ring has been attacked by C. W. Nelson[9] in connection with stress analysis of roller bearings.

[1] *Abhandl. aerodynam. Inst. Tech. Hochschule Aachen*, vol. 7, p. 3, 1927.
[2] *Nachr. Ges. Wiss. Göttingen*, 1891, p. 27.
[3] *Z. Math. u. Physik*, vol. 52, p. 348, 1905.
[4] *Proc. London Math. Soc.*, vol. 31, p. 100, 1899. See also V. Volterra, *Ann. école norm.*, 3d series, vol. 24, pp. 401–517, 1907; A. Föppl, *Tech. Mech.* vol. 5, p. 293, 1907.
[5] *J. Appl. Mechanics*, vol. 4, p. A-115, 1937.
[6] S. Timoshenko, *Bull. Polytech. Inst. Kiev*, 1910; *Phil. Mag.*, vol. 44, p. 1014, 1922.
[7] "Selected Engineering Papers," no. 12, published by the Institution of Civil Engineers, London, 1924.
[8] *Jahrbuch wiss. Ges. Luftfahrt*, p. 126, 1928.
[9] *J. Applied Mechanics*, Paper 50-A-16, presented at the meeting of the American Society of Mechanical Engineers, December, 1950.

Theory of Elasticity during the Period 1900–1950 407

Stress concentrations which occur at grooves and fillets and holes of various shapes are of great practical importance, and rigorous solutions of the two-dimensional elasticity problems have been found in a number of instances. The stress distribution around a circular hole in a plate subjected to uniform tension in one direction was analyzed by G. Kirsch.[1] This solution shows that the maximum stress occurs at the boundary of the hole (at the ends of the diameter perpendicular to the direction of the applied tensile stress) and that it is three times larger than the applied stress. In this way, it was shown, probably for the first time, how essential it is to study local irregularities in stress distribution produced by holes. Since that time engineers have subjected stress concentration problems to intensive theoretical and experimental study. R. C. J. Howland[2] considered the plate of finite width with a circular hole on the axis of symmetry. The case when the forces are applied to the boundary of the hole was treated by W. G. Bickley.[3] The elliptic hole was investigated by G. V. Kolosoff[4] and later by C. E. Inglis,[5] and the reduction of stresses by reinforcing the edge of a circular hole by a bead was investigated by the writer.[6] Problems of two and more circular holes have been discussed by several authors.[7]

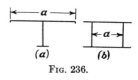

Fig. 236.

Many important two-dimensional problems of elasticity have been solved by using the functions of a complex variable. The method was evolved principally by G. V. Kolosoff[8] and his pupil N. I. Muschelišvili. A bibliography, containing the principal Russian publications, can be found in Muschelišvili's book.[9]

The analysis of stresses in thin-walled structures, such as airplanes, very often requires investigation of bending of beams having cross sections like those depicted in Fig. 236. The width a of the flange and the distance a between the webs is not small in comparison with the span of the beam, and questions of *effective width* and of *shear lag* have to be considered. Approximate solutions of this kind of problem have been offered

---

[1] *VDI*, vol. 42, 1898.
[2] *Trans. Roy. Soc. (London)*, (A), vol. 229, p. 49, 1930; vol. 232, pp. 155–222, 1932.
[3] *Trans. Roy. Soc. (London)*, (A), vol. 227, p. 383, 1928.
[4] Dissertation, St. Petersburg, 1910.
[5] *Trans. Inst. Naval Architects*, London, 1913.
[6] S. Timoshenko, *J. Franklin Inst.*, vol. 197, p. 505, 1924.
[7] C. Weber, *ZAMM*, vol. 2, p. 267, 1922; M. A. Sadowsky, *ZAMM*, vol. 8, p. 107, 1928; R. C. J. Howland, *Proc. Roy. Soc. (London)*, (A), vol. 148, p. 471, 1935.
[8] See his previously mentioned Dissertation and also the paper in *Z. Math. u. Physik*, vol. 62, pp. 383–409, 1914.
[9] "Some Fundamental Problems of the Mathematical Theory of Elasticity" (Russian), Moscow, 1949.

by several writers, and the appropriate bibliography can be found in the papers of E. Chwalla,[1] E. Reissner,[2] and J. Hadji-Argyris and H. L. Cox.[3] The use of wood and other nonisotropic materials in various structures has focused engineers' attention on the theory of anisotropic bodies. Some progress has been made in investigating two-dimensional problems in which it is assumed that the elastic properties of a plate can be defined by the equations

$$\epsilon_x = \frac{1}{E_1} \sigma_x - \frac{\mu_1}{E_2} \sigma_y$$

$$\epsilon_y = -\frac{\mu_2}{E_1} \sigma_x + \frac{1}{E_2} \sigma_y$$

$$\gamma_{xy} = \frac{1}{G} \tau_{xy}$$

where $E_1\mu_1 = E_2\mu_2$. Using these relations, a differential equation for a stress function, similar to that used under conditions of isotropy, can be derived. S. Lechnitski[4] obtained solutions of this equation for rectangular beams, circular rings, and circular and elliptic holes, similar to those previously mentioned for isotropic plates.

84. Bending of Plates and Shells

The wide use of comparatively thin plates, slabs, and shells in modern structures has brought about advances in the theory of plates and shells. Although the relevant equations were derived by Kirchhoff and some applications to acoustics were shown, wide use of the theory of plates in engineering began in the twentieth century. The rectangular plate which is simply supported along two opposite sides and which has any boundary conditions along the other two sides was investigated by M. Lévy and E. Estanave (see page 333). It was of great practical importance and various particular cases of loading were studied by engineers who compiled tables for maximum deflections and maximum bending moments. Other forms of plates, elliptical, triangular, and plates in the form of a sector, have been studied and books dealing principally with bending of plates have appeared.[5] Where no rigorous solution of a problem has been available, or the solutions were given by series unsuitable for practical application, engineers have fallen back on approximate anal-

[1] *Stahlbau*, vol. 9, p. 73, 1936.
[2] *Quart. Applied. Math.*, vol. 4, p. 268, 1946.
[3] *Aeronaut. Research Council (Brit.)*, Rept. 1969, 1944.
[4] S. Lechnitski, "Anisotropic Plates" (Russian), Moscow, 1947. A complete bibliography on the subject is given in this book.
[5] S. Timoshenko, "Theory of Elasticity" (Russian), vol. 2, St. Petersburg, 1916; A. Nadai, "Elastische Platten," Berlin, 1925; B. G. Galerkin, "Elastic Thin Plates," Moscow, 1933.

ysis. Ritz's method has been widely used in the theory of plates and it has led to many useful results. In some complicated cases, finite difference equations have been used and the necessary information obtained by numerical calculations.[1]

The rectangular plate with all four sides built-in is encountered in many engineering problems, but a mathematical treatment of this case presents many difficulties. The first solution of the problem that was suitable for numerical computations was given by B. M. Kojalovich.[2] It was somewhat simplified by I. G. Boobnov[3] who calculated tables of maximum deflections and maximum bending moments for plates of various proportions. More elaborate tables, based on the writer's solution,[4] were calculated by T. H. Evans.[5]

A concentrated force acting on a plate leads to the study of local stresses at the point of application of the load. This cannot be done with elementary plate theory. This problem has been discussed by A. Nadai[6] and by S. Woinowsky-Krieger.[7] The action of a concentrated force on a rectangular plate with simply supported edges has been investigated by several authors.[8] A concentrated force acting on a plate with clamped edges was discussed by the writer in the previously mentioned paper presented at the Fifth International Congress and also by D. Young.[9]

A plate supported by rows of equidistant columns is of a great practical importance in reinforced concrete design. The first approximate solution of the problem was given by F. Grashof[10] and further work was done by V. Lewe.[11] It is also discussed in the previously mentioned books of A. Nadai and B. G. Galerkin. A more recent discussion of the same subject is given in the paper by S. Woinowsky-Krieger.[12]

In connection with the analysis of stresses in slabs used in highway construction, the plate on an elastic foundation has received atten-

[1] See the book by H. Marcus, "Die Theorie elastischer Gewebe," 2d ed., Berlin, 1932. See also the paper by D. L. Holl, *J. Applied Mechanics*, vol. 3, p. 81, 1936.
[2] See his doctoral dissertation, St. Petersburg, 1902.
[3] See Boobnov's book "Theory of Structure of Ships," vol. 2, p. 465, St. Petersburg, 1914.
[4] See *Proc. 5th Intern. Congr. Applied Mechanics*, Cambridge, Mass., 1938.
[5] *J. Applied Mechanics*, vol. 6, p. A-7, 1939.
[6] See his "Elastische Platten," p. 308.
[7] *Ing.-Arch.*, vol. 4, p. 305, 1933.
[8] See A. Nadai, *Bauing.*, 1921, p. 11; S. Timoshenko, *Bauing*, 1922, p. 51. See also Galerkin's paper in *Messenger Math.*, vol. 55, p. 26, 1925.
[9] *J. Applied Mechanics*, vol. 6, p. A-114, 1939.
[10] See his book "Theorie d. Elasticität u. Festigkeit," 2d ed., p. 358, 1878.
[11] See his book "Pilzdecken," 2d ed., Berlin, 1926. The book contains a complete bibliography of the subject.
[12] *ZAMM*, vol. 14, p. 13, 1934.

tion, especially in the works of H. M. Westergaard.[1] Experiments with such plates were made by J. Vint and W. N. Elgood[2] and G. Murphy.[3]

The many applications of wooden plates and reinforced concrete slabs have given rise to the theory of bending of anisotropic plates. Although the first work of this type was done by Gehring,[4] the solutions which allowed practical application were mostly contributed by M. T. Huber. His many publications on this subject are collected in the book "Probleme der Statik technisch wichtiger orthotroper Platten" (Warsaw, 1929). Reference may also be had to his recently published book on theory of elasticity.[5] Further work in this field is mainly that of Russian engineers and the results obtained are compiled in the previously mentioned book by Lechnitski.

It is assumed in the elementary theory of plates that deflections are small in comparison with thickness. For larger deflections, the stretching of the middle plane of the plate must be taken into consideration; these equations were derived by Kirchhoff[6] and Clebsch (see page 258). The equations are not linear and are difficult to handle, and Kirchhoff used them only in the simplest case in which stretching of the middle plane is uniform. Progress in this field was accomplished by engineers, principally in connection with the analysis of stresses in ships' hulls. Considering bending of a long uniformly loaded rectangular plate, Boobnov[7] reduced the problem to that of bending of a strip and solved it for several edge conditions such as are encountered in ship construction. He also prepared tables, which greatly simplify stress analysis and are now widely used in the shipbuilding industry. Large deflections of circular plates by couples uniformly distributed along the boundary were discussed by the writer,[8] who investigated the limit of accuracy of the elementary linear theory. A further study of the problem has been made by S. Way,[9] who investigates bending of a uniformly loaded circular plate

[1] See *Ingeniøren*, vol. 32, p. 513, 1923; see also his papers in *J. Public Roads*, 1926, 1929, 1933.

[2] *Phil. Mag.*, 7th series, vol. 19, p. 1, 1935.

[3] *Iowa Eng. Exp. Sta. Bull.*, vol. 135, 1937.

[4] See his doctoral thesis, Berlin, 1860.

[5] "Théorie de L'élasticité" (Polish), Cracow, vol. 1, 1948; vol. 2, 1950.

[6] See his "Vorlesungen über mathematische Physik, Mechanik," 2d ed., 1877; also F. Gehring's dissertation in which Kirchhoff's suggestions, made in *Crelle's J.*, vol. 56, were used.

[7] The English translation of the paper was published in *Trans. Inst. Naval Architects.*, vol. 44, p. 15.

[8] *Mem. Inst. Engrs. Ways of Communication*, vol. 89, 1915.

[9] *Trans. ASME*, vol. 56, p. 627, 1934. See also K. Federhofer, *Luftfahrt-Forsch.*, vol. 21, p. 1, 1944, *Sitzber. Akad. Wiss. Wien, Math.-naturw. Klasse*, Abt. IIa, vol. 155, p. 15, 1946.

with clamped edges, theoretically and experimentally. He also analyzed[1] the uniformly loaded rectangular plate and showed that, when the ratio of the sides a/b is greater than 2, the maximum stress does not differ much from that obtained by Boobnov for an infinitely long plate.

The general equations for the large deflections of very thin plates were simplified by A. Föppl by the use of the stress function for the stresses acting in the middle plane of the plate.[2] The requirement that the plate be "very thin" was removed by von Kármán[3] whose equations were used by A. Nadai in his book (previously mentioned) and by Samuel Levy[4] in his investigation of large deflections of rectangular plates.

In the derivation of the equations of the elementary theory of plates, it is assumed that each thin layer of the plate parallel to the middle plane xy is in a state of plane stress by which only the stress components σ_x, σ_y, and τ_{xy} are different from zero. In thicker plates, it is useful to have a complete solution of the problem in which all six stress components are considered. Some solutions of this kind are given by Saint-Venant in his translation of Clebsch's book.[5] Some elementary rigorous solutions for circular plates have been found by A. Korobov[6] and a general discussion of the rigorous theory of plates has been given by J. H. Michell[7] and was further developed by A. E. H. Love in his book on elasticity.[8] In recent times, the rigorous theory of plates has attracted the interest of engineers and several problems have been completely solved. Special mention should be made of the papers by S. Woinowsky-Krieger[9] and by B. G. Galerkin.[10]

As a result of the increasing popularity of thin-walled structures, the theory of shells has received much attention in recent times. In many problems of thin shells, a satisfactory solution can be obtained by neglecting bending and assuming that stresses are uniformly distributed through the thickness of the structures. These *membrane stresses* have been investigated for several surfaces of revolution, especially for spherical and cylindrical forms.[11] The general theory of bending of shells was offered

---

[1] See the paper in *Proc. 5th Intern. Congr. Applied Mechanics*, Cambridge, Mass., 1938.

[2] See his "Technische Mechanik," vol. 5, p. 132, 1907.

[3] See his article "Festigkeit im Maschinenbau," *Encykl. Math. Wiss.*, vol. IV₄, p. 311, 1910.

[4] See *Natl. Advisory Comm. Aeronaut. Tech. Notes*, 846, 847, 853.

[5] See p. 337 of that translation.

[6] See S. Timoshenko, "Theory of Elasticity," p. 315, 1934.

[7] *Proc. London. Math. Soc.*, vol. 31, p. 100, 1900.

[8] See 4th ed., p. 473, 1927.

[9] *Ing.-Arch.*, vol. 4, pp. 203, 305, 1933.

[10] *Compt. rend.*, vol. 190, p. 1047; vol. 193, p. 568; vol. 194, p. 1440.

[11] For a bibliography, see W. Flügge, "Statik und Dynamik der Schalen," Berlin, 1934; K. Girkmann, "Flächentragwerke," 2d ed., Vienna, 1948.

by H. Aron[1] and A. E. H. Love[2] and the first engineering application that this theory found was in A. Stodola's analysis[3] of stresses in conical shells of constant thickness. The symmetrically loaded spherical shell of a constant thickness has been investigated by H. Reissner.[4] He shows that bending stresses are produced principally by reactive forces acting along the fixed edge of the structure and derives the differential equations for calculating these forces. These equations were used by E. Meissner, who, together with his pupils succeeded in getting rigorous solutions for several problems of practical importance.[5] The applicability of these rigorous solutions depends on the rapidity of convergence of the series involved and this becomes slower for thinner shells. To overcome such difficulties, the approximate methods devised by O. Blumenthal[6] and J. W. Geckeler[7] can be successfully applied. Nonsymmetrical load distributions on spherical shells have been studied by A. Havers.[8] The theory of cylindrical shells has received the attention of several authors[9] in connection with the design of boilers and pipe lines.

85. Elastic Stability

The wide use that is being made of steel and high-strength alloys in engineering structures (especially in bridges, ships, and aircraft) has made elastic instability a problem of great importance.[10] Urgent practical requirements have given rise in recent years to extensive investigations, both theoretical and experimental, of the conditions governing the stability of various structural elements. Originally the problem of lateral buckling of a compressed member was the only one which concerned engineers, but now it has become necessary to answer a variety of other important questions. We have already mentioned the problem of lateral buckling of beams, discussed by L. Prandtl for a narrow rectangular cross section and by the writer in the case of I beams.

[1] *J. reine u. ang. Math.*, vol. 78, p. 136, 1874.
[2] *Phil. Trans.*, (A), vol. 179, p. 491, 1888.
[3] "Die Dampfturbinen," 4th ed., p. 597, Berlin, 1910; see also H. Keller, *Schweiz. Bauztg.*, 1913, p. 111, and the book "Berechnung gewölbter Böden," Berlin, 1922.
[4] "Müller-Breslau Festschrift," p. 192, 1912.
[5] E. Meissner, *Physik. Z.*, vol. 14, p. 343, 1913; L. Bolle, *Schweiz. Bauztg.*, vol. 66, p. 105, 1915; F. Dubois, *Promotionsarb.*, Zürich, 1916; E. Honegger, *ZAMM*, vol. 7, p. 120, 1927.
[6] *Z. Math. Physik*, vol. 62, p. 343, 1914.
[7] *Forschungsarb.*, no. 276, Berlin, 1926. See also *Ing.-Arch.*, vol. 1, p. 255, 1930.
[8] *Ing.-Arch.*, vol. 6, p. 282, 1935.
[9] A bibliography on theory of shells can be found in the article by J. W. Geckeler, "Handbuch der Physik," vol. 6, 1928; W. Flügge, "Statik und Dynamik der Schalen," Berlin, 1934; S. Timoshenko, "Theory of Plates and Shells," 1940.
[10] A general discussion of the stability criterion in theory of elasticity is given by H. Ziegler. See *ZAMM*, vol. 20, p. 49, 1952.

Theory of Elasticity during the Period 1900–1950

In connection with the latter problem, the question of torsion of thin-walled opened sections became of practical importance. The simplest case of torsional buckling (Fig. 237) of an angle section has already been treated.[1] A general investigation of torsional buckling of compressed struts of thin-walled sections, such as those used in airplane construction, has been made by H. Wagner[2] and the theory given a more rigorous basis by R. Kappus.[3] Since that time, the theory of lateral buckling of beams and torsional buckling of struts has been studied by many engineers and wide applications have been found for it, not only in airplane construction, but also in bridgebuilding. Mention should be made of the works of Goodier[4] who has investigated not only the stability of an isolated strut under various conditions, but also that of a strut rigidly connected to elastic plates. By using the theory of large deformation, he has given a rigorous proof that the assumption upon which Wagner's theory of torsional buckling was founded is in fact a valid one.[5] H. Nylander[6] has contributed to the theory of lateral buckling of I beams and has made an extensive experimental study of the problem. E. Chwalla[7] investigated lateral buckling of beams of nonsymmetrical cross section and gives general equations from which those for an I beam are obtainable as a particular case.

FIG. 237.

The general theory of bending, torsion, and buckling of thin-walled members of open cross section was discussed by the writer.[8] In his book, V. Z. Vlasov[9] develops a different method of approach to the theory of torsional buckling and indicates that Saint-Venant's principle does not hold for thin-walled members and that it is possible, for example, to produce torsion of a bar of Z section by applying bending couples to the flanges at the ends.

As a result of some structural problems which arose in the Russian navy, the writer investigated the elastic stability of rectangular plates[10]

[1] See S. Timoshenko, *Bull. Polytech. Inst. Kiev*, 1907. See also H. J. Plass, *J. Applied Mechanics*, vol. 18, p. 285, 1951.

[2] See his paper in "Festschrift Fünfundzwanzig Jahre Technische Hochschule Danzig," p. 329, 1929.

[3] See *Jahrbuch der deutschen Luftfahrtforschung*, 1937; and *Luftfahrt-Forsch.*, vol. 14, p. 444, 1938.

[4] *Cornell Univ. Eng. Exp. Sta. Bulls.* 27, 1941, and 28, 1942.

[5] *J. Applied Mechanics*, vol. 17, p. 383, 1950.

[6] *Proc. Roy. Swed. Inst. Eng. Research*, no. 174, 1943.

[7] See *Sitzsber. Akad. Wiss. Wien*, Abt. IIa, vol. 153, pp. 25–60, 1944.

[8] *J. Franklin Inst.*, vol. 239, p. 201, 1945.

[9] V. Z. Vlasov, "Thin Walled Elastic Bars," Moscow, 1940.

[10] *Bull. Polytech. Inst. Kiev*, 1907. See also German translation in *Z. Math. Physik*, vol. 58, p. 357, 1910.

subjected to forces in the middle plane. The simplest case of a uniformly compressed rectangular plate with simply supported edges had already been solved by G. H. Bryan (see page 299), but in shipbuilding the engineer usually encounters other kinds of edge conditions and the determination of critical values of stresses requires more elaborate calculations. The problem was solved at that time for many particular cases and tables of critical values of stresses were prepared.

The investigation of elastic stability problems often leads to differential equations, the rigorous solution of which presents much difficulty, and it is necessary to have recourse to some approximate method for calculating the critical forces. A method based on consideration of the energy of a system, similar to that used by Rayleigh for approximate calculation of frequencies of vibration of elastic systems, has been developed by the author[1] and applied in solving many stability problems. Considering a compressed column, for example, we can assume some expression for the deflection curve of the buckled column, such that the end conditions are satisfied, and calculate the critical value of the load from the condition that the increase in strain energy of the column during buckling must be equal to the work done by the compressive force. In this way, we usually obtain a value for the critical load which is higher than the true one, since working with an assumed curve is equivalent to introducing some additional constraints into the system. These prevent the column from taking any other shape than that prescribed by the assumed curve. By introducing such additional constraints, the critical load can only be made larger. A better approximation for critical loads can be had by taking expressions for the assumed curves with several parameters by the variation of which the shape of the curve may be somewhat changed. By adjusting these parameters so as to make the expression for the critical load a minimum, we get the value of this load with better accuracy. The method was used for several cases, including that of a system of bars. In this way, the stability of built-up columns, of compressed chords of open bridges, and of several kinds of frame systems was investigated.

The problem of reinforcing plates by various kinds of stiffeners readily submits to the approximate method of attack. In shipbuilding we often have uniformly compressed rectangular plates which have to be reinforced by a system of longitudinal or transverse stiffeners. The critical values of compressive stresses for such strengthened plates have been determined by using the energy method, and tables have been computed which simplify selection of the proper dimensions of stiffeners. The buckling of a rectangular plate under the action of shearing stresses was also solved by

---

[1] *Bull. Polytech. Inst. Kiev*, 1910. See also French translation in *Ann. ponts et chaussées*, Paris, 1913.

using the same approximate method and the proper selection of stiffeners was investigated.[1] With the introduction of large-span plate girders the question of stiffening the web was solved by the same means.[2]

The elastic stability of compressed curved bars has become of practical importance in connection with the construction of thin arches. The two-hinged circular arch of constant cross section was solved by E. Hurlbrink,[3] and the case of very small initial curvature was discussed by the writer.[4] The three-hinged arch was investigated by R. Mayer,[5] and experiments with such structures were conducted by E. Gaber.[6] The uniformly compressed circular arch of constant cross section with clamped ends was taken up by E. Nicolai,[7] while arches of variable cross section have been considered by E. Steuermann,[8] A. N. Dinnik,[9] and K. Federhofer.[10] A very complete study of stability was made on helical springs by J. A. Haringx.[11]

The elastic stability of thin shells is of prime importance in modern airplane construction. The existing work deals mainly with cylindrical shells, and the case of axial compression has been treated by several authors.[12] It has been shown that there are possible not only symmetrical, but also unsymmetrical forms of buckling. In studying the elastic stability of submarine hulls, the problem of a cylindrical shell under combined axial and lateral pressure was investigated by R. von Mises.[13] Axial compression of curved sheet panels was treated by the writer[14] while buckling of a curved sheet panel under shearing forces was investigated

---

[1] See S. Timoshenko's paper in *Mem. Inst. Engrs. Ways of Communication*, vol. 89, p. 23, 1915. The case of an infinitely long rectangular plate was independently investigated by R. V. Southwell, *Phil. Mag.*, vol. 48, p. 540, 1924.

[2] See S. Timoshenko's papers in *Eisenbau*, vol. 12, p. 147, 1921; *Engineering*, vol. 138, p. 207, 1934. For a further study of buckling of plates reinforced by stiffeners, see R. Barbré, *Bauing.*, vol. 17, 1936; E. Chwalla, *Bauing.*, vol. 17, 1936.

[3] *Schiffbau*, vol. 9, p. 517, 1908.

[4] *J. Applied Mechanics*, vol. 2, p. 17, 1935.

[5] *Eisenbau*, vol. 4, p. 361, 1913.

[6] *Bautech.*, 1934, p. 646.

[7] *Bull. Polytech. Inst. St. Petersburg*, vol. 27, 1918; *ZAMM*, vol. 3, p. 227, 1923.

[8] *Ing.-Archiv.*, vol. 1, p. 301, 1930.

[9] *Vestnik Inzhenerov i Tech.*, nos. 6, 12, 1933.

[10] *Bauing.*, vol. 22, p. 340, 1941.

[11] *Philips Research Repts.*, vol. 3, 1948; vol. 4, 1949.

[12] R. Lorenz, *VDI*, vol. 52, p. 1766, 1908; *Physik. Z.*, vol. 13, p. 241, 1911; S. Timoshenko, *Z. Math. u. Physik*, vol. 58, p. 378, 1910; *Bull. Electrotech. Inst. St. Petersburg*, vol. 11, 1914. See also R. V. Southwell, *Phil. Mag.*, vol. 25, p. 687, 1913; *Trans. Roy. Soc. (London)*, (A), vol. 213, p. 187, 1914.

[13] *VDI*, vol. 58, p. 750, 1914.

[14] See S. Timoshenko, "Theory of Elasticity" (Russian), vol. 2, p. 395, 1916.

by D. M. A. Leggett.[1] The buckling of cylindrical shells subjected to torsion has been studied by E. Schwerin[2] and L. H. Donnell.[3] The stability of thin cylindrical tubes subjected to pure bending has been investigated by W. Flügge.[4] He shows that the critical compressive stress is about 30 per cent higher than that for a symmetrically buckled axially compressed cylindrical shell. Various methods of stiffening cylindrical shells have been tried theoretically and experimentally in connection with airplane construction. If a cylindrical shell is stiffened by equidistant longitudinal and circumferential ribs, the problem can be reduced to that of investigating the buckling of a nonisotropic shell. The differential equations for this have been established by W. Flügge[4] and some calculations made by Dji-Djüän Dschou.[5]

Uniformly compressed spherical shells have received attention from R. Zoelly,[6] who considered symmetrical forms of buckling, and by Van der Neut,[7] who carried out a more general investigation of the matter. The buckling problem for a segment of a spherical shell of small curvature was investigated by C. B. Biezeno.[8]

The theoretical formulas for critical stresses in plates and shells have been obtained by using linear differential equations for deflections, derived on the assumption that deflections are small. To examine the deformations of buckled plates and shells when the loads are greater than their critical values, the theory of large deflection must be used. H. Wagner[9] showed that a plate girder with a very thin web can carry a large additional load after buckling of the web has occurred and he gave an approximate formula for calculating the ultimate load. The ultimate load for a compressed rectangular plate was discussed by Theodore von Kármán[10] and an approximate formula for the *effective width* was given. Further study of the same subject was made by the writer[11] and E. Trefftz and K. Marguerre.[12] Experiments made on the buckling of thin shells showed that buckling usually begins at loads much smaller than those predicted by theory. The explanation of this phenomenon was given by

[1] *Proc. Roy. Soc. (London)*, (A), vol. 162, pp. 62–83, 1937.
[2] *Repts. Intern. Cong. Applied Mechanics*, Delft, 1924; *ZAMM*, vol. 5, p. 235, 1925.
[3] *Natl. Advisory Comm. Aeronautic. Rept.* 479, 1933.
[4] *Ing.-Arch.*, vol. 3, p. 463, 1932.
[5] *Luftfahrt-Forsch.*, vol. 11, p. 223, 1935.
[6] Dissertation, Zürich, 1915.
[7] Dissertation, Delft, 1932.
[8] *ZAMM*, vol. 15, p. 10, 1935.
[9] *Z. Flugtech. u. Motorluftschiffahrt*, vol. 20, p. 200, 1929.
[10] *Trans. ASME*, vol. 54, p. 53, 1932.
[11] "Theory of Elastic Stability," p. 390, 1936.
[12] *ZAMM*, vol. 16, p. 353, 1936; vol. 17, p. 121, 1937.

von Kármán and Tsien[1] who showed, by using the theory of large deflections, that there exist forms of stable equilibrium which require smaller values of load than those obtained from the classical theory. The problem was further studied by L. H. Donnell[2] and K. O. Friedrichs.[3]

The problem of elastic stability usually arises in bodies one or two of whose dimensions are small (*i.e.*, in thin bars or thin plates and shells). But in such materials as rubber, which may have large deformations within the elastic range, the question of stability may crop up in bodies which have all their dimensions of the same order. The first problems of this kind were discussed by R. V. Southwell.[4] Further study of the general theory of elastic stability was made by C. B. Biezeno and H. Hencky[5] and by M. A. Biot[6] and E. Trefftz.[7]

86. Vibrations and Impact

Modern machinery frequently presents problems which require the analysis of stresses produced by dynamical causes. Such problems of practical significance as the torsional vibrations of shafts, vibration of turbine blades and turbine disks, the whirling of rotating shafts, the vibrations of railway tracks and bridges under rolling loads, and the vibration of foundations can only be thoroughly understood through the theory of vibration. Only by using this theory, can favorable design proportions be found which will remove working conditions of a machine as far as possible from the critical conditions of resonance (where heavy vibrations may occur).

One of the first problems in which the importance of studying vibration was recognized by engineers was that of the torsional vibrations in propeller shafts of steamships. Frahm[8] was probably the first to investigate this problem theoretically and experimentally and to show that, at resonance, torsional stresses may become very large so as to lead to sudden fracture due to fatigue. Since the time of Frahm's famous paper, torsional vibrations have been studied by many engineers, not only with regard to propeller shafts, but also in more complicated cases of engine crankshafts with many rotating masses. The problem can be idealized and reduced to the solution of a system of linear differential equations

---

[1] *J. Aeronaut. Sci.*, vol. 7, p. 43, 1939; vol. 8, p. 303, 1941.
[2] *J. Applied Mechanics*, vol. 17, p. 73, 1950.
[3] Theodore von Kármán Anniversary Volume, p. 258, 1941.
[4] *Trans. Roy. Soc. (London)*, (A), vol. 213, pp. 187–244, 1913.
[5] *Proc. Acad. Sci. Amsterdam*, vol. 31, pp. 569–592, 1928.
[6] *Proc. 5th Intern. Congr. Applied Mechanics*, Cambridge, Mass., 1938; and *Phil. Mag.* (7), vol. 27, pp. 468–489, 1939.
[7] *ZAMM*, vol. 13, pp. 160–165, 1933.
[8] *VDI*, 1902, p. 797.

with constant coefficients. From these, the frequency equation can be derived and the evaluation of the coefficients of this equation and the numerical solution of the equation itself become more and more involved as the number of rotating masses increases. In practical applications we often have systems with many masses and the amount of work required to evaluate the coefficients and to solve the frequency equation becomes prohibitive. Then use must be made of some approximate method of calculating frequencies without deriving the frequency equation. In calculation of the lowest frequency, the Rayleigh method usually gives satisfactory results. For calculating this and also the higher frequencies, a method of successive approximation can be applied. The method was suggested by L. Vianello,[1] who used it for calculating buckling loads for struts. Its application to calculating principal frequencies of shafts was made by A. Stodola in his book on steam turbines. A complete discussion of the method is given in the book by C. B. Biezeno and R. Grammel,[2] and a very complete review of the various methods devised for investigating torsional problems (with a comparison of them) is made by K. Klotter.[3]

Lateral vibrations of shafts and beams are also of great practical importance. The simplest cases of vibration of prismatical bars were investigated during the eighteenth century and their solutions included in books on acoustics. In applying these solutions to practical beams in which the cross-sectional dimensions are not very small in comparison with the length, or to cases where higher modes of vibration are of importance, it has become necessary to derive a more complete differential equation, in which the effect of shearing forces on deflections is considered.[4] Very often the cross-sectional dimensions vary along the length of the beam. Rigorous analysis of the vibration of such beams can only be accomplished in the simplest cases,[5] and usually recourse has to be had to one of the approximate methods of integrating differential equations. These methods were brought to the fore in connection with the frequencies of lateral vibration of ships.[6] They are usually based on the above-mentioned idea of successive approximation, originated by L. Vianello. The vibration of prismatical bars with masses attached at the

[1] *VDI*, vol. 42, p. 1436, 1898.

[2] "Technische Dynamik," Berlin, 1939.

[3] *Ing.-Arch.*, vol. 17, p. 1, 1949.

[4] See S. Timoshenko, "Theory of Elasticity" (Russian), vol. 2, p. 206, 1916. See also *Phil. Mag.* (6), vol. 41, p. 744; vol. 43, p. 125.

[5] See P. F. Ward, *Phil. Mag.*, vol. 25, pp. 85–106, 1913; and A. N. Dinnik, Dissertation, Ekaterinoslav, 1915.

[6] A very complete review of literature on vibration of ships and comparison of various methods is given by P. F. Papkovich, *Appl. Math. Mechanics* (Russian), vol. 1, pp. 97–124, 1933.

ends is also of practical interest and the fundamental modes of the motions were investigated.[1] The problem of lateral vibration of bridges, under traveling loads, has continued to interest engineers. In 1905, A. N. Kryloff gave[2] a complete solution of the problem, neglecting the mass of the rolling load and assuming that a constant force moves across a prismatical beam with constant speed. The case of a pulsating load, such as that produced by an imperfectly balanced locomotive passing over the bridge was also considered.[3] The investigation showed that a pulsating force may give rise to considerable vibration under the condition of resonance. The problem was again taken up by C. E. Inglis,[4] who considers the effect of the rolling mass with some simplifying assumptions. A general discussion of the effect of rolling masses was made by A. Schallenkamp,[5] who did experiments also with a small model and found that the experimental deflection curve agreed satisfactorily with theoretical calculations. A study of vibration produced by a moving force in a continuous two-span beam has been carried on theoretically and experimentally at Stanford University by R. S. Ayre, George Ford, and L. S. Jacobsen.[6]

Several important vibration problems have been studied by engineers in connection with advances in steam-turbine design. The whirling of a shaft carrying a disk was first investigated theoretically and experimentally by A. Föppl.[7] Further work on the problem was given in A. Stodola's book on steam turbines. Consideration of the hysteresis effect on shaft whirling was covered by A. L. Kimball.[8] Vibrations of turbine blades may sometimes become excessive and lead to fractures. This problem has been studied by many engineers. We have here the vibration of a bar of variable cross section under the combined action of lateral and axial forces. In calculating the fundamental frequency, the Rayleigh method is usually applied.[9] To reduce this effect, the blades are often bound in groups and the vibration of such complicated systems has been discussed by E. Schwerin[10] and by R. P. Kroon.[11]

Vibration of rotating disks presents another important problem in

[1] See S. Timoshenko, *Z. Math. u. Physik*, vol. 59, 1911.
[2] *Math. Ann.*, vol. 61, 1905.
[3] S. Timoshenko, *Bull. Polytech. Inst. Kiev*, 1908; *Phil. Mag.*, vol. 43, p. 1018, 1922.
[4] "A Mathematical Treatise on Vibrations in Railway Bridges," Cambridge, 1934.
[5] *Ing.-Arch.*, vol. 8, p. 182, 1937.
[6] *J. Applied Mechanics*, vol. 17, pp. 1, 283.
[7] *Civiling.*, vol. 41, p. 333, 1895.
[8] *Phys. Rev.*, 1923; *Phil. Mag.*, 6th series, vol. 49, p. 724, 1925.
[9] See E. Sörensen, *Ing.-Arch.*, vol. 8, p. 381, 1937; G. Mesmer, *Ing.-Arch.*, vol. 8, p. 396, 1937.
[10] *Z. tech. Physik*, vol. 8, p. 312, 1927.
[11] *Trans. ASME*, vol. 56, p. 109, 1934; *Mech. Eng.*, vol. 62, p. 531, 1940.

turbine design. It was studied theoretically, using the Rayleigh method, by A. Stodola[1] and experimentally by Wilfred Campbell,[2] who describes a method of observing disk vibrations under service conditions. A very complete experimental study of vibration of circular disks was made by Mary D. Waller,[3] who developed a new powerful method for obtaining *Chladni lines*. The study of vibration of rectangular plates has also progressed, mainly as a result of the epoch-making work of Walter Ritz (see page 339). Dana Young[4] recently applied the Ritz method to frequency calculations for rectangular plates with various edge conditions.

Problems of vibration of circular rings and portions of rings are encountered in investigations of the vibration of circular frames, rotating electrical machinery, and arches. The bending vibration of a ring of circular cross section in its plane was studied by R. Hoppe[5] and at right angles to that plane by J. H. Michell.[6] Torsional vibration of the same kind of rings was investigated by A. B. Basset.[7] Vibration of a portion of a ring with various end conditions has been discussed by several authors, and a very complete study of the problem was made by K. Federhofer,[8] who recently collected his numerous publications on this subject and published them in book form.

The study of impact of elastic bodies has been continued in the twentieth century. Experiments on the impact of steel balls have shown[9] the validity of Hertz's theory. Saint-Venant's theory of longitudinal impact of cylindrical bars has been found to be in agreement with experiments by J. E. Sears,[10] who used bars with spherical ends in his tests. The study of impact of balls beyond the elastic limit has been made by D. Tabor.[11]

The lateral impact of a ball on a beam was studied theoretically by the writer.[12] Combining the Hertz theory for the deformation at the surface of contact with the theory of lateral vibration of the beam, it was found possible to calculate the duration of impact and to show that several interruptions of the contact between the ball and the beam usually occur

[1] *Schweiz. Bauztg.*, vol. 63, p. 112, 1914.
[2] *Trans. ASME*, vol. 46, p. 31, 1924.
[3] *Proc. Phys. Soc. (Brit.)*, vol. 50, p. 70, 1938.
[4] *J. Applied Mechanics*, vol. 17, p. 448, 1950.
[5] *J. Math.* (Crelle), vol. 73, 1871.
[6] *Messenger Math.*, vol. 19, 1890.
[7] *Proc. London Math. Soc.*, vol. 23, 1892.
[8] "Dynamik des Bogenträgers und Kreisringes," Vienna, 1950.
[9] A. Dinnik, *J. Russian Phys. Soc.*, vol. 38, p. 242, 1906.
[10] *Proc. Cambridge Phil. Soc.*, vol. 14, p. 257, 1908; see also J. E. P. Wagstaff, *Proc. Roy. Soc. (London)*, (A), vol. 105, p. 544, 1924; W. A. Prowse, *Phil. Mag.* (7), vol. 22, p. 209, 1936; R. M. Davies, *Trans. Roy. Soc. (London)*, (A), vol. 240, pp. 375–457, 1948.
[11] *Engineering*, vol. 167, p. 145, 1949.
[12] *Z. Math. u. Physik*, vol. 62, p. 198, 1914.

during impact. This result was verified by the experiments of H. L. Mason.[1] Further study of lateral impact has been made by many investigators.[2] Plastic deformation of bars and elastic and plastic deformation of various structures under conditions of impact have recently received a great deal of attention in connection with some military problems.

[1] *J. Applied Mechanics*, vol. 58, p. A-55, 1936.
[2] See R. N. Arnold, *Proc. Inst. Mech. Engrs. (London)*, vol. 137, p. 217, 1937; E. H. Lee, *J. Applied Mechanics*, vol. 62, p. A-129, 1940; C. Zener, *J. Applied Mechanics*, vol. 61, p. A-67, 1939; Zirô Tuzi and Masataka Nizida, *Bull. Inst. Phys. Chem. Research (Tokyo)*, vol. 15, pp. 905–922, 1936.

CHAPTER XIV

Theory of Structures during the Period 1900–1950

87. New Methods of Solving Statically Indeterminate Systems

The theory of structures was developed in the nineteenth century principally for the analysis of trusses. Satisfactory solutions for these can be obtained by assuming that joints are hinged so that all members are subjected only to axial forces. With the introduction of reinforced concrete into structural engineering, various kinds of frame structures found wide application. These structures are usually highly indeterminate systems the members of which work essentially in bending. Previously developed methods were soon found to be unsuitable for analyzing such systems and some new methods, based on considerations of deformation, were introduced.

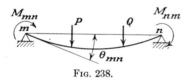

Fig. 238.

We have already seen (page 322) that, when analyzing secondary stresses, it is to our advantage to take the angles of rotation of rigid joints as the unknown quantities. A systematic use of angles of rotation in analysis of frame structures was introduced by Axel Bendixen,[1] who developed the so-called *slope-deflection method*. Considering frame structures in which the displacements of joints can be neglected,[2] the well-known equation

$$M_{mn} = 2k_{mn}(2\theta_{mn} + \theta_{nm}) + M_{mn}{}^0 \qquad (a)$$

is used for the bending moment M_{mn} acting on end m of the bar mn (Fig. 238). θ_{mn} and θ_{nm} are the angles of rotation of the ends, taken positive in the clockwise sense; $k_{mn} = EI_{mn}/l_{mn}$ is the so-called *stiffness factor;* and $M_{mn}{}^0$ is the *fixed-end moment*, that is, the moment which would act on the end m of the bar if its ends were fixed and only lateral loads P and Q were acting. These moments are taken positive if they act on

---

[1] See his book "Die Methode der Alpha-Gleichungen zur Berechnung von Rahmenkonstruktionen," Berlin, 1914.

[2] In his paper Bendixen also considers more general cases.

the bar in the clockwise direction. Taking, as an example, the system shown in Fig. 239 and considering the end moments acting on the bars ab, ac, ad, and ae as unknowns, we should have to derive and solve seven equations; but the problem reduces to that of one equation, if we take the angle of rotation θ_a of the rigid joint a as the unknown. This equation is obtained if we put the summation of the four end moments of the bars jointed at a equal to zero. Using Eq. (a) for end moments and observing that the ends b, c, d, and e are fixed, we obtain

$$4\theta_a(k_{ab} + k_{ac} + k_{ad} + k_{ae}) = -(M_{ab}{}^0 + M_{ad}{}^0) \qquad (b)$$

Solving this equation for θ_a and substituting into Eq. (a), we obtain all the end moments of the bars of the system. In more complicated structures where we have to consider several joints (like joint a in the above example), we write an equation, equivalent to Eq. (b), for each joint. The number of these equations is equal to the number of unknown angles of rotation, and when these angles are calculated, all end moments can be found from Eqs. (a).

A further step in the simplification of framed structure analysis was made by K. A. Čališev.[1] He uses the method of successive approximations and reduces the problem to that of solving only one equation with one unknown at each step. Considering a framed system (shown in

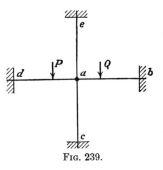

Fig. 239.

Fig. 240) where any lateral movement is prevented by introducing the horizontal bar ab, he assumes first that all joints are fixed and that they do not rotate during application of the loads P_1, P_2. Under this assumption, he calculates the *fixed-end moments* for all loaded bars and sums these for each joint. This summation, with a negative sign, represents the *unbalanced moment* acting on the joint. The final values of the end moments are obtained by adding (algebraically) the end moments produced by the *unbalanced moments* acting at the joints to the previously calculated *fixed-end moments*. Čališev solves this last problem by successive approximations. Denoting the unbalanced moments of the system (Fig. 240) by M_c, M_d, M_e, he unlocks one joint at a time, keeping all others fixed, and computes first approximations for the angles of rotation. For example, unlocking joint d in Fig. 240[2] and considering

[1] *Tehnički List*, nos. 1–2, 1922; nos. 17–21, 1923.

[2] Čališev recommends beginning with the joint at which the unbalanced moment is the largest.

the rest of the joints as fixed, he writes an equation similar to Eq. (b) for the first approximation θ'_d of the angle of rotation θ_d; that is,

$$4\theta_d'(k_{dc} + k_{de} + k_{dd_1}) = -(M_{de}{}^0 + M_{dc}{}^0) = M_d$$

from which

$$\theta_d' = \frac{M_d}{4\Sigma k_{mn}} \qquad (c)$$

The corresponding first approximation M_{mn}' of the end moments is obtained from the angles of rotation of joints by means of the well-known equation

$$M_{mn}' = 2k_{mn}(2\theta_{mn}' + \theta_{nm}') \qquad (d)$$

Since the values θ_{mn}' and θ_{nm}' in this equation are not exact, the summation $-\Sigma M_{mn}'$ will not be equal to the unbalanced moment M_d at the

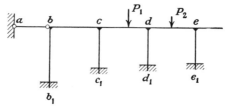

Fig. 240.

joint d. To get a second approximation, Čališev introduces the difference $\Delta' M_d = M_d + \Sigma M_{mn}'$ and by using it in Eq. (c), instead of M_d, he obtains a correction $\Delta\theta_d'$. Having such corrections for all joints and using them in Eq. (d), the corresponding corrections $\Delta M_{mn}'$ can be computed. Adding these corrections to the first approximations of the end moments, better approximations of these moments are obtained. These are usually accurate enough for practical purposes. If a closer approximation is required, he proceeds with second approximations M_{mn}'' in the same way as described above and obtains the third approximation, and so on.

Instead of getting successive approximations by using Eqs. (c) and (d), we can accomplish the same end by directly distributing the unbalanced moments while unlocking one joint at a time and considering all other joints as fixed. This procedure of successive approximations was proposed by Hardy Cross[1] and found wide applications in America.

In selecting proper cross-sectional dimensions of steel structures, it is sometimes necessary to consider not only loads at which yielding of the material sets in but also those loads under which complete collapse of the structure is brought about. Analysis shows that, if two structures are

[1] *Trans. ASCE*, vol. 96, 1932.

designed with the same factor of safety with respect to yielding, they may have very different factors of safety with respect to complete failure. For example, considering pure bending of beams and assuming that structural steel follows Hooke's law up to the yield point and that beyond that limit it stretches without strain hardening, we obtain the stress distributions, shown in Figs. 241a and 241b for the two limiting conditions: (1) the beginning of yielding and (2) complete collapse. The corresponding bending moments for a rectangular cross section (Fig. 241c) are

$$M_{yp} = \sigma_{yp} \frac{bh^2}{6}, \qquad M_{ult} = \sigma_{yp} \frac{bh^2}{4}$$

so that

$$M_{ult} : M_{yp} = 1.5$$

For an I beam this ratio is much smaller and depends on the proportions of the cross-sectional dimensions (Fig. 241d). From this it follows that,

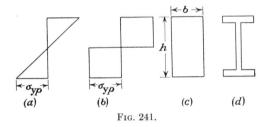

Fig. 241.

if a rectangular beam and an I beam are both designed with the same factor of safety with respect to yielding, the factor of safety of the I beam with respect to complete collapse is smaller than that of a rectangular beam.

To allow for this fact some engineers have recommended[1] that cross-sectional dimensions of structural members should be chosen on the basis of ultimate strength. They have shown that, if we take the stress distribution shown in Fig. 241b for the ultimate strength condition, the corresponding ultimate loads can be readily calculated. For instance from a beam with built-in ends loaded at the middle Fig. 242a, we conclude that complete collapse of the beam occurs when the bending moment reaches its ultimate value M_{ult} at the three cross sections, a, b, and c. For any further loading, the condition will be the same as for two-hinged bars (Fig. 242b). The magnitude of the ultimate load is then

[1] See the paper by N. C. Kist, *Eisenbau*, 1920, p. 425, and the theoretical analysis of the problem by M. Grüning in his book "Die Tragfähigkeit statisch unbestimmter Tragwerke . . . ," Berlin, 1926.

obtainable from the corresponding bending-moment diagram (Fig. 242a), from which we have

$$\frac{P_{ult}l}{4} = 2M_{ult}$$

If we proceed in a similar manner for such highly statically indeterminate systems as the structures of office buildings, and put hinges at all those cross sections at which bending moments reach their ultimate values, the problem of determining ultimate loads can be reduced to that of simple statics of rigid bodies, and the entire stress analysis of the system can be greatly simplified.

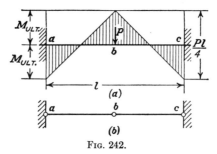

Fig. 242.

In analyzing systems of slender bars which carry axial and lateral forces simultaneously, beam-column problems become of practical importance. The first systematic investigation of such problems was made by A. P. Van der Fleet.[1] He did not confine himself to a theoretical study of the subject, but also calculated tables for simplifying such analyses. With the development of airplane structures, the subject of beam-column analysis has come to the forefront and further, more elaborate tables have been prepared by several authors[2] to simplify this analysis.

88. Arches and Suspension Bridges

With the introduction of reinforced concrete into structural engineering practice, arches have found a wide application, especially in bridge construction, and methods of analyzing stresses in arches have been the object of research. In an arch without hinges, we have a system with three statically indeterminate quantities and analysis can be greatly simplified by proper selection of those quantities. K. Culmann[3] intro-

---

[1] *Mem. Inst. Engrs. Ways of Communication*, 1900–1903; *Bull. Polytech. Inst. St. Petersburg*, 1904.

[2] Very complete tables were calculated by A. S. Niles and given in the book "Aeroplane Structures" by A. S. Niles and J. S. Newell, 2d ed., 1938.

[3] See the book "Anwendungen der graphischen Statik" by W. Ritter, vol. 4, p. 197, Zürich, 1906.

duced the so-called *elastic center* and showed that if the reactive forces acting at an arch abutment are represented by a force applied at its elastic center and by a couple, the three unknowns (two components of the force and the couple) may be determined, each from one equation with one unknown. For determining the position of the elastic center and for calculating the reactions, he used a graphic method in which fictitious forces, similar to those used in calculating the deflection of beams, are applied. Various methods of arch analysis have been presented by E. Mörsch,[1] J. Melan,[2] and by A. Strassner.[3] Numerous tables for thrust and bending moments in arches of various proportions are given in their books.

It is advantageous to make the center line of an arch coincide with the funicular curve of permanent loads. If this condition is fulfilled in a three-hinged arch, the resultant force on each cross section is tangential to the center line, producing only uniformly distributed compressive stresses. In a hingeless arch, internal forces depend on the deformation of the structure, and we cannot entirely eliminate bending stresses. If the center line of such an arch coincides with the funicular curve for the dead load initially, there will be a certain contraction of the axis of the arch due to compression of the material, shrinkage of the concrete, and lowering of temperature. The thrust of the arch, calculated as for the three-hinged structure, will be somewhat diminished. As a result of this, the thrust line of a hingeless arch is displaced upward at the crown and downward at the abutments from the assumed center line (coinciding with the funicular curve for the dead load). This displacement becomes particularly great in thick and flat arches where we may expect the appearance of undesirable tensile stresses at the abutments. Various methods have recently been developed to reduce these displacements in large arch structures. Sometimes temporary hinges are put at the crown and at the abutments; these permit free rotation during decentering. In this way, displacements of the thrust line, which occur as a result of the compression of the center line by the dead load, are eliminated. By filling up the gaps at the temporary hinges with concrete after decentering, we obtain a hingeless arch, the center line of which coincides with the funicular curve for the dead load. To allow for the thrust-line displacements due to shrinkage and lowering of temperature, the temporary hinges must be placed with proper eccentricities such that, when the gaps have been filled after decentering, bending moments are produced at the

[1] *Schweiz. Bauztg.*, vol. 47, p. 83, 1906; see also E. Mörsch, "Der Eisenbetonbau," vol. 2, part 3, 1935.

[2] J. Melan and T. Gesteschi, "Bogenbrücken, Handbuch für Eisenbetonbau," vol. 11, 1931.

[3] A. Strassner, "Der Bogen und das Brückengewölbe," 3d ed., vol. 2, Berlin, 1927.

crown and the abutments, the signs of which are opposite to those expected later from shrinkage and temperature drop. E. Freyssinet[1] suggested putting jacks in the temporary gap at the crown between the two halves of the arch. By using them, the most favorable position of the thrust line at the crown can be ascertained before filling up the gap with concrete. In the case of an arch with a tie beam, F. Dischinger[2] proposed the use of tie cables; by stretching these, the bending moments resulting from dead-load action and shrinkage of concrete can be greatly reduced.[3] The idea of *prestressing* in structures is now widely used. For example, by this means, Dischinger succeeded in considerably reducing the weight of recently designed reinforced concrete bridges in Germany.

The use of prestressing in concrete work has brought about a considerable amount of experimental and theoretical work in that field.[4] By using strong steel wires, which are kept under high tensile stress during hardening of the concrete, reinforced beams can be fabricated which have considerable initial compressive stresses in the concrete. Such prestressed beams can withstand much greater loads than is otherwise the case, before the first cracks appear in concrete. To simplify prestressing, H. Lossier has suggested the use of a special kind of concrete,[5] which expands while hardening. As a result of this, tensile stresses are produced in the steel wires and compression is set up in the concrete, so that the special devices for producing prestressing are rendered unnecessary.

Designers of arches usually assume that the displacements which result from elastic deformations are small and consider the initial undeformed center line of the arch in their analysis. But in large-span arches, the effects of deflections on the magnitude of the redundant forces may become considerable.[6] If this is so, calculations made on the basis of the initial undeformed center line can be taken only as a first approximation; using them, better accuracy can be obtained by taking into account the deflections indicated by the first approximation.

With the increased use of arch dam construction, engineers have had to solve a very complicated problem of stress analysis. An approximate

[1] *Génie civil*, vol. 79, p. 97, 1921; vol. 93, p. 254, 1928.

[2] *Bauing.*, vol. 24, p. 193, 1949.

[3] *Génie civil*, 1944.

[4] Methods of calculating initial stresses and descriptions of experiments can be found in the following books: M. Ritter and P. Lardy, "Vorgespannter Beton," Zürich, 1946; and Gustave Magnel, "Prestressed Concrete," Ghent, 1948, English translation, London, 1950.

[5] This kind of prestressing was used by Dischinger in the construction of the Saale Bridge near Alsleben in 1927–1928 and also in the construction of many hangars during the war.

[6] See articles by J. Melan in "Handbuch der Ingenieurwissenschaften," vol. 2, 1906, and in *Bauing.*, 1925. See also the papers by S. Kasarnowsky, "Stahlbau," 1931, and by B. Fritz, "Theorie und Berechnung vollwandiger Bogenträger," Berlin, 1934.

method was devised in connection with the large arch dams in the United States. A first approximation is obtained by replacing the dam by a system of horizontal arches plus a system of vertical cantilevers. The horizontal water pressure is divided by trial into two radial components, the one acting on the arches and the other on the cantilevers. The desired division of the load is that giving the arches and the cantilevers common radial components of deflection at all points. This method was proposed by the engineers of the U.S. Bureau of Reclamation.[1] To get a better approximation, the effects of twisting moments in the horizontal and vertical sections and of shearing forces acting on horizontal sections along the center lines of the arches and the corresponding vertical shearing forces in the radial sections were introduced in the analysis.[2] To check this theory, experiments were conducted with models in several important cases. In connection with the construction of Hoover Dam, a plastercelite model was tested, using mercury for loading, and the deformations were found to be in good agreement with the design analysis. Subsequent measurements on the actual dam tallied satisfactorily with the predictions of the calculations and model tests.

Recently considerable progress has been made[3] in the analysis and construction of suspension bridges. Those that were erected at the beginning of the nineteenth century did not fulfil expectations, for they were too flexible and many of them collapsed due to the excessive vibration produced by moving loads or wind. This undesirable flexibility was overcome in later structures by introducing stiffening trusses. It was also found that vibrations produced by moving loads diminish with increase of the span and weight of bridges so that, in very large bridges, satisfactory conditions are obtained without stiffening trusses. In the early stress analysis of suspension bridges with stiffening trusses, it was usually assumed that deformations were small and the same methods were applied as for rigid trusses. The first attempt to allow for deflections of the stiffening trusses was made by W. Ritter, a professor at the Polytechnical Institute of Riga.[4] Further advances were made in this respect by several investigators and in a form suitable for practical use it is presented by J. Melan.[5] This theory was used in the design of large suspension bridges in the United States. The theory covers a uniformly

[1] See the paper by C. H. Howell and A. C. Jaquith, *Trans. ASCE*, vol. 93, p. 1191, 1929.
[2] See the paper by H. M. Westergaard in *Proc. 3d Intern. Cong. Applied Mechanics*, 1931, Stockholm, vol. 2, p. 366.
[3] A very complete history of suspension bridges in bibliographical form is given by A. A. Jakkula; see *Bull. Agr. Mech. Coll. Texas*, 4th series, vol. 12, 1941.
[4] See *Z. Bauwesen*, 1877, p. 189.
[5] See his book "Theorie der eisernen Bogenbrücken und der Hängebrücken," 2d ed. Berlin, 1888.

distributed dead load and also a uniformly distributed live load on a portion of the span.

To select the most unfavorable load distribution, Godard recommended[1] the use of influence lines in a somewhat restricted sense (restricted by the fact that the principle of superposition cannot be applied rigorously). The writer applied[2] trigonometric series to the analysis of the bending suffered by stiffening trusses. By using this method, the effect of a concentrated force on the bending of stiffening trusses can be investigated as required in the construction of influence lines. The series method has been extended by H. H. Bleich,[3] who used it for several kinds of end conditions of stiffening trusses and by E. Steurman,[4] who applied it to a stiffening truss of variable cross section. As demonstrated by F. Stüssi,[5] finite difference equations can be used in analyzing bending of stiffening trusses of variable cross section. An investigation of suspension bridges with continuous three span stiffening trusses was made by the writer in collaboration with S. Way.[6]

The theory of suspension bridges continues to attract the interest of engineers and recently several important publications have appeared in this field.[7] The problem of self-induced vibration of suspension bridges has become very important, and several papers on the subject have been published recently. Mention should be made here of the report on the failure of the Tacoma Narrows Bridge by O. H. Amman, Theodore von Kármán, L. G. Dunn, and G. B. Woodruff, the paper by H. Reissner,[8] and that of K. Klöppel and K. H. Lie.[9]

89. Stresses in Railway Tracks

It appears that the analysis of stresses produced in rails by moving loads has attracted engineer's attention ever since the first railroads were

[1] *Ann. ponts et chaussées*, vol. 8, pp. 105–189, 1894.
[2] *Proc. ASCE*, vol. 54, p. 1464, 1928. The use of trigonometric series in analyzing bending of stiffening trusses was shown independently by Martin, *Engineering*, vol. 125, p. 1, 1928. A comparison of the series method with that of Melan was made by A. A. Jakkula, *Pub. Intern. Assoc. Bridge Structural Eng.*, vol. 4, p. 333, 1936, Zürich.
[3] See his book "Die Berechnung verankerter Hängebrücken," Vienna, 1935.
[4] *Trans. ASCE*, vol. 94, p. 377, 1930.
[5] *Publ. Intern. Assoc. Bridge Structural Eng.*, vol. 4, p. 531, 1936. See also his articles in *Schweiz. Bauztg.*, vol. 116, 1940; vol. 117, 1941.
[6] *Publ. Intern. Assoc. Bridge Structural Eng.*, vol. 2, p. 452, 1934.
[7] Reference should be made to the following papers: R. I. Atkinson and R. V. Southwell, *J. Inst. Civil Engrs.* (*London*), 1939, pp. 289–312; K. Klöppel and K. H. Lie, *Forschungshefte Stahlbau*, no. 5, Berlin, 1942; H. Granholm, *Trans. Chalmers Univ. Technol., Göthenburg*, 1943; S. O. Asplund, *Proc. Roy. Swed. Inst. Eng. Research*, nr. 184, Stockholm, 1945; A. Selberg, "Design of Suspension Bridges," Trondheim, 1946; R. Gran Olsson, *Kgl. Norske Videnskab. Selskabs.*, vol. 17, no. 2.
[8] *J. Applied Mechanics*, vol. 10, pp. A23–A32, 1943.
[9] *Ing.-Arch.*, vol. 13, pp. 211–266, 1942.

built. Barlow (see page 100) investigated the bending strength of rails of various cross sections by considering them as beams on two supports. Thus, for a load P and distance l between the supports, the maximum bending moment is $0.250Pl$. Winkler[1] took the rail as a continuous beam on rigid supports and found the value $0.189Pl$ for the maximum bending moment.

Further progress in the subject was made by H. Zimmermann,[2] who prepared tables for simplifying the Winkler analysis of a beam on an elastic foundation and applied the theory in calculating the deflection of ties. The rail that he considers is a continuous beam on elastic supports. With conventional distances between the ties, the vertical rail load is distributed over several ties so that the isolated elastic supports can be replaced by an equivalent continuous elastic foundation. Thus the theory of beams on elastic foundations can be applied in analyzing stresses and deflections of rails.[3] Denoting the modulus of the foundation by K, the differential equation of the deflection curve (Fig. 243) becomes

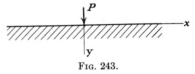

Fig. 243.

$$EI \frac{d^4y}{dx^4} = -Ky \quad (a)$$

Considering the rail as an infinitely long beam and integrating Eq. (a), we find that the maximum bending moment is

$$M_{max} = \frac{P}{4} \sqrt[4]{\frac{4EI}{K}} \quad (b)$$

and that corresponding maximum stress is

$$\sigma_{max} = \frac{M_{max}}{S} = \frac{P}{4} \cdot \frac{\sqrt[4]{I}}{S} \sqrt[4]{\frac{4E}{K}} \quad (c)$$

where S denotes the section modulus of the rail. The second factor in this formula has the dimensions of in.$^{-2}$ and it can be concluded that, for

[1] E. Winkler, "Vorträge über Eisenbahnbau," 3d ed., Prague, 1875.

[2] H. Zimmermann, "Die Berechnung des Eisenbahn-Oberbaues," Berlin, 1888; 2d ed., Berlin, 1930. See also his article "Berechnung des Oberbaues" in *Handbuch Ingenieurwiss.*, vol. 2, pp. 1–68, 1906.

[3] This simplified theory was extended by the writer for the Russian railroad system. See *Mem. Inst. Engrs. Ways of Communication*, St. Petersburg, 1915. It was used on the Polish railroad system by A. Wasiutyński; see *Ann. acad. sci. tech. Varsovie*, vol. 4, pp. 1–136, 1937. In Germany it was used by Dr. Saller; see *Organ Fortschritte Eisenbahnw.*, vol. 87, p. 14, 1932. See also the paper by E. Czitary in the same periodical, vol. 91, 1936.

geometrically similar cross sections of rails and for constant values of the quantity E/K, the value of σ_{max} remains unchanged if the load P increases in the same proportion as the weight of the rail per unit length.

Using the theory of beams on an elastic foundation, the bending-moment diagram can be readily obtained for a system of loads acting on the rail, by superposition. For example, Fig. 244 represents the bending-moment diagram for four equal loads with $I = 44$ in.[4] and $K = 1,500$ lb per in.[2] The value of M_{max} for an isolated load P is taken from Eq. (b) as unity for the moment scale. It is seen from the figure that the maximum moment occurs under the first and the last loads and that it is equal to 75 per cent of the moment produced by an isolated load P. Between the loads, the rail is bent convex upward and the maximum numerical value of the corresponding bending moment is about 25 per

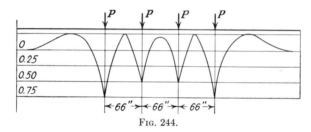

Fig. 244.

cent of that given by Eq. (b). From this it follows that, during the passage of a locomotive, the stresses in a cross section of a rail change in magnitude and sign several times. This explains fatigue failures of rails.

A considerable amount of experimental work has been done in measuring railroad-track deformation under moving loads. But earlier investigations, in which mechanical devices were used for the purpose, were not reliable. A. Wasiutyński devised an optical method and succeeded in getting photographic records of bending strains and deflections in a rail under the wheels of a moving locomotive.[1]

An extensive investigation of track stresses was carried out in the United States by the Westinghouse Electric Corporation. In these tests magnetic strain gauges were used for strain measurements.[2] These

---

[1] A complete description of these important tests is given in Wasiutyński's dissertation presented at the Institute of Engineers of Ways of Communication, St. Petersburg, 1899. See also *Verhandl. Internat. Eisenbahn Kong.*, 1895–1898, and Wasiutyński's book "Beobachtungen über die elastischen Formänderungen des Eisenbahngleises," Wiesbaden, 1899.

[2] The original gauge was invented by J. G. Ritter. Its further development is due to B. F. Langer and J. P. Shamberger, Westinghouse engineers, who made track stress experiments. A complete description of the instrument is given by B. F. Langer in "Handbook of Experimental Stress Analysis," p. 238.

experiments showed that the theory of beams on elastic foundations is accurate enough for this purpose. They also showed that lateral forces as well as vertical loads produce bending and also torsion of rails. These effects cannot be neglected.[1] To select the best method of calculating those forces from field tests, preliminary laboratory experiments were conducted on bending of an elastically supported rail and local stresses at the point of application of a concentrated load were studied.[2] The proper location of strain gauges in field tests was established by this investigation.

Field experiments showed the great influence of the dynamical factor on track stresses produced by moving wheels. Wasiutyński, in the previously mentioned dissertation, indicates that some freight-car wheels with flat spots produce larger deflections of rails than heavier locomotive wheels with smooth surfaces. A theoretical investigation, apparently the first of the dynamical effect of flat spots on wheels and of low spots on rails, was made by N. P. Petrov,[3] the originator of the hydrodynamical theory of friction in bearings. Neglecting the mass of the rail in his derivation, and considering it as a beam on equidistant elastic supports, he derives the differential equation, similar to that of Willis' (see page 175). An arithmetical step-by-step method is used for the integration of that equation. In this way the pressure of the wheel on the rail is calculated by taking into consideration not only the bending of rails but also elastic vertical movements of the supports and vertical movements of the wheel due to low spots.

A simplification of this analysis was accomplished by considering a beam on an elastic foundation instead of a beam on isolated elastic supports.[4] In this way, the dynamical effect of low spots on the deflection of rails can be readily analyzed. Assume, for example, that the profile of the low spot (of length l and of depth δ) is given by the equation

$$y = \frac{\delta}{2}\left(1 - \cos\frac{2\pi x}{l}\right) \qquad (d)$$

It can be shown that the additional dynamic deflection, due to the low spot, is proportional to δ and depends upon the magnitude of the ratio

[1] See paper by S. Timoshenko, *Proc. 2d Intern. Cong. Applied Mechanics*, Zürich, 1926, p. 407.
[2] See the paper by S. Timoshenko and B. F. Langer, *Trans. ASME*, vol. 54, p. 277, 1932.
[3] See the *Bull. Russian Imp. Tech. Soc.*, 1903, and his subsequent publications in the *J. Depart. Ways of Communication*.
[4] See S. Timoshenko's paper mentioned on p. 431; see also his paper in *Génie civil*, 1921, p. 551, Paris. Some further study of dynamic action of the wheel on the rail is given by B. K. Hovey in his doctoral dissertation, Göttingen, 1933.

T_1/T, in which T is the period of vertical vibration of the wheel, considering the rail as a spring, and T_1 is the time it takes the wheel to cross the low spot. The maximum additional deflection, equal to 1.47δ, occurs at the speed corresponding to $T_1/T = \frac{2}{3}$. Thus it can be concluded that the additional dynamic pressure, due to the low spot, is about equal to the load which produces a static deflection of the rail equal to 1.5δ. We see that a comparatively small low spot produces a very appreciable dynamic effect at certain speeds. In this analysis, the mass of the rail was neglected in comparison with the mass of the wheel. This is permissible if the time T_1 is long in comparison with the period of vibration of the rail on its elastic foundation.

In the derivation of Eq. (a), it is assumed that the deflection at any cross section of the beam is proportional to the pressure on the foundation at that cross section. But if we consider the foundation as a semi-infinite elastic body, the deflection at any cross section of the beam is a function of the pressure distribution along the length of the beam and the problem of deflections becomes more involved. This more elaborate investigation was made by M. A. Biot[1] and by K. Marguerre.[2] It was shown, by the latter, that the more accurate theory indicates a maximum bending stress in the beam that is about 20 per cent higher than that which follows from the elementary formula.

90. Theory of Ship Structures

Great progress has been made in the application of stress analysis to the structural design of ships during the twentieth century; this is particularly true of men-of-war. Designers have been confronted with many new problems owing to the rapid increase in size of these ships, the tendency to minimize hull weight so as to allow the installation of heavier guns and armor protection, increased speeds, etc. To meet these demands, they have turned to theoretical analysis.

Thomas Young considered a ship as a girder and suggested the construction of curves of buoyancy and weight. The difference between the ordinates of these curves was the load acting on the "ship girder." This analysis of the longitudinal strength of ships was generally accepted and its accuracy was checked by direct tests. Experiments made on the bending of the destroyer *Wolf*,[3] and later on the destroyers *Preston* and *Bruce*[4] showed that the measured deflections and stresses could be made to agree

---

[1] *J. Applied Mechanics*, vol. 4, pp. 1–7, 1937.

[2] *ZAMM*, vol. 17, pp. 224–231, 1937.

[3] See J. H. Biles, *Trans. Inst. Naval Architects.*, vol. 1, 1905; G. H. Hoffmann, *Trans. Inst. Naval Architects*, 1925.

[4] See W. Hovgaard's paper, *Trans. Soc. Naval Architects Marine Engrs.*, 1931, p. 25; see also the paper presented at a meeting of the Society of Naval Architects and Marine Engineers, November, 1931, by C. O. Kell.

with the beam theory if, in calculating the flexural rigidities of the ships, the fact is taken into consideration that the effectiveness of some plates is reduced by buckling. Failure in the *Preston* and *Bruce* took place by buckling of compressed plates and stringers.

Experience has showed that, in addition to general considerations of bending strength of ship girders, local stresses at points of abrupt changes of cross section should be investigated and necessary reinforcements made. We have such a condition where a deckhouse is built on a structurally important (or *strength*) deck. There are many cases on record where cracks have developed in the deck, and in the wall of the deckhouse, at the corners. We have another important instance of stress concentration at the corners of hatches in a strength deck. There have been cases where cracks have started at these points and spread progressively through the deck plating, so that the safety of the ship was seriously endangered.[1] The stress concentrations produced by the holes in decks were investigated by C. E. Inglis[2] and the reinforcement of circular holes has been studied by the writer.[3]

The problem of transverse strength of ships has also been studied by naval architects and various methods of analyzing the deformation of frames have been developed. This problem is especially important in torpedo vessels and submarines. The frames of these vessels resemble closed rings, and the curved-bar theory and methods used in theory of arches have been applied[4] for their analysis.

Fig. 245. A. N. Krylov.

The work of A. N. Krylov (1863–1945) has had an important influence on the development of rational methods of stress analysis and on their application in structural design of ships. A brief outline will now be given of the principal achievements of this great engineer and scientist.[5]

---

[1] The case of the S.S. *Majestic* was described by E. Ellsberg, "Marine Engineering," 1925. A similar case with the *Leviathan* was discussed by J. Lyell Wilson, *Trans. Soc. Naval Architects Marine Engrs.*, 1930.

[2] *Trans. Inst. Naval Architects*, London, 1913.

[3] *J. Franklin Inst.*, vol. 197, p. 505, 1924.

[4] See the paper by M. Marbec, *Bull. assoc. tech. maritime*, 1908.

[5] See "My Reminiscences" by A. N. Krylov, publication of the Russian Academy of Sciences, 1945. See also the obituary of Krylov, *J. Applied Math. Mechanics* (Russian), vol. 10, p. 3, 1946. A list of Krylov's papers and books is added to this obituary.

While he still was in the military marine school, Krylov's great mathematical ability was noticed. He usually spent his free time reading mathematical books and, by the time of his graduation from the school (1884), he already had a broad mathematical background extending far beyond the established school programs. In 1888, after some practical experience in the Russian Navy, Krylov entered the naval academy where he was particularly interested in the theory of ships and in their structural design. As the most outstanding student, Krylov was invited to stay at the school in the capacity of instructor in mathematics, after graduating (1890). By 1891 he had become the lecturer in the theory of ships. As an introduction to this course, the young man gave a series of lectures on approximate calculations and showed how the methods of calculation devised by mathematicians and astronomers could be used to advantage by engineers. Based on these lectures he later prepared and published (1906) a book on approximate calculations. This is still one of the most important books in its field.

Very soon, Krylov became interested in the theory of motion of ships on waves. While the problem of ship rolling had been treated by Froude, Krylov occupied himself with the more complicated question of pitching. In 1896 he successfully solved the problem and published his results in English[1] and in French.[2] Next, he went on to investigate the general motion of a ship on waves and presented his conclusions in a paper, "General Theory of the Oscillations of a Ship on Waves."[3] These publications placed Krylov in the first rank of authorities on the theory of ships and the Society of English Shipbuilding Engineers awarded him the gold medal of the Society, an honor bestowed on no foreigner before him. Working on oscillatory motion of ships, Krylov paid attention to the stresses produced in hulls by the inertia forces and offered a valid method of calculating these stresses.[4] He also made experimental investigations of dynamical stresses on several ships of the Russian navy by using a sensitive extensometer of his own design.

In 1900, A. N. Krylov was put in charge of the model basin of the Russian Navy in St. Petersburg. He reorganized the experimental work with models at that institution, and coordinated that work with the sea testing of newly constructed ships. At the same time, he helped to reorganize the department of ship construction at the Polytechnical Institute in St. Petersburg and prepared a course of lectures on ship vibration for that school. This was published later in book form.[5]

[1] A new theory of the pitching motion of ships on waves . . . , *Trans. Inst. Naval Architects*, 1896, pp. 326–359, London.
[2] *Bull. assoc. tech. maritime*, vol. 8, pp. 1–31, 1897.
[3] *Trans. Inst. Naval Architects*, vol. 40, pp. 135–190, 1898, London.
[4] See *Engineering*, 1896, pp. 522–524; *Trans. Inst. Naval Architects*, vol. 40, pp. 197–209, 1898.
[5] "Vibration of Ships" (Russian), 1936.

What were possibly the first investigations of vibration of ships were made by O. Schlick[1] who made a special instrument for recording these motions[2] and determined frequencies of various modes of vibration experimentally. Krylov gives a theoretical analysis of free vibrations of ships in the above-mentioned course. In it, the ship is considered as a beam of variable cross section and Adams' approximate method[3] of integrating ordinary differential equations is employed. About the same time Krylov became interested in vibration of bridges and published the previously mentioned paper (see page 419) on forced vibration of beams produced by moving loads. The method used in this paper was applied later to investigation of longitudinal vibration of cylinders and to the measurement of gas pressure in guns.[4]

In connection with the problem of motion of ships on waves, Krylov studied gyroscopic devices for stabilization purposes[5] and later published a book on gyroscopes.[6]

This scientist wrote the article on theory of ships[7] in the "Encyklopädie der mathematischen Wissenschaften." He did this at the request of Klein (see page 391) who, as we have seen, undertook to collect material for the volume on mechanics.

In 1908, Krylov was put in charge of all ship construction for the Russian Navy. Russia was then making efforts to rebuild her fleet after the severe losses inflicted upon it by Japan in the Russo-Japanese War. Some battleships of a completely new type (dreadnoughts) had just been built in England and the construction of similar ships was put on the Russian program. Many new problems appeared in the design of these new ships and Krylov had ample scope to use scientific methods as far as possible in solving them. His idea of replacing rules of thumb by mathematical analysis proved very successful and, under Krylov's leadership, many structural problems were satisfactorily solved. He participated in this work himself and developed a new method of analysing beams on elastic foundations, which greatly simplified calculations, especially those involving beams of variable cross section.[8] He also made a very complete

[1] See *Trans. Inst. Naval Architects*, 1884, p. 24.
[2] See *Trans. Inst. Naval Architects*, 1893, p. 167.
[3] See the book by F. Bashford, "An Attempt to Test the Theories of Capillary Action with an Explanation of the Method of Integration Employed by J. C. Adams," Cambridge, 1883.
[4] *Bull. Russian Acad. Sci.*, 6th series, 1909, pp. 623–654.
[5] See his paper in *Bull. assoc. tech. maritime*, 1909, pp. 109–139.
[6] "General Theory of Gyroscopes and Some of Their Technical Applications," publication of the Russian Academy of Sciences, 1932, p. 394.
[7] See "Encyklopädie der mathematischen Wissenschaften," vol. 4, part 3, pp. 517–562, 1906.
[8] Details of this investigation were later published in book form, see "Analysis of Beams on Elastic Foundation," a publication of the Russian Academy of Sciences, 2d ed., p. 154, 1931.

analysis of the elastic curve of a buckled strut whose deflections are large.[1]

As a result of his teaching, Krylov published several books on applied mathematics and mechanics. The one on the partial differential equations of engineering[2] attracted so much attention from engineers and physicists, that the first edition was sold out in a few days. His book on the numerical integration of ordinary differential equations was translated into French.[3] In his "Reminiscences,"[4] Krylov states that, after tiring days in his office, he read classical works in astronomy and mathematics for relaxation. In this way, his important work "On Determination of Trajectories of Comets and Planets from a Small Number of Observations" was published.[5] He also undertook the huge task of translating Newton's "Principia" into Russian; more than 200 notes were added by Krylov in this work. In the latter part of his life, Krylov also translated Euler's book on the new theory of motion of the moon.[6] A complete collection of Krylov's works has been edited and published in eight volumes by the Russian Academy of Sciences, 1936–1943.

I. G. Boobnov, who was Krylov's pupil and collaborator and who designed the first Russian dreadnoughts and submarines, made very important contributions to the theory of ship structures. The first to apply the theory of bending of plates in the structural design of ships, he showed that deflections of plates under hydrostatic pressure are not usually small, so that not only bending but also stretching of the middle plane of plates must be considered. He derived the general solution of the problem and also prepared numerical tables to simplify its application. This work commanded wide attention both in Russia and in other countries. An early paper on this subject was translated into English[7] and further results are included in Boobnov's "Theory of Structure of Ships."[8]

The theory of interconnected longitudinal and transverse beams is of great importance in the design of ships, and Boobnov contributed much

[1] *Bull. Russian Acad. Sci.*, 7th series, 1931, pp. 963–1012.
[2] Publication of the Naval Academy, 2d ed., 1913; a publication of the Russian Academy of Sciences, 472 pp., 1932.
[3] "Sur l'intégration numerique approchée des équations differentielles avec application au calcul des trajectoires des projectiles," Paris, 1927.
[4] See p. 217.
[5] See *Publs. Naval Academy*, 1911, pp. 1–161.
[6] See *Publs. Russian Acad. Sci.*, 1934.
[7] *Trans. Inst. Naval Architects*, vol. 44, p. 15, 1902.
[8] The first two volumes of this important book appeared in print in 1912 (St. Petersburg). The third volume, dealing with the applications of the theory of structures in ship design, remained in mimeographed form. See also S. Timoshenko, "Theory of Plates and Shells," pp. 1–33, 1940; and S. Way's paper presented at the National Meeting of Applied Mechanics, June, 1932.

to this theory. Considering a system of parallel equidistant longitudinal beams supported by a crossbeam, Boobnov showed that this support can be treated as a beam on an elastic foundation and prepared tables simplifying the analysis of this beam. Later Boobnov extended his method to the case of several crossbeams.[1]

The analysis of the elastic stability of plates under various kinds of loading and edge conditions was first introduced in ship design[2] in the Russian dreadnoughts. The condition that a battleship is docked on its center keel only presents considerable difficulty regarding the strength and elastic stability of transverse bulkheads. The previously mentioned theory of buckling of stiffened plates (see page 415) was developed, and a series of tests on 15 by 7 ft models made, as a result of this problem. In the analysis of bending of interconnected longitudinal and transverse beams, the Rayleigh-Ritz method was applied[3] and sufficiently accurate solutions were obtained in this way.

All these refined methods of the stress analysis of ship structures are critically discussed and extended by P. F. Papkovich (1887–1946) in his "Theory of Structure of Ships," vol. 1, pp. 1–816 (Theory of Frames and Interconnected Beams), Moscow, 1947, and vol. 2, pp. 1–960 (Theory of Bending and Buckling of Plates and Shells), 1941. These two volumes constitute the most complete and up-to-date treatise on theory of ship structures.

[1] A further study of this kind of problem and a bibliography of this subject can be found in M. Hetényi, "Beams on Elastic Foundation," The University of Michigan Press, 1946. See also the book by J. M. Hlitčijev, "Theory of Elasticity," Beograd, 1950.

[2] In the capacity of a consulting engineer (1912–1917), the writer participated in this work.

[3] See S. Timoshenko, "Strength of Materials," p. 321, 1911; "Theory of Elasticity" (Russian), vol. 2, p. 72, 1916. See also S. Timoshenko, $ZAMM$, vol. 13, p. 153, 1933.

Name Index

Page numbers followed by asterisks denote biographical material

A

Adams, F. D., 362
Adams, J. C., 437
Airy, G. B., 223, 224*, 226, 274
Airy, Wilfrid, 224
Amman, O. H., 430
Amsler, Laffon, 282
Anderson, J., 197
Andrews, E. S., 289
Anthes, H., 393
Arago, 69
Archimedes, 1
Arnold, R. N., 421
Aron, H., 412
Asimont, 321
Asplund, S. O., 430
Atkinson, R. I., 430
Auday, 84
Auerbach, F., 326
Ayre, R. S., 419

B

Babbage, C., 222
Bach, C., 281, 283, 356, 365, 392
Bailey, R. W., 373, 375, 377
Baillet, 115
Bairstow, L., 387
Banks, 102
Barba, 282
Barbré, R., 415
Barlow, P. W., 85, 99, 200, 223, 431
Barton, M. V., 404
Bashford, F., 437
Basset, A. B., 420
Baumann, R., 279
Bauschinger, J., 279, 287, 294, 297, 301, 303, 377, 381
Bautz, W., 398
Bazain, 114
Beare, Thomas H., 197
Becker, M., 178

Belajev, N. M., 350
Belanger, 144, 146
Belelubsky, N. A., 282
Belidor, 41, 60, 64, 72
Beltrami, E., 369
Bendixen, A., 422
Berliner, S, 395
Bernoulli, Daniel, 27*, 32
Bernoulli, Jacob, 25*, 71
Bernoulli, Jacques, 119*
Bernoulli, John, 25*, 27
Bernoulli, Nicolas, 28
Berthollet, 107
Bertot, 145
Bertrand, J. L. F., 17, 80, 214, 230
Bessel, 247
Bétancourt, A., 114*
Betti, E., 320
Bickley, W. G., 407
Biezeno, C. B., 397, 416, 418
Biles, J. H., 434
Biot, M. A., 69, 398, 403, 417, 434
Birkbech, G., 198
Blasius, 39
Bleich, H. H., 430
Blood, W. B., 186
Blumenthal, O., 412
Boas, W., 355, 360, 362, 363
Boistard, L. C., 65, 66
Boker, R., 369
Bolle, L., 412
Boltzmann, L., 252
Boobnov, I. G., 409, 410, 438
Borchardt, C. W., 248, 347
Boscovich, 104
Bossut, C., 50, 69
Boussinesq, J., 181, 229, 239, 328*, 334, 404
Boyle, R., 16, 17
Breguet, 251
Bresse, J. A. C., 144, 145, 146*, 211, 241, 333

Brewster, D., 73, 223, 249
Brolling, 102
Brown, D. M., 31
Brunel, 162, 192
Brunner, J., 183
Bryan, G. H., 299, 414
Buchholtz, H., 387
Buckley, 403
Buffon, 55*
Buckwalter, T. V., 383
Bühler, H., 387
Bülfinger, G. B., 59, 355
Bunsen, 252
Burg, von, 135
Burkhardt, H., 104
Burr, 192
Busby, 17

C

Čališev, K. A., 423
Campbell, L., 268
Campbell, W., 420
Camus, 69
Cardan, 7
Carnot, Lazare, 69
Carnot, S. N. L., 262
Cassini, 17
Castigliano, A., 289*, 311, 316, 320
Cauchy, A., 69, 106, 107*, 115, 142, 220, 233, 248, 262, 328
Chales, 230
Challis, 226
Charpy, 360
Chevandier, E., 219
Chevenard, P., 372
Chezy, 64
Chladni, E. F. F., 119
Chree, C., 344
Chwalla, E., 393, 408, 413, 415
Clapeyron, B. P. E., 84, 114*, 144, 160, 194, 205, 213, 262, 288
Clark, E., 157
Clausen, 39
Clebsch, A., 104, 238, 248, 255*, 266, 297, 316, 389, 406
Coker, E. G., 384
Collignon, E., 144
Colonnetti, G., 289
Condorcet, 28, 30
Considère, 297, 326
Contamin, V., 133
Coriolis, 230, 242
Cornu, 136
Coulomb, C. A., 47*, 61, 70, 71, 100, 141, 287, 325
Cox, H. L., 408

Cox, Homersham, 177, 178
Craggs, J. W., 404
Cranz, H., 398
Cremona, L., 196, 304
Crew, H., 6
Cross, H., 424
Culmann, K., 190*, 426
Cyran, A., 319
Czitary, E., 431

D

D'Alembert, 35
Dalton, J., 127
Danizy, 65
Darby, Abraham, 72
Dartein, de, 42
Darwin, G. H., 265
Davidenkov, N. N., 361, 367, 387
Delambre, J. B. J., 27, 37, 47
Den Hartog, J. P., 397
Desargues, 63
Des Billettes, 44
Descartes, 16
Dickenson, J. H. S., 372
Dietrich, 385
Dji-Djüän Dschou, 416
Dirichlet, 389
Dinnik, A. N., 415, 418, 420
Dischinger, F., 428
Dixon, 269
Donnell, L. H., 416, 417
Dorey, S. F., 382
Dove, 247
Doyn, 186
Dubois, F., 412
Duguet, C., 369
Duhamel, J. M. C., 228, 242*, 347
Duhem, P., 7
Duleau, A., 81, 99, 234
Dunn, L. G., 430
Dupin, F. P. C., 80

E

Eggenschwyler, A., 401
Eichinger, A., 371, 377
Elam, C. F., 355, 362, 363
Elgood, W. N., 410
Ellis, C. A., 31
Ellsberg, E., 435
Engesser, F., 292, 297*, 322
Escher, G., 124, 132
Estanave, E., 333
Ettinghausen, 232
Euler, L., 27, 28*, 98, 119, 120, 288

Name Index

Evans, T. H., 409
Everett, F. L., 375
Ewing, J. A., 364, 378
Eytelwein, J. A., 101*

F

Fahie, J. J., 7
Fairbairn, W., 100, 123*, 156, 157, 166, 192, 223
Faraday, M., 275
Favre, H., 385
Federhofer, K., 410, 415, 420
Ferrers, N. M., 217
Filon, L. N. G., 344, 384, 404–406
Finley, J., 73
Flamant, 181, 229, 239, 240, 332, 333*, 384
Flügge, W., 397, 411, 416
Fontana, 1
Föppl, August, 283, 299*, 308, 349, 361, 369, 381, 382, 392, 397, 403, 404, 411, 419
Föppl, L., 350, 369, 397, 403, 404
Föppl, O., 358, 378, 383
Forbes, 197, 268, 269
Ford, G., 419
Fourier, 142, 242, 246, 260, 262, 328
Frahm, H., 417
Fränkel, W., 322
French, H. J., 380
Fresnel, A., 249
Frézier, 65
Freyssinet, E., 428
Friedrichs, K. O., 417
Fritz, B., 428
Frocht, M. M., 385
Fromm, H., 350, 368
Fuchs, S., 387
Fuss, P. H., 99

G

Gaber, E., 415
Galerkin, B. G., 408, 409, 411
Galileo, 7*, 139
Galton, D., 165, 173
Garnett, W., 268
Gassendi, 16
Gauss, C. F., 120, 389
Gauthey, 57*, 59, 66, 71, 182, 183
Gautier, H., 182
Gay-Lussac, 69
Geckeler, J. W., 401, 412
Gehring, F., 410
Gerber, W., 377
Germain, Sophie, 120*
Gerstner, F. J., 101*, 211

Gesteschi, T., 427
Gibb, A., 73
Gibbons, C. H., 114
Gilbert, D., 85
Girard, P. S., 7, 42, 58, 99
Girkmann, K., 411
Glazebrook, R., 225, 337, 354
Godard, T. 430
Golovin, H., 351, 406
Goodier, J. N., 402, 403, 405, 413
Gordon, Lewis, 209
Gough, H. J., 377, 378, 380, 381
Grammel, R., 397, 418
Granholm, H., 430
Grashof, F., 133*, 146, 238, 409
Green, G., 217*, 262, 288
Gregory, 102
Griffith, A. A., 358, 393
Grubenmann, J. U., 182
Grüneisen, E., 355
Grüning, M., 425
Guest, J. J., 369
Guidi, C., 289

H

Hadji-Argyris, J., 408
Haigh, P. B., 380
Hanson, D., 378
Harcourt, W. V., 223
Haringx, J. A., 415
Harsanyi, Z., 7
Hartmann, L., 287
Haupt, H., 185
Havers, A., 412
Helmholtz, H., 267, 336, 337, 347, 354
Hencky, H., 403, 417
Henneberg, L., 307
Herbert, H., 395
Herschel, J. F., 222
Hertz, H. R., 347*
Hetényi, M., 384, 385, 439
Heun, E., 281
Heydecamp, G. S., 378
Hill, R., 396
Hire, de la, 21
Hlitčijev, J. M., 439
Hodge, P. R., 165
Hodgkinson, 100, 123, 126*, 157, **162**, 209, 223, 277, 356
Hoffmann, G. H., 434
Holl, D. L., 409
Hollister, S. C., 47
Honegger, E., 412
Hooke, R., 12, 16, 17*
Hôpital (see l'Hôpital)
Hopkins, W., 226, 273

Hopkinson, B., 378
Hoppe, R., 420
Horger, O. J., 382, 383
Hort, H., 395
Houbotte, 162
Hovey, B. K., 433
Hovgaard, W., 434
Howe, 184
Howell, C. H., 429
Howland, R. C. J., 407
Huber, M. T., 369, 410
Hugoniot, 181, 239
Hugueny, M. F., 348
Hülsenkamp, F., 326
Humber, W., 183
Humfrey, J. C. W., 378
Hurlbrink, E., 415
Huyghens, 23

I

Inglis, C. E., 407, 419, 435

J

Jacobsen, L. S., 398, 419
Jacoby, C. G. J., 247
Jakkula, A. A., 73, 429
James, H., 165, 173
Jaquith, A. C., 429
Jasinsky, F. S., 295*, 298, 301
Jenkin, C. F., 365, 379
Jenkin, Fleming, 202
Jenny, K., 281
Joffe, A. F., 360
Joukowski, N. E., 306
Jourawski, D. J., 141*, 150, 161, 186, 304

K

Kalakoutzky, N., 386
Kappus, R., 413
Kármán, Theodore von, 369, 395, 397, 401, 406, 411, 416, 417, 430
Karmarsch, K., 135
Kasarnowsky, S., 428
Kaul, H. W., 380
Kell, C. O., 434
Kelland, 92
Keller, H., 412
Kelvin, Lord (William Thomson), 97, 113, 249, 260*, 275, 329, 333, 335, 338, 341, 356
Kick, F., 362
Kimball, A. L., 419
Kirchhoff, G. R., 113, 222, 248, 252*, 256, 266, 332, 345, 347
Kirkaldy, 114, 276*
Kirpitchev, V. L., 321

Kirsch, G., 407
Kist, N. C., 425
Klein, F., 256, 389*, 394
Klöppel, K., 430
Klotter, K., 418
Kojalovich, B. M., 409
Kolossoff, G. V., 407
Kommers, J. B., 377, 379
Köpcke, 146
Korobov, A., 404, 411
Köster, W., 366
Kroon, R. P., 419
Krylov, A. N., 419, 435*
Kupffer, A. T., 220
Kurdjumoff, V. J., 327

L

Lacroix, S. F., 222
Lagerhjelm, P., 102, 114, 281
Lagrange, J., 37*, 68, 107, 111, 120
Lahire, 63
Laisle, F., 145
Lamarle, E., 208
Lamb, H., 340*, 405
Lamblardie, J. E., 58*
Lamé, G., 83, 84, 114*, 194, 205, 213, 233, 282, 288, 328, 329, 368
Langer, B. F., 380, 384, 432, 433
Laplace, P. S., 104, 107, 111, 328
Lardy, P., 428
Larmor, J., 225, 249, 274, 344
Laue, 355
Lechnitski, S., 408
Lee, E. H., 421
Legendre, 120
Leger, A., 1
Leggett, D. M. A., 416
Lehr, 385
Leon, A., 279, 382
Leonardo da Vinci, 2*
Lesage, 64, 65
Lévy, M., 266, 326, 333*
Levy, S., 411
Lewe, V., 409
Lexell, 99
Lie, K. H., 430
Liebmann, H., 399
Lillie, J., 124
Liouville, 262
Lode, W., 370
Lodge, O., 354
Long, S. H., 184, 191
l'Hôpital, 41
Lorenz, R., 415
Lossier, H., 428
Love, A. E. H., 251, 342*, 404, 411, 412

Name Index

Love, G. H., 123
Lüders, W., 135, 287
Ludwik, P., 366, 368
Lyons, H., 16

M

McAdam, D. J., 380
McConnell, J. E., 163
McGregor, C. W., 367
Macneil, J. B., 197
McVetty, P. G., 373, 374
Maillart, R., 401
Malus, 69
Manderla, H., 322
Manjoine, M. J., 366
Marbec, M., 435
Marcus, H., 409
Marguerre, K., 416, 434
Mariotte, E., 17, 21*, 71, 139
Martens, A., 281
Martin, H. M., 430
Mason, H. L., 421
Matschoss, C., 279
Maupertuis, 30
Maxwell, J. C., 202, 228, 249, 268*, 304, 311, 320, 335, 336, 349, 368, 384
Mayer, R., 415
Mayniel, K., 60
Mechel, Chrétien de, 183
Mehmke, R., 356
Mehrtens, G. C., 73, 114, 184
Meier, J. H., 388
Meinesz, V., 397
Meissner, E., 412
Melan, J., 427, 429
Ménabréa, L. F., 289
Merian, P., 25
Mersenne, 16
Mesmer, G., 419
Mesnager, A., 367, 384, 405
Meyer, O. E., 251
Michell, A. G. M., 393
Michell, J. H., 334, 353*, 403, 404, 406, 411, 420
Michon, 186, 194
Mindlin, R. D., 406
Mises, R. von, 369, 389, 396, 415
Möbius, A. F., 304, 308
Mohr, O., 146, 193, 196, 207, 283*, 305, 310–312, 316, 319, 320, 322, 324, 326
Moigno, 104, 108, 232, 269
Molinos, L., 145, 146
Monge, Gaspard, 67, 68*
Montmor, H., 16
Moore, H. F., 377, 379
Morin, A., 89, 123, 126, 163

Morphy, G., 410
Mörsch, E., 427
Mosley, H., 212*, 269
Müller, C. H., 104
Müller-Breslau, H., 289, 306, 309, 310, 327
Muschelišvili, N. I., 398, 407
Musschenbroek, P., 54*

N

Nádái, A., 340, 366, 370, 396, 397, 404, 408, 409
Napier, R., 276
Navier, 57, 69, 70*, 73, 105, 115, 161, 180, 190, 213, 216, 230, 231, 238, 248, 249, 251, 253
Neifert, H. R., 382
Nelson, C. W., 406
Neuber, H., 405
Neumann, Carl, 256
Neumann, F. E., 221, 246*, 347
Neumann, Luise, 246
Newell, J. S., 426
Newman, Francis, 225
Newton, I., 25, 41, 104, 120, 233
Nichol, J. P., 260
Nicol, William, 268
Nicolai, E., 415
Nicolson, J. T., 362
Niles, A. S., 207, 426
Niven, W. D., 268
Nizida, M., 421
Nylander, H., 413

O

Odqvist, F. K. G., 377
Oldfather, W. A., 31
Olsson, Gran R., 430
Orowan, E., 360
Ostenfeld, A., 294
Ostrogradsky, M. V., 112, 142*, 282

P

Palladio, 182
Papkovich, P. E., 404, 418
Parent, 43*
Parsons, W. B., 3
Pascal, 16, 233
Pauker, 327
Pauli, W., 135
Peacock, G., 90, 222
Pearson, K., 342*, 401
Pernetty, 120
Perrodil, 323
Perronet, J. R., 42, 58, 64
Persy, 136

Peterson, R. E., 360, 381, 382
Petrov, N. P., 433
Pettersson, O., 393
Pinet, 67
Phillips, E., 177, 241, 244*
Picard, E., 83, 328
Piobert, 233
Pirlet, J., 319
Plass, H. J., 413
Plucker, 389
Pochhammer, L., 252, 350
Poinsot, L., 69, 115
Poisson, S. D., 69, 104, 111*, 121, 142, 216, 242, 248, 249, 251, 253
Pole, W., 124, 157, 161
Pollard, H. V., 381
Pomp, A., 362
Poncelet, J. V., 63, 84, 85, 87*, 131, 162, 194, 210, 214, 233, 368
Pöschl, T., 241
Pothier, F., 214
Potier, 114
Prager, W., 397
Prandtl, L., 364, 368, 371, 391, 392*, 397
Prechtl, J. J., 102
Prieur, 69
Pronnier, C., 145, 146
Prony, G. C. F. M., 62*, 87
Purser, F., 402

Q

Quinney, H., 371

R

Ramsauer, C., 346
Rankin, A. W., 404
Rankine, W. J. M., 163, 197*, 238, 326, 332, 368
Rayleigh, Lord, 97, 98, 225, 228, 320, 334*, 399
Réaumur, 54
Rebhann, G., 144, 188, 326
Redtenbacher, F., 132*
Regnault, H. V., 219, 262
Reissner, E., 401, 408
Reissner, H., 406, 412, 430
Renaudot, 177
Rennie, J., 124
Ribière, C., 405
Richardson, L. F., 399
Rider, 193
Riemann, 389
Ritter, A., 189, 304
Ritter, J. G., 432
Ritter, M., 428
Ritter, W., 426, 429
Ritz, W., 339, 399, 409, 420

Roberval, 16, 23
Robison, J., 18, 98, 223
Römer, 17
Rondelet, 58, 66
Roś, M., 371, 377
Rosenhain, W., 364
Rouse Ball, W. W., 222
Routh, 273, 334
Rowett, 358
Runge, C., 391, 398, 404

S

Saalschütz, L., 248, 347
Sachs, G., 367, 371, 387
Sadowsky, M. A., 405, 407
Saller, 431
Saint-Venant, Barré de, 89, 104, 108, 109, 131, 135, 179, 216, 228, 229*, 248, 249, 252, 256, 257, 332, 334, 343, 353, 368, 395
Salvio, A, 6
Savart, 88, 347
Saverin, M. M., 350
Saviotti, C., 308
Schallenkamp, A., 419
Scheffler, H., 134, 146, 155, 210, 213
Scheu, 366
Schimmack, 389
Schlick, O., 437
Schmid, E., 355, 360, 362, 363
Schnuse, 89
Schübler, A., 145
Schüle, W., 356
Schultz, F. H., 387
Schur, 308
Schwedler, J. W., 189, 304, 309
Schwerin, E., 416
Sears, J. E., 346, 429
Sébert, 181, 239
Seebeck, 249
Seegar, M., 401
Séguin, M., 83
Selberg, A., 430
Shamberger, J. P., 432
Shaw, 337
Siebel, E., 362
Siemens, W., 354
Simon, H. T., 391
Smeaton, John, 72
Smekal, A., 358
Smiles, 73, 124
Smith, G. V., 372
Sobko, P. I., 282*
Soderberg, C. R., 377
Sokolovsky, W. W., 396
Sopwich, D. C., 380

Name Index

Sörensen, E., 419
Southwell, R. V., 341, 344, 399, 403, 405, 415, 430
Stark, J., 354
Steinhardt, O., 297
Stephenson, G., 156
Stephenson, R., 156*, 164, 165
Sternberg, E., 405
Steuerman, E., 430
Stodola, A., 355, 412, 418-420
Stokes, G. G., 176, 224, 225*, 269, 275, 335, 352
Strassner, A., 427
Strehlke, 254
Sturm, 262
Stüssi, F., 430
Styffe, K., 102, 282

T

Tabor, D., 420
Tabor, P., 7
Tait, P. G., 197, 265, 273, 336
Tapsel, H. J., 372
Taylor, G. I., 344, 364, 371, 393
Taylor, W. P., 202, 304
Tedone, O., 401
Telford, 73, 85, 100
Tetmajer, L., 281, 283, 297
Thompson, S. P., 260
Thomson, James, 260
Thum, A., 398
Timpe, A., 104, 393, 401, 402, 406
Tipper, C. F., 368
Todhunter, I., 223
Torricelli, 15
Town, 184
Tranter, C. J., 404
Tredgold, Thomas, 100
Trefftz, E., 299, 400, 401, 403, 416, 417
Tresca, H., 232, 242
Tsien, 417
Tuzi, Z., 385, 421
Tyndall, 337

V

Valson, C. A., 107
Van der Neut, 416
Van Swinden, 48
Varignon, 45, 194
Venske, O., 406
Vianello, L., 299, 418
Vicat, 83, 287
Villarceau, Y., 200, 214*
Vinci, Leonardo da, 2*
Vint, J., 410
Vitruvius, 1
Viviani, 15

Vlasov, V. Z., 402, 413
Voigt, W., 241, 246, 248, 249, 344*, 394

W

Wagner, H., 413, 416
Wagstaff, J. E. P., 346
Wahl, A. M., 381
Waller, M. D., 420
Wallis, J., 15
Walter, Gaspar, 182
Wangerin, A., 246, 248, 347
Ward, P. E., 418
Warren, 186
Wasiutyński, A., 431-433
Watt, James, 123
Way, S., 410, 430, 438
Weale, J., 160, 183, 212
Weber, C., 401-404, 407
Weibel, E. E., 397, 398
Weibull, W., 360
Weisbach, J., 123, 131*, 133
Weiss, E. C., 247
Werder, L., 135, 279
Wertheim, G., 219*, 345
Westergaard, H. M., 293, 410, 429
Whewell, W., 223*
Whipple, S., 184, 193, 304
Wiebeking, 114, 183
Wiedemann, G., 301
Wieghardt, K., 398
Willers, F. A., 382, 403
Williams, G. T., 378
Williot, 314
Willis, R., 99, 173, 174, 269
Wilson, Carus, 352
Wilson, J. Lyell, 435
Winkler, E., 134, 151, 152*, 196, 238, 285, 310, 313, 316, 322, 323, 326, 431
Wöhler, A., 167*, 276, 303, 377
Woinowsky-Krieger, S., 409, 411
Wolf, K., 402
Woodruff, G. B., 430
Wren, C., 16
Wyss, T., 132

Y

Young, D., 409, 420
Young, Thomas, 74, 85, 90*, 141, 147, 263, 269, 403, 434

Z

Zener, C., 358, 421
Ziegler, H., 412
Zimmermann, H., 431
Zoelly, R., 416
Zwicky, F., 360

Subject Index

A

Academies of Science, 15
Adiabatic strain, 357
Aelotropy (anisotropy), 106, 408
Aftereffect, elastic, 356
Analogies in stress analysis, 397
Anticlastic surface, 136
Applied mechanics at Göttingen, 389
Arches, center line of, 328
    experiments on, 65, 153, 323
    geostatic, 200
    hydrostatic, 200
    pressure line of, 323
    prestressing of, 428
    theory of, Bress's, 147
        Coulomb's, 65
        Lahire's, 63
        Mosley's, 212
        Villarceau's, 214
Area-moment method, 137
Axes, principal, 109
Axles, railway, 163, 168, 383

B

Ballistic pendulum, 21
Baltimore lectures, 338
Bauschinger effect, 280, 365
Beams, built-up, 82, 143
    cast-iron, 127
    continuous, 77, 144, 151, 161
    dynamic test of, 127, 173
    of equal strength, 14
    not following Hooke's law, 127, 137
    rolling load on, 173
    strength of, 12, 21, 45, 49
Bending moment, 189
    diagram of, 195
Bridges, cast-iron, 72
    deflection of, 322
        dynamical, 173
    early history of, 181
    suspension, 73, 86, 429
Bridges, tubular, 156
Brittle coating, 385
Brittle material, fracture of, 358
Buckling, of columns, 413
    of curved bars, 415
    of I beams, 393
    of plates and shells, 415

C

Cast iron, beams of, 127
    crushing of, 127
Castigliano theorem, 288
Catenary, in arches, 211
    in suspension bridges, 85
Center of shear, 401
Chains links, 153
Circle of stress, 195, 286
Circular arch, 63
Circular ring, 140
Clapeyron's theorem, 118
Coefficients, elastic, 110, 216, 248, 345, 356
Columns, built-up, 298
    design formulas for, 208
    Euler's theory of, 33
    experiments with, 56, 128, 294
    torsional buckling of, 413
Complementary energy, 292
Components of strain, 110
Compression strength of cast iron, 129
Compression test of materials, 57, 128, 362
Concentration of stresses, 358, 382, 405
Constants, elastic (see Coefficients)
Continuous beams, 77, 144, 151, 161
Coordinates, curvilinear, 117
Core of a cross section, 147
Corrosion fatigue, 380
Crane, tubular, 160
Creep of metals, 372
Crushing of cast-iron, 129
Crystalline materials, 364

Crystals, elastic constants of, 248, 345
  lattice structure of, 362
  tests of single, 362
Cup and cone fracture, 368
Curvature in bending of bars, 27
Curved bars, bending of, 34, 77, 147, 152
  of double curvature, 140
Curves, elastic (see Elastic curves)
Curvilinear coordinates, 117
Cylinder, hollow, 116

D

Damping capacity, 378
Deflection of beams, 32, 75
  due, to moving load, 173
  to shear, 89, 201, 270
  produced by impact, 94, 127, 179
Deflection curve as funicular curve, 282
Diagram, of bending moment, 195
  of tensile test, 88
Differences equations in stress analysis, 398
Disk, loaded in its plane, 257, 349, 353, 406
  rotating, 155, 270, 341
Dislocation, 364
Distortion energy, 370
Ductile materials, testing of, 362

E

Earth, stability of loose, 201, 332
École Polytechnique, 67
Elastic after effect, 356
Elastic constants (see Coefficients)
  experimental determination of, 216
Elastic curves, 33
  equation of, 27, 32
Elastic limit, 280
Elastic line (see Elastic curves)
Elastic moduli (see Modulus)
Elasticity, general equations of, 106, 109
Ellipsoid, Lamé's, 115
Endurance limit, 169
Energy, distortion, 370
  potential, 288
  strain, 288
Energy method, 399, 414

F

Fatigue of metals, early work on, 162, 167
  effect of corrosion on, 380

Fatigue of metals, new investigation in, 377
  size effect on, 382
Flexural rigidity, 113
Foundation depth, 326
Fracture, brittle, 368
  ductile, 368
Friction, internal, 221, 357, 378
Fringe method in photo-elasticity, 385
Funicular curve as deflection curve, 282
Funicular polygon, 194

G

Glass, tensile strength of, 358
  tubes and globes of, 126

H

Hardness, 349
Hardening, 364
Heat, 357
Helix, 140
Hooke's law, 20
  generalized, 110, 218
Hysteresis, 356

I

Impact, lateral, on a beam, 94, 127, 179
  longitudinal, on a bar, 93, 179, 180, 239, 252, 345, 420
  of solid spheres, 348
Influence lines, 152, 285
Initial stresses, 171, 251, 365, 386 252, 345, 420
Internal friction, 221, 357, 378
Isothermal strain, 357
Isotropy, 106

J

Joints, rigid, in trusses, 321
  testing of riveted, 125

L

Lateral contraction, calculated, 112
  measured, 220, 222
Light, 249, 271, 351
Limit, elastic, 280
  proportional, 280
Limit design, 424
Limiting stress, 169
Live load, 173

Subject Index

M

Maxwell-Mohr method, 208
Membrane analogy, in bending, 397
    in torsion, 393
Membranes, vibration of, 36
Metals, creep of, 372
    fatigue of (*see* Fatigue of metals)
Modulus, in shear, 216
    static and kinetic, 219, 264
    tangent, 298
    in tension, 74
Molecular forces, 104
Moment-distribution method, 423
Multiconstancy, 218

N

Necking, 367
Neutral axis, 23, 26, 46, 50, 74, 136
Nodal lines, 119

O

Obelisk, erection of, 4
Open sections, thin-walled, 401, 402
Optical stress analysis, 249, 271, 351

P

Photoelasticity, 249, 271, 351
Plane strain, stress, 257, 350, 405
Plasticity, 242, 395, 396
Plate girders, 162
Plates, bending of, 72, 119, 253, 266, 408
    boundary conditions of, 112, 254, 266, 340
    large deflections of, 254, 258, 410
    transverse vibrations of, 119, 254, 420
Potential energy, 288
Principal strain, 110
Principle, of least work, 292
    of Saint-Venant, 139
    of superposition, 149
Prisms, torsion of (*see* Torsion)
Proportional limit, natural, 377

R

Rails, dynamic stresses in, 433
    shapes of cross sections of, 100
    static tests of, 433
Range of stress, 170, 377
Rariconstancy, 218
Reciprocal figures, 203
Reciprocity theorem, 207
Relaxation, 357
Relaxation method, 399
Residual stresses, 171, 251, 365, 386
Resilience, 93, 127
Resistance line in arches, 212
Retaining walls, 60, 201, 326
Rigidity, flexural, of beams, 113
    of plates, 254
    torsional, 237
Rings, circular, 140
Ritz method, 399
Rivets, stresses in, 143
Rollers, 349
Rolling load on beams, 173
Rotating disks, 155, 270, 341

S

Saint-Venant's principle, 139
Schools of engineering, in France, 41, 61
    in Germany, 100, 129
    in Russia, 114
Semi-inverse method, 233
Shafts in torsion, cylindrical, 234
    of variable cross section, 403
Shear center, 401
Shear fracture, 368
Shear modulus, 216
Shear strength, 49
Shearing force, 189
Shearing stress in torsion, 92, 236
    in beams, 142
Shells, buckling of, 415
    theory of, 342, 411
Slip bands, 364
Slope-deflection method, 422
Spheres, impact of, 348
Springs, built-up of laminae, 244
    helical, 140
    spiral, of watches, 40, 245
Stability, elastic, of plates and shells, 413
    of structures, 293, 414
Stiffeners, 159, 161, 414
Strain, adiabatic, 357
    components of, 110
    isothermal, 357
    principal, 110
Strain energy, 288
Strength theories, 287, 346, 368
Stress, analysis of, analogies in, 397
    differences equations in, 398
    circle of, 195, 286
    components of, 109
    concentration of, 358, 382, 405
    principal, 109
    range of, 170, 377
    residual, 171, 251, 365, 386

Stress, secondary, 322
 thermal, 242, 250
Stress function, 225, 353
Strings, vibration of, 243
Struts (*see* Columns)
Superposition, principle of, 149
Suspension bridges, 73, 86, 429

T

Thermal stresses, 242, 250
Thermoelasticity, 263
Tensile test of ductile materials, 362
Tension, eccentric, 95
Testing laboratories, 276, 354
Torsion, experiments on, 52, 82, 220
 Saint-Venant's theory on, 233
 Young's theory on, 92, 402
Torsional vibration, 52, 417
 damping of, 53, 221
Trusses, deflection of, 311
 history of, 181
 statically determinate, 304
 statically indeterminate, 316
Tubes, bursting of, 24
 collapse of, 126, 134, 151
 thick, 154, 161
 thin-walled, 162
Tubular bridges, 156

U

Ultimate load, 74
Ultimate stress, 74

V

Vibration, of bars, lateral, 35, 241, 255, 418
 longitudinal, 88, 178
 torsional, 417
 of disks, 419
 of membranes, 36
 of plates, transverse, 119, 254, 420
 of shells, 339, 340
 of ships, 418
 of spheres, 243, 257
 of strings, 243
 torsional (*see* Torsional vibration)

W

Whirling of shafts, 419
Width, effective, 407, 416
Williot diagram, 314

Y

Yield point, 280, 365
Young's modulus, 92

A CATALOG OF SELECTED
DOVER BOOKS
IN ALL FIELDS OF INTEREST

A CATALOG OF SELECTED DOVER BOOKS IN ALL FIELDS OF INTEREST

DRAWINGS OF REMBRANDT, edited by Seymour Slive. Updated Lippmann, Hofstede de Groot edition, with definitive scholarly apparatus. All portraits, biblical sketches, landscapes, nudes. Oriental figures, classical studies, together with selection of work by followers. 550 illustrations. Total of 630pp. 9⅛ × 12¼.
21485-0, 21486-9 Pa., Two-vol. set $25.00

GHOST AND HORROR STORIES OF AMBROSE BIERCE, Ambrose Bierce. 24 tales vividly imagined, strangely prophetic, and decades ahead of their time in technical skill: "The Damned Thing," "An Inhabitant of Carcosa," "The Eyes of the Panther," "Moxon's Master," and 20 more. 199pp. 5⅜ × 8½. 20767-6 Pa. $3.95

ETHICAL WRITINGS OF MAIMONIDES, Maimonides. Most significant ethical works of great medieval sage, newly translated for utmost precision, readability. Laws Concerning Character Traits, Eight Chapters, more. 192pp. 5⅜ × 8½.
24522-5 Pa. $4.50

THE EXPLORATION OF THE COLORADO RIVER AND ITS CANYONS, J. W. Powell. Full text of Powell's 1,000-mile expedition down the fabled Colorado in 1869. Superb account of terrain, geology, vegetation, Indians, famine, mutiny, treacherous rapids, mighty canyons, during exploration of last unknown part of continental U.S. 400pp. 5⅜ × 8½. 20094-9 Pa. $6.95

HISTORY OF PHILOSOPHY, Julián Marías. Clearest one-volume history on the market. Every major philosopher and dozens of others, to Existentialism and later. 505pp. 5⅜ × 8½. 21739-6 Pa. $8.50

ALL ABOUT LIGHTNING, Martin A. Uman. Highly readable non-technical survey of nature and causes of lightning, thunderstorms, ball lightning, St. Elmo's Fire, much more. Illustrated. 192pp. 5⅜ × 8½. 25237-X Pa. $5.95

SAILING ALONE AROUND THE WORLD, Captain Joshua Slocum. First man to sail around the world, alone, in small boat. One of great feats of seamanship told in delightful manner. 67 illustrations. 294pp. 5⅜ × 8½. 20326-3 Pa. $4.50

LETTERS AND NOTES ON THE MANNERS, CUSTOMS AND CONDITIONS OF THE NORTH AMERICAN INDIANS, George Catlin. Classic account of life among Plains Indians: ceremonies, hunt, warfare, etc. 312 plates. 572pp. of text. 6⅛ × 9¼. 22118-0, 22119-9 Pa. Two-vol. set $15.90

ALASKA: The Harriman Expedition, 1899, John Burroughs, John Muir, et al. Informative, engrossing accounts of two-month, 9,000-mile expedition. Native peoples, wildlife, forests, geography, salmon industry, glaciers, more. Profusely illustrated. 240 black-and-white line drawings. 124 black-and-white photographs. 3 maps. Index. 576pp. 5⅜ × 8½. 25109-8 Pa. $11.95

CATALOG OF DOVER BOOKS

THE BOOK OF BEASTS: Being a Translation from a Latin Bestiary of the Twelfth Century, T. H. White. Wonderful catalog real and fanciful beasts: manticore, griffin, phoenix, amphivius, jaculus, many more. White's witty erudite commentary on scientific, historical aspects. Fascinating glimpse of medieval mind. Illustrated. 296pp. 5⅜ × 8¼. (Available in U.S. only) 24609-4 Pa. $5.95

FRANK LLOYD WRIGHT: ARCHITECTURE AND NATURE With 160 Illustrations, Donald Hoffmann. Profusely illustrated study of influence of nature—especially prairie—on Wright's designs for Fallingwater, Robie House, Guggenheim Museum, other masterpieces. 96pp. 9¼ × 10¾. 25098-9 Pa. $7.95

FRANK LLOYD WRIGHT'S FALLINGWATER, Donald Hoffmann. Wright's famous waterfall house: planning and construction of organic idea. History of site, owners, Wright's personal involvement. Photographs of various stages of building. Preface by Edgar Kaufmann, Jr. 100 illustrations. 112pp. 9¼ × 10. 23671-4 Pa. $7.95

YEARS WITH FRANK LLOYD WRIGHT: Apprentice to Genius, Edgar Tafel. Insightful memoir by a former apprentice presents a revealing portrait of Wright the man, the inspired teacher, the greatest American architect. 372 black-and-white illustrations. Preface. Index. vi + 228pp. 8¼ × 11. 24801-1 Pa. $9.95

THE STORY OF KING ARTHUR AND HIS KNIGHTS, Howard Pyle. Enchanting version of King Arthur fable has delighted generations with imaginative narratives of exciting adventures and unforgettable illustrations by the author. 41 illustrations. xviii + 313pp. 6⅛ × 9¼. 21445-1 Pa. $5.95

THE GODS OF THE EGYPTIANS, E. A. Wallis Budge. Thorough coverage of numerous gods of ancient Egypt by foremost Egyptologist. Information on evolution of cults, rites and gods; the cult of Osiris; the Book of the Dead and its rites; the sacred animals and birds; Heaven and Hell; and more. 956pp. 6⅛ × 9¼. 22055-9, 22056-7 Pa., Two-vol. set $20.00

A THEOLOGICO-POLITICAL TREATISE, Benedict Spinoza. Also contains unfinished *Political Treatise*. Great classic on religious liberty, theory of government on common consent. R. Elwes translation. Total of 421pp. 5⅜ × 8½. 20249-6 Pa. $6.95

INCIDENTS OF TRAVEL IN CENTRAL AMERICA, CHIAPAS, AND YUCATAN, John L. Stephens. Almost single-handed discovery of Maya culture; exploration of ruined cities, monuments, temples; customs of Indians. 115 drawings. 892pp. 5⅜ × 8½. 22404-X, 22405-8 Pa., Two-vol. set $15.90

LOS CAPRICHOS, Francisco Goya. 80 plates of wild, grotesque monsters and caricatures. Prado manuscript included. 183pp. 6⅞ × 9⅞. 22384-1 Pa. $4.95

AUTOBIOGRAPHY: The Story of My Experiments with Truth, Mohandas K. Gandhi. Not hagiography, but Gandhi in his own words. Boyhood, legal studies, purification, the growth of the Satyagraha (nonviolent protest) movement. Critical, inspiring work of the man who freed India. 480pp. 5⅜ × 8½. (Available in U.S. only) 24593-4 Pa. $6.95

ILLUSTRATED DICTIONARY OF HISTORIC ARCHITECTURE, edited by Cyril M. Harris. Extraordinary compendium of clear, concise definitions for over 5,000 important architectural terms complemented by over 2,000 line drawings. Covers full spectrum of architecture from ancient ruins to 20th-century Modernism. Preface. 592pp. 7½ × 9⅝. 24444-X Pa. $14.95

THE NIGHT BEFORE CHRISTMAS, Clement Moore. Full text, and woodcuts from original 1848 book. Also critical, historical material. 19 illustrations. 40pp. 4⅝ × 6. 22797-9 Pa. $2.25

THE LESSON OF JAPANESE ARCHITECTURE: 165 Photographs, Jiro Harada. Memorable gallery of 165 photographs taken in the 1930's of exquisite Japanese homes of the well-to-do and historic buildings. 13 line diagrams. 192pp. 8⅝ × 11¼. 24778-3 Pa. $8.95

THE AUTOBIOGRAPHY OF CHARLES DARWIN AND SELECTED LETTERS, edited by Francis Darwin. The fascinating life of eccentric genius composed of an intimate memoir by Darwin (intended for his children); commentary by his son, Francis; hundreds of fragments from notebooks, journals, papers; and letters to and from Lyell, Hooker, Huxley, Wallace and Henslow. xi + 365pp. 5⅜ × 8. 20479-0 Pa. $5.95

WONDERS OF THE SKY: Observing Rainbows, Comets, Eclipses, the Stars and Other Phenomena, Fred Schaaf. Charming, easy-to-read poetic guide to all manner of celestial events visible to the naked eye. Mock suns, glories, Belt of Venus, more. Illustrated. 299pp. 5¼ × 8¼. 24402-4 Pa. $7.95

BURNHAM'S CELESTIAL HANDBOOK, Robert Burnham, Jr. Thorough guide to the stars beyond our solar system. Exhaustive treatment. Alphabetical by constellation: Andromeda to Cetus in Vol. 1; Chamaeleon to Orion in Vol. 2; and Pavo to Vulpecula in Vol. 3. Hundreds of illustrations. Index in Vol. 3. 2,000pp. 6⅛ × 9¼. 23567-X, 23568-8, 23673-0 Pa., Three-vol. set $36.85

STAR NAMES: Their Lore and Meaning, Richard Hinckley Allen. Fascinating history of names various cultures have given to constellations and literary and folkloristic uses that have been made of stars. Indexes to subjects. Arabic and Greek names. Biblical references. Bibliography. 563pp. 5⅜ × 8½. 21079-0 Pa. $7.95

THIRTY YEARS THAT SHOOK PHYSICS: The Story of Quantum Theory, George Gamow. Lucid, accessible introduction to influential theory of energy and matter. Careful explanations of Dirac's anti-particles, Bohr's model of the atom, much more. 12 plates. Numerous drawings. 240pp. 5⅜ × 8½. 24895-X Pa. $4.95

CHINESE DOMESTIC FURNITURE IN PHOTOGRAPHS AND MEASURED DRAWINGS, Gustav Ecke. A rare volume, now affordably priced for antique collectors, furniture buffs and art historians. Detailed review of styles ranging from early Shang to late Ming. Unabridged republication. 161 black-and-white drawings, photos. Total of 224pp. 8⅜ × 11¼. (Available in U.S. only) 25171-3 Pa. $12.95

VINCENT VAN GOGH: A Biography, Julius Meier-Graefe. Dynamic, penetrating study of artist's life, relationship with brother, Theo, painting techniques, travels, more. Readable, engrossing. 160pp. 5⅜ × 8½. (Available in U.S. only) 25253-1 Pa. $3.95

HOW TO WRITE, Gertrude Stein. Gertrude Stein claimed anyone could understand her unconventional writing—here are clues to help. Fascinating improvisations, language experiments, explanations illuminate Stein's craft and the art of writing. Total of 414pp. 4⅜ × 6⅜. 23144-5 Pa. $5.95

ADVENTURES AT SEA IN THE GREAT AGE OF SAIL: Five Firsthand Narratives, edited by Elliot Snow. Rare true accounts of exploration, whaling, shipwreck, fierce natives, trade, shipboard life, more. 33 illustrations. Introduction. 353pp. 5⅜ × 8½. 25177-2 Pa. $7.95

THE HERBAL OR GENERAL HISTORY OF PLANTS, John Gerard. Classic descriptions of about 2,850 plants—with over 2,700 illustrations—includes Latin and English names, physical descriptions, varieties, time and place of growth, more. 2,706 illustrations. xlv + 1,678pp. 8½ × 12¼. 23147-X Cloth. $75.00

DOROTHY AND THE WIZARD IN OZ, L. Frank Baum. Dorothy and the Wizard visit the center of the Earth, where people are vegetables, glass houses grow and Oz characters reappear. Classic sequel to *Wizard of Oz*. 256pp. 5⅜ × 8. 24714-7 Pa. $4.95

SONGS OF EXPERIENCE: Facsimile Reproduction with 26 Plates in Full Color, William Blake. This facsimile of Blake's original "Illuminated Book" reproduces 26 full-color plates from a rare 1826 edition. Includes "The Tyger," "London," "Holy Thursday," and other immortal poems. 26 color plates. Printed text of poems. 48pp. 5¼ × 7. 24636-1 Pa. $3.50

SONGS OF INNOCENCE, William Blake. The first and most popular of Blake's famous "Illuminated Books," in a facsimile edition reproducing all 31 brightly colored plates. Additional printed text of each poem. 64pp. 5¼ × 7. 22764-2 Pa. $3.50

PRECIOUS STONES, Max Bauer. Classic, thorough study of diamonds, rubies, emeralds, garnets, etc.: physical character, occurrence, properties, use, similar topics. 20 plates, 8 in color. 94 figures. 659pp. 6⅛ × 9¼. 21910-0, 21911-9 Pa., Two-vol. set $14.90

ENCYCLOPEDIA OF VICTORIAN NEEDLEWORK, S. F. A. Caulfeild and Blanche Saward. Full, precise descriptions of stitches, techniques for dozens of needlecrafts—most exhaustive reference of its kind. Over 800 figures. Total of 679pp. 8⅛ × 11. Two volumes. Vol. 1 22800-2 Pa. $10.95
Vol. 2 22801-0 Pa. $10.95

THE MARVELOUS LAND OF OZ, L. Frank Baum. Second Oz book, the Scarecrow and Tin Woodman are back with hero named Tip, Oz magic. 136 illustrations. 287pp. 5⅜ × 8½. 20692-0 Pa. $5.95

WILD FOWL DECOYS, Joel Barber. Basic book on the subject, by foremost authority and collector. Reveals history of decoy making and rigging, place in American culture, different kinds of decoys, how to make them, and how to use them. 140 plates. 156pp. 7⅞ × 10¾. 20011-6 Pa. $7.95

HISTORY OF LACE, Mrs. Bury Palliser. Definitive, profusely illustrated chronicle of lace from earliest times to late 19th century. Laces of Italy, Greece, England, France, Belgium, etc. Landmark of needlework scholarship. 266 illustrations. 672pp. 6⅛ × 9¼. 24742-2 Pa. $14.95

CATALOG OF DOVER BOOKS

ILLUSTRATED GUIDE TO SHAKER FURNITURE, Robert Meader. All furniture and appurtenances, with much on unknown local styles. 235 photos. 146pp. 9 × 12. 22819-3 Pa. $7.95

WHALE SHIPS AND WHALING: A Pictorial Survey, George Francis Dow. Over 200 vintage engravings, drawings, photographs of barks, brigs, cutters, other vessels. Also harpoons, lances, whaling guns, many other artifacts. Comprehensive text by foremost authority. 207 black-and-white illustrations. 288pp. 6 × 9.
24808-9 Pa. $8.95

THE BERTRAMS, Anthony Trollope. Powerful portrayal of blind self-will and thwarted ambition includes one of Trollope's most heartrending love stories. 497pp. 5⅜ × 8½. 25119-5 Pa. $8.95

ADVENTURES WITH A HAND LENS, Richard Headstrom. Clearly written guide to observing and studying flowers and grasses, fish scales, moth and insect wings, egg cases, buds, feathers, seeds, leaf scars, moss, molds, ferns, common crystals, etc.—all with an ordinary, inexpensive magnifying glass. 209 exact line drawings aid in your discoveries. 220pp. 5⅜ × 8½. 23330-8 Pa. $3.95

RODIN ON ART AND ARTISTS, Auguste Rodin. Great sculptor's candid, wide-ranging comments on meaning of art; great artists; relation of sculpture to poetry, painting, music; philosophy of life, more. 76 superb black-and-white illustrations of Rodin's sculpture, drawings and prints. 119pp. 8⅜ × 11¼. 24487-3 Pa. $6.95

FIFTY CLASSIC FRENCH FILMS, 1912–1982: A Pictorial Record, Anthony Slide. Memorable stills from Grand Illusion, Beauty and the Beast, Hiroshima, Mon Amour, many more. Credits, plot synopses, reviews, etc. 160pp. 8¼ × 11.
25256-6 Pa. $11.95

THE PRINCIPLES OF PSYCHOLOGY, William James. Famous long course complete, unabridged. Stream of thought, time perception, memory, experimental methods; great work decades ahead of its time. 94 figures. 1,391pp. 5⅜ × 8½.
20381-6, 20382-4 Pa., Two-vol. set $19.90

BODIES IN A BOOKSHOP, R. T. Campbell. Challenging mystery of blackmail and murder with ingenious plot and superbly drawn characters. In the best tradition of British suspense fiction. 192pp. 5⅜ × 8½. 24720-1 Pa. $3.95

CALLAS: PORTRAIT OF A PRIMA DONNA, George Jellinek. Renowned commentator on the musical scene chronicles incredible career and life of the most controversial, fascinating, influential operatic personality of our time. 64 black-and-white photographs. 416pp. 5⅜ × 8¼. 25047-4 Pa. $7.95

GEOMETRY, RELATIVITY AND THE FOURTH DIMENSION, Rudolph Rucker. Exposition of fourth dimension, concepts of relativity as Flatland characters continue adventures. Popular, easily followed yet accurate, profound. 141 illustrations. 133pp. 5⅜ × 8½. 23400-2 Pa. $3.50

HOUSEHOLD STORIES BY THE BROTHERS GRIMM, with pictures by Walter Crane. 53 classic stories—Rumpelstiltskin, Rapunzel, Hansel and Gretel, the Fisherman and his Wife, Snow White, Tom Thumb, Sleeping Beauty, Cinderella, and so much more—lavishly illustrated with original 19th century drawings. 114 illustrations. x + 269pp. 5⅜ × 8½. 21080-4 Pa. $4.50

CATALOG OF DOVER BOOKS

SUNDIALS, Albert Waugh. Far and away the best, most thorough coverage of ideas, mathematics concerned, types, construction, adjusting anywhere. Over 100 illustrations. 230pp. 5⅜ × 8½. 22947-5 Pa. $4.00

PICTURE HISTORY OF THE NORMANDIE: With 190 Illustrations, Frank O. Braynard. Full story of legendary French ocean liner: Art Deco interiors, design innovations, furnishings, celebrities, maiden voyage, tragic fire, much more. Extensive text. 144pp. 8⅜ × 11¼. 25257-4 Pa. $9.95

THE FIRST AMERICAN COOKBOOK: A Facsimile of "American Cookery," 1796, Amelia Simmons. Facsimile of the first American-written cookbook published in the United States contains authentic recipes for colonial favorites—pumpkin pudding, winter squash pudding, spruce beer, Indian slapjacks, and more. Introductory Essay and Glossary of colonial cooking terms. 80pp. 5⅜ × 8½. 24710-4 Pa. $3.50

101 PUZZLES IN THOUGHT AND LOGIC, C. R. Wylie, Jr. Solve murders and robberies, find out which fishermen are liars, how a blind man could possibly identify a color—purely by your own reasoning! 107pp. 5⅜ × 8½. 20367-0 Pa. $2.00

THE BOOK OF WORLD-FAMOUS MUSIC—CLASSICAL, POPULAR AND FOLK, James J. Fuld. Revised and enlarged republication of landmark work in musico-bibliography. Full information about nearly 1,000 songs and compositions including first lines of music and lyrics. New supplement. Index. 800pp. 5⅜ × 8¼. 24857-7 Pa. $14.95

ANTHROPOLOGY AND MODERN LIFE, Franz Boas. Great anthropologist's classic treatise on race and culture. Introduction by Ruth Bunzel. Only inexpensive paperback edition. 255pp. 5⅜ × 8½. 25245-0 Pa. $5.95

THE TALE OF PETER RABBIT, Beatrix Potter. The inimitable Peter's terrifying adventure in Mr. McGregor's garden, with all 27 wonderful, full-color Potter illustrations. 55pp. 4¼ × 5½. (Available in U.S. only) 22827-4 Pa. $1.75

THREE PROPHETIC SCIENCE FICTION NOVELS, H. G. Wells. *When the Sleeper Wakes, A Story of the Days to Come* and *The Time Machine* (full version). 335pp. 5⅜ × 8½. (Available in U.S. only) 20605-X Pa. $5.95

APICIUS COOKERY AND DINING IN IMPERIAL ROME, edited and translated by Joseph Dommers Vehling. Oldest known cookbook in existence offers readers a clear picture of what foods Romans ate, how they prepared them, etc. 49 illustrations. 301pp. 6⅛ × 9¼. 23563-7 Pa. $6.00

SHAKESPEARE LEXICON AND QUOTATION DICTIONARY, Alexander Schmidt. Full definitions, locations, shades of meaning of every word in plays and poems. More than 50,000 exact quotations. 1,485pp. 6½ × 9¼. 22726-X, 22727-8 Pa., Two-vol. set $27.90

THE WORLD'S GREAT SPEECHES, edited by Lewis Copeland and Lawrence W. Lamm. Vast collection of 278 speeches from Greeks to 1970. Powerful and effective models; unique look at history. 842pp. 5⅜ × 8½. 20468-5 Pa. $10.95

CATALOG OF DOVER BOOKS

THE BLUE FAIRY BOOK, Andrew Lang. The first, most famous collection, with many familiar tales: Little Red Riding Hood, Aladdin and the Wonderful Lamp, Puss in Boots, Sleeping Beauty, Hansel and Gretel, Rumpelstiltskin; 37 in all. 138 illustrations. 390pp. 5⅜ × 8½. 21437-0 Pa. $5.95

THE STORY OF THE CHAMPIONS OF THE ROUND TABLE, Howard Pyle. Sir Launcelot, Sir Tristram and Sir Percival in spirited adventures of love and triumph retold in Pyle's inimitable style. 50 drawings, 31 full-page. xviii + 329pp. 6½ × 9¼. 21883-X Pa. $6.95

AUDUBON AND HIS JOURNALS, Maria Audubon. Unmatched two-volume portrait of the great artist, naturalist and author contains his journals, an excellent biography by his granddaughter, expert annotations by the noted ornithologist, Dr. Elliott Coues, and 37 superb illustrations. Total of 1,200pp. 5⅜ × 8.
Vol. I 25143-8 Pa. $8.95
Vol. II 25144-6 Pa. $8.95

GREAT DINOSAUR HUNTERS AND THEIR DISCOVERIES, Edwin H. Colbert. Fascinating, lavishly illustrated chronicle of dinosaur research, 1820's to 1960. Achievements of Cope, Marsh, Brown, Buckland, Mantell, Huxley, many others. 384pp. 5¼ × 8¼. 24701-5 Pa. $6.95

THE TASTEMAKERS, Russell Lynes. Informal, illustrated social history of American taste 1850's–1950's. First popularized categories Highbrow, Lowbrow, Middlebrow. 129 illustrations. New (1979) afterword. 384pp. 6 × 9.
23993-4 Pa. $6.95

DOUBLE CROSS PURPOSES, Ronald A. Knox. A treasure hunt in the Scottish Highlands, an old map, unidentified corpse, surprise discoveries keep reader guessing in this cleverly intricate tale of financial skullduggery. 2 black-and-white maps. 320pp. 5⅜ × 8½. (Available in U.S. only) 25032-6 Pa. $5.95

AUTHENTIC VICTORIAN DECORATION AND ORNAMENTATION IN FULL COLOR: 46 Plates from "Studies in Design," Christopher Dresser. Superb full-color lithographs reproduced from rare original portfolio of a major Victorian designer. 48pp. 9¼ × 12¼. 25083-0 Pa. $7.95

PRIMITIVE ART, Franz Boas. Remains the best text ever prepared on subject, thoroughly discussing Indian, African, Asian, Australian, and, especially, Northern American primitive art. Over 950 illustrations show ceramics, masks, totem poles, weapons, textiles, paintings, much more. 376pp. 5⅜ × 8. 20025-6 Pa. $6.95

SIDELIGHTS ON RELATIVITY, Albert Einstein. Unabridged republication of two lectures delivered by the great physicist in 1920-21. *Ether and Relativity* and *Geometry and Experience*. Elegant ideas in non-mathematical form, accessible to intelligent layman. vi + 56pp. 5⅜ × 8½. 24511-X Pa. $2.95

THE WIT AND HUMOR OF OSCAR WILDE, edited by Alvin Redman. More than 1,000 ripostes, paradoxes, wisecracks: Work is the curse of the drinking classes, I can resist everything except temptation, etc. 258pp. 5⅜ × 8½. 20602-5 Pa. $3.95

ADVENTURES WITH A MICROSCOPE, Richard Headstrom. 59 adventures with clothing fibers, protozoa, ferns and lichens, roots and leaves, much more. 142 illustrations. 232pp. 5⅜ × 8½. 23471-1 Pa. $3.95

CATALOG OF DOVER BOOKS

PLANTS OF THE BIBLE, Harold N. Moldenke and Alma L. Moldenke. Standard reference to all 230 plants mentioned in Scriptures. Latin name, biblical reference, uses, modern identity, much more. Unsurpassed encyclopedic resource for scholars, botanists, nature lovers, students of Bible. Bibliography. Indexes. 123 black-and-white illustrations. 384pp. 6 × 9. 25069-5 Pa. $8.95

FAMOUS AMERICAN WOMEN: A Biographical Dictionary from Colonial Times to the Present, Robert McHenry, ed. From Pocahontas to Rosa Parks, 1,035 distinguished American women documented in separate biographical entries. Accurate, up-to-date data, numerous categories, spans 400 years. Indices. 493pp. 6½ × 9¼. 24523-3 Pa. $9.95

THE FABULOUS INTERIORS OF THE GREAT OCEAN LINERS IN HISTORIC PHOTOGRAPHS, William H. Miller, Jr. Some 200 superb photographs capture exquisite interiors of world's great "floating palaces"—1890's to 1980's: Titanic, Ile de France, Queen Elizabeth, United States, Europa, more. Approx. 200 black-and-white photographs. Captions. Text. Introduction. 160pp. 8⅜ × 11¼. 24756-2 Pa. $9.95

THE GREAT LUXURY LINERS, 1927-1954: A Photographic Record, William H. Miller, Jr. Nostalgic tribute to heyday of ocean liners. 186 photos of Ile de France, Normandie, Leviathan, Queen Elizabeth, United States, many others. Interior and exterior views. Introduction. Captions. 160pp. 9 × 12. 24056-8 Pa. $9.95

A NATURAL HISTORY OF THE DUCKS, John Charles Phillips. Great landmark of ornithology offers complete detailed coverage of nearly 200 species and subspecies of ducks: gadwall, sheldrake, merganser, pintail, many more. 74 full-color plates, 102 black-and-white. Bibliography. Total of 1,920pp. 8⅜ × 11¼. 25141-1, 25142-X Cloth. Two-vol. set $100.00

THE SEAWEED HANDBOOK: An Illustrated Guide to Seaweeds from North Carolina to Canada, Thomas F. Lee. Concise reference covers 78 species. Scientific and common names, habitat, distribution, more. Finding keys for easy identification. 224pp. 5⅜ × 8½. 25215-9 Pa. $5.95

THE TEN BOOKS OF ARCHITECTURE: The 1755 Leoni Edition, Leon Battista Alberti. Rare classic helped introduce the glories of ancient architecture to the Renaissance. 68 black-and-white plates. 336pp. 8⅜ × 11¼. 25239-6 Pa. $14.95

MISS MACKENZIE, Anthony Trollope. Minor masterpieces by Victorian master unmasks many truths about life in 19th-century England. First inexpensive edition in years. 392pp. 5⅜ × 8½. 25201-9 Pa. $7.95

THE RIME OF THE ANCIENT MARINER, Gustave Doré, Samuel Taylor Coleridge. Dramatic engravings considered by many to be his greatest work. The terrifying space of the open sea, the storms and whirlpools of an unknown ocean, the ice of Antarctica, more—all rendered in a powerful, chilling manner. Full text. 38 plates. 77pp. 9¼ × 12. 22305-1 Pa. $4.95

THE EXPEDITIONS OF ZEBULON MONTGOMERY PIKE, Zebulon Montgomery Pike. Fascinating first-hand accounts (1805-6) of exploration of Mississippi River, Indian wars, capture by Spanish dragoons, much more. 1,088pp. 5⅜ × 8½. 25254-X, 25255-8 Pa. Two-vol. set $23.90

CATALOG OF DOVER BOOKS

A CONCISE HISTORY OF PHOTOGRAPHY: Third Revised Edition, Helmut Gernsheim. Best one-volume history—camera obscura, photochemistry, daguerreotypes, evolution of cameras, film, more. Also artistic aspects—landscape, portraits, fine art, etc. 281 black-and-white photographs. 26 in color. 176pp. 8⅜ × 11¼. 25128-4 Pa. $12.95

THE DORÉ BIBLE ILLUSTRATIONS, Gustave Doré. 241 detailed plates from the Bible: the Creation scenes, Adam and Eve, Flood, Babylon, battle sequences, life of Jesus, etc. Each plate is accompanied by the verses from the King James version of the Bible. 241pp. 9 × 12. 23004-X Pa. $8.95

HUGGER-MUGGER IN THE LOUVRE, Elliot Paul. Second Homer Evans mystery-comedy. Theft at the Louvre involves sleuth in hilarious, madcap caper. "A knockout."—Books. 336pp. 5⅜ × 8½. 25185-3 Pa. $5.95

FLATLAND, E. A. Abbott. Intriguing and enormously popular science-fiction classic explores the complexities of trying to survive as a two-dimensional being in a three-dimensional world. Amusingly illustrated by the author. 16 illustrations. 103pp. 5⅜ × 8½. 20001-9 Pa. $2.00

THE HISTORY OF THE LEWIS AND CLARK EXPEDITION, Meriwether Lewis and William Clark, edited by Elliott Coues. Classic edition of Lewis and Clark's day-by-day journals that later became the basis for U.S. claims to Oregon and the West. Accurate and invaluable geographical, botanical, biological, meteorological and anthropological material. Total of 1,508pp. 5⅜ × 8½.
21268-8, 21269-6, 21270-X Pa. Three-vol. set $25.50

LANGUAGE, TRUTH AND LOGIC, Alfred J. Ayer. Famous, clear introduction to Vienna, Cambridge schools of Logical Positivism. Role of philosophy, elimination of metaphysics, nature of analysis, etc. 160pp. 5⅜ × 8½. (Available in U.S. and Canada only) 20010-8 Pa. $2.95

MATHEMATICS FOR THE NONMATHEMATICIAN, Morris Kline. Detailed, college-level treatment of mathematics in cultural and historical context, with numerous exercises. For liberal arts students. Preface. Recommended Reading Lists. Tables. Index. Numerous black-and-white figures. xvi + 641pp. 5⅜ × 8½.
24823-2 Pa. $11.95

28 SCIENCE FICTION STORIES, H. G. Wells. Novels, *Star Begotten* and *Men Like Gods*, plus 26 short stories: "Empire of the Ants," "A Story of the Stone Age," "The Stolen Bacillus," "In the Abyss," etc. 915pp. 5⅜ × 8½. (Available in U.S. only)
20265-8 Cloth. $10.95

HANDBOOK OF PICTORIAL SYMBOLS, Rudolph Modley. 3,250 signs and symbols, many systems in full; official or heavy commercial use. Arranged by subject. Most in Pictorial Archive series. 143pp. 8⅛ × 11. 23357-X Pa. $5.95

INCIDENTS OF TRAVEL IN YUCATAN, John L. Stephens. Classic (1843) exploration of jungles of Yucatan, looking for evidences of Maya civilization. Travel adventures, Mexican and Indian culture, etc. Total of 669pp. 5⅜ × 8½.
20926-1, 20927-X Pa., Two-vol. set $9.90

CATALOG OF DOVER BOOKS

DEGAS: An Intimate Portrait, Ambroise Vollard. Charming, anecdotal memoir by famous art dealer of one of the greatest 19th-century French painters. 14 black-and-white illustrations. Introduction by Harold L. Van Doren. 96pp. 5⅜ × 8½.
25131-4 Pa. $3.95

PERSONAL NARRATIVE OF A PILGRIMAGE TO ALMANDINAH AND MECCAH, Richard Burton. Great travel classic by remarkably colorful personality. Burton, disguised as a Moroccan, visited sacred shrines of Islam, narrowly escaping death. 47 illustrations. 959pp. 5⅜ × 8½. 21217-3, 21218-1 Pa., Two-vol. set $17.90

PHRASE AND WORD ORIGINS, A. H. Holt. Entertaining, reliable, modern study of more than 1,200 colorful words, phrases, origins and histories. Much unexpected information. 254pp. 5⅜ × 8½.
20758-7 Pa. $4.95

THE RED THUMB MARK, R. Austin Freeman. In this first Dr. Thorndyke case, the great scientific detective draws fascinating conclusions from the nature of a single fingerprint. Exciting story, authentic science. 320pp. 5⅜ × 8½. (Available in U.S. only)
25210-8 Pa. $5.95

AN EGYPTIAN HIEROGLYPHIC DICTIONARY, E. A. Wallis Budge. Monumental work containing about 25,000 words or terms that occur in texts ranging from 3000 B.C. to 600 A.D. Each entry consists of a transliteration of the word, the word in hieroglyphs, and the meaning in English. 1,314pp. 6⅞ × 10.
23615-3, 23616-1 Pa., Two-vol. set $27.90

THE COMPLEAT STRATEGYST: Being a Primer on the Theory of Games of Strategy, J. D. Williams. Highly entertaining classic describes, with many illustrated examples, how to select best strategies in conflict situations. Prefaces. Appendices. xvi + 268pp. 5⅜ × 8½.
25101-2 Pa. $5.95

THE ROAD TO OZ, L. Frank Baum. Dorothy meets the Shaggy Man, little Button-Bright and the Rainbow's beautiful daughter in this delightful trip to the magical Land of Oz. 272pp. 5⅜ × 8.
25208-6 Pa. $4.95

POINT AND LINE TO PLANE, Wassily Kandinsky. Seminal exposition of role of point, line, other elements in non-objective painting. Essential to understanding 20th-century art. 127 illustrations. 192pp. 6½ × 9¼.
23808-3 Pa. $4.50

LADY ANNA, Anthony Trollope. Moving chronicle of Countess Lovel's bitter struggle to win for herself and daughter Anna their rightful rank and fortune—perhaps at cost of sanity itself. 384pp. 5⅜ × 8½.
24669-8 Pa. $6.95

EGYPTIAN MAGIC, E. A. Wallis Budge. Sums up all that is known about magic in Ancient Egypt: the role of magic in controlling the gods, powerful amulets that warded off evil spirits, scarabs of immortality, use of wax images, formulas and spells, the secret name, much more. 253pp. 5⅜ × 8½.
22681-6 Pa. $4.00

THE DANCE OF SIVA, Ananda Coomaraswamy. Preeminent authority unfolds the vast metaphysic of India: the revelation of her art, conception of the universe, social organization, etc. 27 reproductions of art masterpieces. 192pp. 5⅜ × 8½.
24817-8 Pa. $5.95

CATALOG OF DOVER BOOKS

CHRISTMAS CUSTOMS AND TRADITIONS, Clement A. Miles. Origin, evolution, significance of religious, secular practices. Caroling, gifts, yule logs, much more. Full, scholarly yet fascinating; non-sectarian. 400pp. 5⅜ × 8½.
23354-5 Pa. $6.50

THE HUMAN FIGURE IN MOTION, Eadweard Muybridge. More than 4,500 stopped-action photos, in action series, showing undraped men, women, children jumping, lying down, throwing, sitting, wrestling, carrying, etc. 390pp. 7⅞ × 10⅝.
20204-6 Cloth. $19.95

THE MAN WHO WAS THURSDAY, Gilbert Keith Chesterton. Witty, fast-paced novel about a club of anarchists in turn-of-the-century London. Brilliant social, religious, philosophical speculations. 128pp. 5⅜ × 8½.
25121-7 Pa. $3.95

A CEZANNE SKETCHBOOK: Figures, Portraits, Landscapes and Still Lifes, Paul Cezanne. Great artist experiments with tonal effects, light, mass, other qualities in over 100 drawings. A revealing view of developing master painter, precursor of Cubism. 102 black-and-white illustrations. 144pp. 8¾ × 6⅛.
24790-2 Pa. $5.95

AN ENCYCLOPEDIA OF BATTLES: Accounts of Over 1,560 Battles from 1479 B.C. to the Present, David Eggenberger. Presents essential details of every major battle in recorded history, from the first battle of Megiddo in 1479 B.C. to Grenada in 1984. List of Battle Maps. New Appendix covering the years 1967-1984. Index. 99 illustrations. 544pp. 6½ × 9¼.
24913-1 Pa. $14.95

AN ETYMOLOGICAL DICTIONARY OF MODERN ENGLISH, Ernest Weekley. Richest, fullest work, by foremost British lexicographer. Detailed word histories. Inexhaustible. Total of 856pp. 6½ × 9¼.
21873-2, 21874-0 Pa., Two-vol. set $17.00

WEBSTER'S AMERICAN MILITARY BIOGRAPHIES, edited by Robert McHenry. Over 1,000 figures who shaped 3 centuries of American military history. Detailed biographies of Nathan Hale, Douglas MacArthur, Mary Hallaren, others. Chronologies of engagements, more. Introduction. Addenda. 1,033 entries in alphabetical order. xi + 548pp. 6½ × 9¼. (Available in U.S. only)
24758-9 Pa. $11.95

LIFE IN ANCIENT EGYPT, Adolf Erman. Detailed older account, with much not in more recent books: domestic life, religion, magic, medicine, commerce, and whatever else needed for complete picture. Many illustrations. 597pp. 5⅜ × 8½.
22632-8 Pa. $8.50

HISTORIC COSTUME IN PICTURES, Braun & Schneider. Over 1,450 costumed figures shown, covering a wide variety of peoples: kings, emperors, nobles, priests, servants, soldiers, scholars, townsfolk, peasants, merchants, courtiers, cavaliers, and more. 256pp. 8⅜ × 11¼.
23150-X Pa. $7.95

THE NOTEBOOKS OF LEONARDO DA VINCI, edited by J. P. Richter. Extracts from manuscripts reveal great genius; on painting, sculpture, anatomy, sciences, geography, etc. Both Italian and English. 186 ms. pages reproduced, plus 500 additional drawings, including studies for *Last Supper, Sforza* monument, etc. 860pp. 7⅞ × 10¾. (Available in U.S. only) 22572-0, 22573-9 Pa., Two-vol. set $25.90

CATALOG OF DOVER BOOKS

AMERICAN CLIPPER SHIPS: 1833-1858, Octavius T. Howe & Frederick C. Matthews. Fully-illustrated, encyclopedic review of 352 clipper ships from the period of America's greatest maritime supremacy. Introduction. 109 halftones. 5 black-and-white line illustrations. Index. Total of 928pp. 5⅜ × 8½.
25115-2, 25116-0 Pa., Two-vol. set $17.90

TOWARDS A NEW ARCHITECTURE, Le Corbusier. Pioneering manifesto by great architect, near legendary founder of "International School." Technical and aesthetic theories, views on industry, economics, relation of form to function, "mass-production spirit," much more. Profusely illustrated. Unabridged translation of 13th French edition. Introduction by Frederick Etchells. 320pp. 6⅛ × 9¼. (Available in U.S. only)
25023-7 Pa. $8.95

THE BOOK OF KELLS, edited by Blanche Cirker. Inexpensive collection of 32 full-color, full-page plates from the greatest illuminated manuscript of the Middle Ages, painstakingly reproduced from rare facsimile edition. Publisher's Note. Captions. 32pp. 9⅜ × 12¼.
24345-1 Pa. $4.50

BEST SCIENCE FICTION STORIES OF H. G. WELLS, H. G. Wells. Full novel *The Invisible Man,* plus 17 short stories: "The Crystal Egg," "Aepyornis Island," "The Strange Orchid," etc. 303pp. 5⅜ × 8½. (Available in U.S. only)
21531-8 Pa. $4.95

AMERICAN SAILING SHIPS: Their Plans and History, Charles G. Davis. Photos, construction details of schooners, frigates, clippers, other sailcraft of 18th to early 20th centuries—plus entertaining discourse on design, rigging, nautical lore, much more. 137 black-and-white illustrations. 240pp. 6⅛ × 9¼.
24658-2 Pa. $5.95

ENTERTAINING MATHEMATICAL PUZZLES, Martin Gardner. Selection of author's favorite conundrums involving arithmetic, money, speed, etc., with lively commentary. Complete solutions. 112pp. 5⅜ × 8½. 25211-6 Pa. $2.95

THE WILL TO BELIEVE, HUMAN IMMORTALITY, William James. Two books bound together. Effect of irrational on logical, and arguments for human immortality. 402pp. 5⅜ × 8½. 20291-7 Pa. $7.50

THE HAUNTED MONASTERY and THE CHINESE MAZE MURDERS, Robert Van Gulik. 2 full novels by Van Gulik continue adventures of Judge Dee and his companions. An evil Taoist monastery, seemingly supernatural events; overgrown topiary maze that hides strange crimes. Set in 7th-century China. 27 illustrations. 328pp. 5⅜ × 8½. 23502-5 Pa. $5.00

CELEBRATED CASES OF JUDGE DEE (DEE GOONG AN), translated by Robert Van Gulik. Authentic 18th-century Chinese detective novel; Dee and associates solve three interlocked cases. Led to Van Gulik's own stories with same characters. Extensive introduction. 9 illustrations. 237pp. 5⅜ × 8½.
23337-5 Pa. $4.95

Prices subject to change without notice.

Available at your book dealer or write for free catalog to Dept. GI, Dover Publications, Inc., 31 East 2nd St., Mineola, N.Y. 11501. Dover publishes more than 175 books each year on science, elementary and advanced mathematics, biology, music, art, literary history, social sciences and other areas.